Geometry Connections
Version 3.1

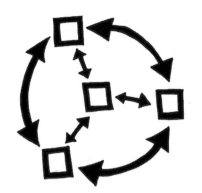

Managing Editor

Leslie Dietiker
Phillip and Sala Burton Academic High School
San Francisco, CA

Contributing Editors

Elizabeth Ellen Coyner
Christian Brothers High School
Sacramento, CA

Lara Lomac
Phillip and Sala Burton Academic
High School, San Francisco, CA

Leslie Nielsen
Isaaquah High School
Issaquah, WA

Barbara Shreve
San Lorenzo High School
San Lorenzo, CA

Lew Douglas
The College Preparatory School
Oakland, CA

Damian Molinari
Phillip and Sala Burton Academic
High School, San Francisco, CA

Karen O'Connell
San Lorenzo High School
San Lorenzo, CA

Michael Titelbaum
University of California
Berkeley, CA

David Gulick
Phillips Exeter Academy
Exeter, NH

Jason Murphy-Thomas
George Washington High School
San Francisco, CA

Ward Quincey
Gideon Hausner Jewish Day School
Palo Alto, CA

Illustrator

Kevin Coffey
San Francisco, CA

Technical Assistants

Erica Andrews
Elizabeth Burke
Carrie Cai
Daniel Cohen

Elizabeth Fong
Rebecca Harlow
Michael Leong
Thomas Leong

Marcos Rojas
Susan Ryan

Program Directors

Leslie Dietiker
Phillip and Sala Burton Academic High School
San Francisco, CA

Brian Hoey
Christian Brothers High School
Sacramento, CA

Judy Kysh, Ph.D.
Departments of Mathematics and Education
San Francisco State University

Tom Sallee, Ph.D.
Department of Mathematics
University of California, Davis

Contributing Editor of the Parent Guide:

Karen Wootten, Managing Editor
Odenton, MD

Technical Manager of Parent Guide and Extra Practice:

Rebecca Harlow
Stanford University
Stanford, CA

Editor of Extra Practice:

Bob Petersen, Managing Editor
Rosemont High School
Sacramento, CA

Assessment Contributors:

John Cooper, Managing Editor
Del Oro High School
Loomis, CA

Leslie Dietiker
Phillip and Sala Burton Academic High School
San Francisco, CA

Lara Lomac
Phillip and Sala Burton Academic High School,
San Francisco, CA

Damian Molinari
Phillip and Sala Burton Academic High School,
San Francisco, CA

Barbara Shreve
San Lorenzo High School
San Lorenzo, CA

2 3 4 5 6 7 8 9   10 09 08 07 06   ISBN-10:   1-931287-60-0

Printed in the United States of America   Version 3.1   ISBN-13: 978-1-931287-60-9

A Note to Students:

Welcome to Geometry! This courses
centers on the study of shapes. As you
study geometry, you will be investigating
new situations, discovering relationships,
and figuring out what strategies can be used
to solve problems. Learning to think this
way is useful in mathematical contexts and
other courses, as well as in situations
outside the classroom.

In meeting the challenges of geometry, you will not be working alone. During this course you will
collaborate with other students as a member of a study team. Working in a team means speaking up
and interacting with others. You will explain your ideas, listen to what others have to say, and ask
questions if there is something you do not understand. In geometry, a single problem can often be
solved more than one way. You will see problems in different ways than your teammates do. Each
of you has something to contribute while you work on the lessons in this course.

Together, your team will complete problems and activities that will help you discover new
mathematical ideas and methods. Your teacher will support you as you work, but will not take away
your opportunity to think and investigate for yourself. The ideas in the course will be revisited
several times and connected to other topics. If something is not clear to you the first time you work
with it, don't worry—you will have more chances to build your understanding as the course
continues.

Learning math this way has a significant advantage: as long as you actively participate, make sure
everyone in your study team is involved, and ask good questions, you will find yourself
understanding mathematics at a deeper level than ever before. By the end of this course, you will
have an understanding of a variety of geometric principles and properties that govern the world
around us. You will see how these principles and properties interweave so that you can use them
together to solve new problems. With your teammates, you will meet mathematical challenges you
would not have known how to approach before.

In addition to the support provided by your teacher and your study team, CPM has also created
online resources to help you, including help with homework, a parent guide, and extra practice. You
will find these resources and more at www.cpm.org.

We wish you well and are confident that you will enjoy learning geometry!

Sincerely,
The CPM Team

Geometry
Connections
Table of Contents

Student Edition

Chapter 1 Shapes and Transformations
                                                                            1

Section 1.1 Course Introduction Problems
1.1.1	Creating a Quilt Using Symmetry	3
1.1.2	Making Predictions and Investigating Results	7
1.1.3	Perimeter and Area of Enlarging Tile Patterns	12
1.1.4	Logical Arguments	16
1.1.5	Building a Kaleidoscope	21

Section 1.2 Transformations and Symmetry
1.2.1	Spatial Visualization and Reflections	26
1.2.2	Rigid Transformations: Rotations and Translations	31
1.2.3	Using Transformations	36
1.2.4	Using Transformations to Create Shapes	40
1.2.5	Symmetry	44

Section 1.3 Shapes and Probability
1.3.1	Attributes and Characteristics of Shapes	49
1.3.2	More Characteristics of Shapes	53
1.3.3	Introduction to Probability	57

Chapter Closure
                                                                            62

Chapter 2 Angles and Measurement
                                                                            71

Section 2.1 Angle Relationships
2.1.1	Complementary, Supplementary, and Vertical Angles	73
2.1.2	Angles Formed by Transversals	78
2.1.3	More Angles Formed by Transversals	83
2.1.4	Angles in a Triangle	89
2.1.5	Applying Angle Relationships	93

Section 2.2 Area
2.2.1	Units of Measure	98
2.2.2	Areas of Triangles and Composite Shapes	102
2.2.3	Areas of Parallelograms and Trapezoids	105
2.2.4	Heights and Area	110

Section 2.3 The Pythagorean Theorem
2.3.1	Squares and Square Roots	113
2.3.2	Triangle Inequality	117
2.3.3	The Pythagorean Theorem	121

Chapter Closure
                                                                            125

Chapter 3 Justification and Similarity 133

Section 3.1 Introduction to Similarity
   3.1.1   Similarity   135
   3.1.2   Proportional Growth and Ratios   139
   3.1.3   Using Ratios of Similarity   144
   3.1.4   Applications and Notation   147

Section 3.2 Triangle Similarity
   3.2.1   Conditions for Triangle Similarity   152
   3.2.2   Creating a Flowchart   157
   3.2.3   Triangle Similarity and Congruence   161
   3.2.4   More Conditions for Triangle Similarity   165
   3.2.5   Determining Similarity   169
   3.2.6   Applying Similarity   173

Chapter Closure   176

Chapter 4 Trigonometry and Probability 185

Section 4.1 The Tangent Ratio
   4.1.1   Constant Ratios in Right Triangles   187
   4.1.2   Connecting Slope Ratios to Specific Angles   192
   4.1.3   Expanding the Trig Table   195
   4.1.4   The Tangent Ratio   198
   4.1.5   Applying the Tangent Ratio   202

Section 4.2 Probability Models
   4.2.1   Introduction to Probability Models   204
   4.2.2   Theoretical and Experimental Probability   209
   4.2.3   Using an Area Model   212
   4.2.4   Choosing a Probability Model   216
   4.2.5   Optional: Applications of Probability Methods   221

Chapter Closure   224

Chapter 5 Trigonometry and Triangle Tool Kit   233

Section 5.1 Trigonometry
5.1.1   Sine and Cosine Ratios   235
5.1.2   Selecting a Trig Tool   239
5.1.3   Inverse Trigonometry   243
5.1.4   Trigonometric Applications   246

Section 5.2 Special Right Triangles
5.2.1   Special Right Triangles   250
5.2.2   Pythagorean Triples   254

Section 5.3 Completing the Triangle Toolkit
5.3.1   Finding Missing Parts of Triangles   257
5.3.2   Law of Sines   262
5.3.3   Law of Cosines   265
5.3.4   Ambiguous Triangles   269
5.3.5   Choosing a Tool   273

Chapter Closure   278

Chapter 6 Congruent Triangles   287

Section 6.1 Congruent Triangles
6.1.1   Congruent Triangles   287
6.1.2   Conditions for Triangle Congruence   293
6.1.3   Flowcharts for Congruence   297
6.1.4   Converses   301

Section 6.2 Closure Activities
6.2.1   Angles on a Pool Table   306
6.2.2   Investigating a Triangle   309
6.2.3   Creating a Mathematical Model   311
6.2.4   Analyzing a Game   315
6.2.5   Using Transformations and Symmetry to Create a Snowflake   318

Chapter Closure   324

Chapter 7 Proof and Quadrilaterals

333

Section 7.1 Introduction to Chapters 7 - 12
7.1.1	Properties of a Circle	335
7.1.2	Building a Tetrahedron	339
7.1.3	Shortest Distance Problems	343
7.1.4	Using Symmetry to Study Polygons	349

Section 7.2 Proof and Quadrilaterals
7.2.1	Special Quadrilaterals and Proof	352
7.2.2	Properties of Rhombi	356
7.2.3	More Proof with Congruent Triangles	359
7.2.4	More Properties of Quadrilaterals	363
7.2.5	Two-Column Proofs	365
7.2.6	Explore-Conjecture-Prove	369

Section 7.3 Coordinate Geometry
7.3.1	Studying Quadrilaterals on a Coordinate Grid	373
7.3.2	Coordinate Geometry and Midpoints	376
7.3.3	Quadrilaterals on a Coordinate Plane	380

Chapter Closure   383

Chapter 8 Polygons and Circles

391

Section 8.1 Angles and Area of a Polygon
8.1.1	Pinwheels and Polygons	393
8.1.2	Interior Angles of a Polygon	398
8.1.3	Angles of Regular Polygons	402
8.1.4	Regular Polygon Angle Connections	405
8.1.5	Finding the Area of Regular Polygons	408

Section 8.2 Ratio of the Area of Similar Figures
8.2.1	Area Ratios of Similar Figures	413
8.2.2	Ratios of Similarity	417

Section 8.3 Area and Circumference of a Circle
8.3.1	A Special Ratio	421
8.3.2	Area and Circumference of Circles	424
8.3.3	Circles in Context	428

Chapter Closure   434

Chapter 9 Solids and Constructions   441

Section 9.1 Solids
9.1.1   Three-Dimensional Solids   443
9.1.2   Volume and Surface Area of Prisms   446
9.1.3   Prisms and Cylinders   450
9.1.4   Volumes of Similar Solids   453
9.1.5   Ratios of Similarity   456

Section 9.2 Construction
9.2.1   Introduction to Construction   459
9.2.2   Constructing Bisectors   463
9.2.3   More Exploration with Constructions   466
9.2.4   Finding a Centroid   470

Chapter Closure   473

Chapter 10 Circles and Expected Value   481

Section 10.1 Relationships Within Circles
10.1.1 Introduction to Chords   483
10.1.2 Angles and Arcs   487
10.1.3 Chords and Angles   492
10.1.4 Tangents and Chords   496
10.1.5 Problem Solving with Circles   499

Section 10.2 Expected Value
10.2.1 Designing Spinners   502
10.2.2 Expected Value   507
10.2.3 More Expected Value   510

Section 10.3 Equation of a Circle
10.3.1 The Equation of a Circle   514

Chapter Closure   518

Chapter 11 Solids and Circles   525

Section 11.1 Pyramids, Cones, Spheres
11.1.1 Platonic Solids   527
11.1.2 Pyramids   533
11.1.3 Volume of a Pyramid   536
11.1.4 Surface Area and Volume of a Cone   541
11.1.5 Surface Area and Volume of a Sphere   545

Section 11.2 More Circle Relationships
11.2.1 Coordinates on a Sphere   548
11.2.2 Tangents and Arcs   553
11.2.3 Secant and Tangent Relationships   556

Chapter Closure   564

Chapter 12 Conics and Closure   571

Section 12.1 Conic Sections
12.1.1 Introduction to Conic Sections   573
12.1.2 Graphing Parabolas Using the Focus and Directrix   576
12.1.3 Circles and Ellipses   580
12.1.4 The Hyperbola   583
12.1.5 Conic Equations and Graphs   587

Section 12.2 Closure Activities
12.2.1 Using Coordinate Geometry and Construction to Explore Shapes   591
12.2.2 Euler's Formula for Polyhedra   594
12.2.3 The Golden Ratio   597
12.2.4 Using Geometry to Find Probability   601

Appendices   607

A   Points, Lines, and Planes   607
B   Euclidean Geometry   612
C   Non-Euclidean Geometry   616

Glossary   621

Index   647

Index to Symbols and Equations   655

CHAPTER 1   Shapes and Transformations

Welcome to Geometry! *Geo* means Earth (*geography* is mapping the Earth, for example) and *metry* means measurement. Geometry applies the arithmetic, algebra and **reasoning** skills you have learned to the objects you see all around you. During this course, you will ask and answer questions such as "How can I describe this shape?", "How can I measure this shape?", "Is this shape symmetrical?", and "How can I convince others that what I think about this shape is true?"

This chapter begins with some activities that will introduce you to the big ideas of the course. Then you will apply motions to triangles and learn how to specify a particular motion. Finally, you will explore attributes of shapes that can be used to categorize and name them and find the probabilities of selecting shapes with certain properties from a "shape bucket."

Guiding Questions

Think about these questions throughout this chapter:

What is geometry?

How can I describe this shape?

How can I move it?

Is this shape symmetrical?

How can I communicate my ideas clearly?

In this chapter, you will:

> ➤ become familiar with basic geometric shapes and learn how to describe each one using its attributes, such as parallel sides or rotation symmetry.

> ➤ investigate three basic rigid transformations: reflection (flip), rotation (turn), and translation (slide).

> ➤ be introduced to probability and learn how to use probability to make predictions.

Chapter Outline

 Section 1.1   Investigations involving quilts, twisted strips of paper, rug designs, precise reasoning, and kaleidoscopes will introduce you to some basic building-blocks of geometry: shapes, motions, measurements, patterns, reasoning and symmetry.

 Section 1.2   You will learn about transformations as you study how to flip, turn, and slide shapes. Then you will learn how to use these motions to build new shapes and to describe symmetry.

 Section. 1.3   A "shape bucket" will introduce you to a variety of basic shapes that you will describe, classify and name according to their attributes. You will also learn about probability.

1.1.1 How can I design it?

Creating a Quilt Using Symmetry

Welcome to Geometry! But what is geometry? At the end of this chapter you will have a better understanding of what geometry is. To start, you will focus on several activities that will hopefully challenge you and introduce you to important concepts in geometry that you will study in this course. While all of the problems are solvable with your current math skills, some will be revisited later in the course so that you may apply new geometric tools to solve and extend them.

Today you will consider an example of how geometry is applied in the world around you. A very popular American tradition is to create quilts by sewing together remnants of cloth in intricate geometric designs. These quilts often integrate geometric shapes in repeated patterns that show symmetry. For centuries, quilts have been designed to tell stories, document special occasions, or decorate homes.

1-1.   DESIGNING A QUILT, Part One

How can you use symmetry to design a quilt? Today you will work with your team to design a patch that will be combined with other team patches to make a class quilt. Before you start, review the team roles, which are outlined on the next page.

a.   Each team member will receive four small squares. With a colored pencil or marker, shade in <u>half</u> of each square (one triangle) as shown at right. Each team member should use a different color.

b.   Next arrange your squares to make a larger 2-by-2 square (as shown at right) with a design that has **reflection symmetry**. A design has reflection symmetry if it can be folded in half so that both sides match perfectly. Make sure that you have arranged your pieces into a different symmetrical pattern than the rest of your team.

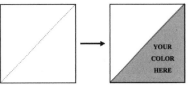

c.   Now, create a 4-by-4 square using the designs created by each team member as shown on the Lesson 1.1.1B Resource Page. Ask your teacher to verify that your designs are all symmetrical and unique. Then glue (or tape) all sixteen pieces carefully to the resource page and cut along the surrounding dashed square so that you have a blank border around your 4-by-4 square.

Problem continues on next page →

1-1.   *Problem continued from previous page.*

d.   Finally, discuss with your team what you all personally have in common. Come up with a team sentence that captures the most interesting facts. Write your names and this sentence in the border so they wrap around your 4-by-4 design.

To help you work together today, each member of your team has a specific job, assigned by your first name (or last name if team members have the same first name).

Team Roles

Resource Manager – If your name comes first alphabetically:

- Make sure the team has all of the necessary materials, such as colored pencils or markers and the Lesson 1.1.1A and 1.1.B Resource Pages.

- Ask the teacher when the **entire** team has a question. You might ask, *"No one has an idea? Should I ask the teacher?"*

- Make sure your team cleans up by delegating tasks. You could say, *"I will put away the _____ while you _____ ."*

Facilitator – If your name comes second alphabetically:

- Start the team's discussion by asking, *"What are some possible designs?"* or *"How can we make sure that all of our designs are symmetrical?"* or *"Are all of our designs different?"*

- Make sure that all of the team members get any necessary help. You don't have to answer all the questions yourself. A good facilitator regularly asks, *"Do you understand what you are supposed to do?"* and *"Who can answer _____ 's question?"*

Recorder/Reporter – If your name comes third alphabetically:

- Coordinate the taping or gluing of the quilt pieces together onto the resource page in the orientation everyone agreed to.

- Take notes for the team. The notes should include phrases like, *"We found that we all had in common …"* and explanations like, *"Each of our designs was found to be unique and symmetrical because …"*

- Help the team agree on a team sentence: *"What do we all have in common?"* and *"How can I write that on our quilt?"*

Task Manager – If your name comes fourth alphabetically:

- Remind the team to stay on task and not to talk to students in other teams. You can suggest, *"Let's try coming up with different symmetrical patterns."*

- Keep track of time. Give your team reminders, such as *"I think we need to decide now so that we will have enough time to …"*

Geometry Connections

1-2.   DESIGNING A QUILT, Part Two

Your teacher will ask the Recorder/Reporters from each team to bring their finished quilt patches up to the board one at a time and tape them to the other patches. Be prepared to explain how you came up with your unique designs and interesting ideas about symmetry. Also be prepared to read your team sentence to the class. As you listen to the presentations, look for relationships between your designs and the other team designs.

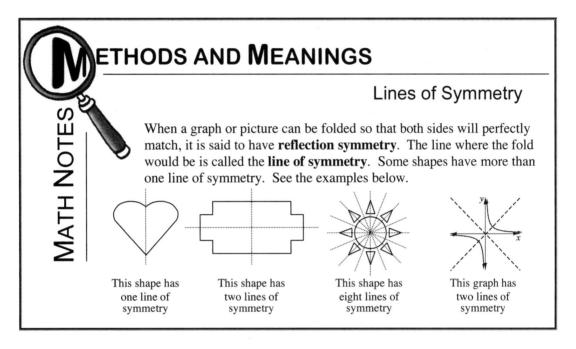

MATH NOTES

METHODS AND MEANINGS

Lines of Symmetry

When a graph or picture can be folded so that both sides will perfectly match, it is said to have **reflection symmetry**. The line where the fold would be is called the **line of symmetry**. Some shapes have more than one line of symmetry. See the examples below.

This shape has one line of symmetry

This shape has two lines of symmetry

This shape has eight lines of symmetry

This graph has two lines of symmetry

1-3.   One focus of this Geometry course is to help you recognize and accurately identify a shape. For example, a **rectangle** is a four-sided shape with four right angles. Which of the shapes below can be called a rectangle? More than one answer is possible.

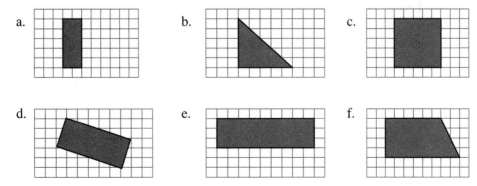

a.

b.

c.

d.

e.

f.

1-4.   Calculate the values of the expressions below. Show all steps in your process. The answers are provided for you to check your result. If you miss two or more of these and cannot find your errors, be sure to seek help from your team or teacher.
[Answers: a: 40, b: –6, c: 7, d: 59]

a. $2 \cdot (3(5+2)-1)$ b. $6-2(4+5)+6$

c. $3 \cdot 8 \div 2^2 +1$ d. $5-2 \cdot 3+6(3^2+1)$

1-5.   Match each table of data on the left with its rule on the right and **briefly explain** why it matches the data.

a.
x	1	0	–4	2	–2	–1
y	4	3	–1	5	1	2

(1) $y = x$

(2) $y = 3x - 1$

b.
x	–1	3	1	0	–2	2
y	–1	–9	–1	0	–4	–4

(3) $y = x + 3$

c.
x	3	–2	1	0	2	–3
y	12	7	4	3	7	12

(4) $y = x^2$

(5) $y = -x^2$

d.
x	–3	4	2	–2	0	–10
y	–10	11	5	–7	–1	–31

(6) $y = x^2 + 3$

1-6.   Simplify the expressions below as much as possible.

a. $2a + 4(7 + 5a)$ b. $4(3x+2) - 5(7x+5)$

c. $x(x+5)$ d. $2x + x(x+6)$

1-7.   **Examine** the graph at right. Then, in a sentence or two, suggest reasons why the graph rises at 11:00 AM and then drops at 1:15 PM.

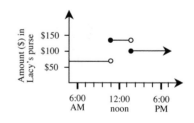

1.1.2 Can you predict the results?

. . . .

Making Predictions and Investigating Results

Today you will **investigate** what happens when you change the attributes of a Möbius strip. As you **investigate**, you will record data in a table. You will then analyze this data and use your results to brainstorm further experiments. As you look back at your data, you may start to consider other related questions that can help you understand a pattern and learn more about what is happening. This Way of Thinking, called **investigating**, includes not only generating new questions, but also rethinking when the results are not what you expected.

1-8.   Working effectively with your study team will be an important part of the learning process throughout this course. Choose a member of your team to read aloud these Study Team Expectations:

STUDY TEAM EXPECTATIONS

Throughout this course you will regularly work with a team of students. This collaboration will allow you to develop new ways of thinking about mathematics, increase your ability to communicate with others about math, and help you strengthen your understanding by having you explain your thinking to someone else. As you work together,

- You are expected to share your ideas and contribute to the team's work.

- You are expected to ask your teammates questions and to offer help to your teammates. Questions can move your team's thinking forward and help others to understand ideas more clearly.

- Remember that a team that functions well works on the same problem together and discusses the problem while it works.

- Remember that one student on the team should not dominate the discussion and thinking process.

- Your team should regularly stop and verify that everyone on the team agrees with a suggestion or a solution.

- Everyone on your team should be consulted before calling on the teacher to answer a question.

1-9.   On a piece of paper provided by your teacher, make a "bracelet" by taping the two ends securely together. Putting tape on both sides of the bracelet will help to make sure the bracelet is secure. In the diagram of the rectangular strip shown at right, you would tape the ends together so that point A would attach to point C, and point B would attach to point D.

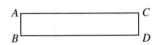

Now predict what you think would happen if you were to cut the bracelet down the middle, as shown in the diagram at right. Record your prediction in a table like the one shown below or on your Lesson 1.1.2 Resource Page.

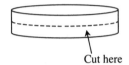

Cut here

	Experiment	Prediction	Result
1-9	Cut bracelet in half as shown in the diagram.		
1-10			
1-11			
1-12a			
1-12b			
1-12c			
1-12d			

Now cut your strip as described above and record your result in the first row of your table. Make sure to include a short description of your result.

1-10.   On a second strip of paper, label a point X in the center of the strip at least one inch away from one end.

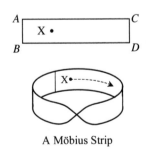

A Möbius Strip

Now turn this strip into a **Möbius strip** by attaching the ends together securely after making one twist. For the strip shown in the diagram at right, the paper would be twisted once so that point A would attach to point D. The result should look like the diagram at right.

Predict what would happen if you were to draw a line down the center of the strip from point X until you ran out of paper. Record your prediction, conduct the experiment, and record your result.

Geometry Connections

1-11.   What do you think would happen if you were to cut your Möbius strip along the central line you drew in problem 1-10? Record your prediction in your table.

Cut just one of your team's Möbius strips. Record your result in your table. Consider the original strip of paper drawn in problem 1-9 to help you explain why cutting the Möbius strip had this result.

1-12.   What else can you learn about Möbius strips? For each experiment below, first record your expectation. Then record your result in your table after conducting the experiment. Use a new Möbius strip for each experiment.

a.   What if the result from problem 1-11 is cut in half down the middle again?

b.   What would happen if the Möbius strip is cut one-third of the way from one of the sides of the strip? Be sure to cut a constant distance from the side of the strip.

c.   What if a strip is formed by 2 twists instead of one? What would happen if it were cut down the middle?

d.   If time allows, make up your own experiment. You might change how many twists you make, where you make your cuts, etc. Try to generalize your findings as you conduct your experiment. Be prepared to share your results with the class.

1-13.   LEARNING REFLECTION

Think over how you and your study team worked today, and what you learned about Möbius strips. What questions did you or your teammates ask that helped move the team forward? What questions do you still have about Möbius strips? What would you like to know more about?

METHODS AND MEANINGS

The Investigative Process

The **investigative process** is a way to study and learn new mathematical ideas. Mathematicians have used this process for many years to make sense of new concepts and to broaden their understanding of older ideas.

In general, this process begins with a **question** that helps you frame what you are looking for. For example, a question such as, *"What if the Möbius strip has 2 half-twists? What will happen when that strip is cut in half down the middle?"* can help start an investigation to find out what happens when the Möbius strip is slightly altered.

Once a question is asked, you can make an educated guess, called a **conjecture**. This is a mathematical statement that has not yet been proven.

Next, **exploration** begins. This part of the process may last awhile as you gather more information about the mathematical concept. For example, you may first have an idea about the diagonals of a rectangle, but as you draw and measure a rectangle on graph paper, you find out that your conjecture was incorrect. When this happens, you just experiment some more until you have a new conjecture to test.

When a conjecture seems to be true, the final step is to **prove** that the conjecture is always true. A proof is a convincing logical argument that uses definitions and previously proven conjectures in an organized sequence.

1-14.   A major focus of this course is learning the **investigative process**, a process you used during the Möbius Strip activity in problems 1-9 through 1-12. One part of this process is asking mathematical questions.

Assume your teacher is thinking of a shape and wants you to figure out what shape it is. Write down three questions you could ask your teacher to determine more about his or her shape.

1-15. The shapes at right are examples of **equilateral triangles**. How can you describe an equilateral triangle? Make at least two statements that seem true for all equilateral triangles. Then trace these equilateral triangles on your paper and draw one more in a different orientation.

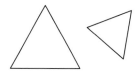

Examples of Equilateral Triangles

1-16. Match each table of data below with the most appropriate graph and **briefly explain** why it matches the data.

    a.   Boiling water cooling down.

Time (min)	0	5	10	15	20	25
Temp (°C)	100	89	80	72	65	59

    b.   Cost of a phone call.

Time (min)	1	2	2.5	3	4	5	5.3	6
Cost (cents)	55	75	75	95	115	135	135	155

    c.   Growth of a baby in the womb.

Age (months)	1	2	3	4	5	6	7	8	9
Length (inches)	0.75	1.5	3	6.4	9.6	12	13.6	15.2	16.8

Graph 1   Graph 2   Graph 3   Graph 4   Graph 5

1-17. Solve for the given variable. Show the steps leading to your solution. Check your solution.

    a. $-11x = 77$

    b. $5c + 1 = 7c - 8$

    c. $\frac{x}{8} = 2$

    d. $-12 = 3k + 9$

1-18. Calculate the values of the expressions below. Show all steps in your process.

    a. $\frac{3(2+6)}{2}$

    b. $\frac{1}{2}(14)(5)$

    c. $7^2 - 5^2$

    d. $17 - 6 \cdot 2 + 4 \div 2$

1.1.3 How can I predict the area?

Perimeter and Area of Enlarging Tile Patterns

One of the core ideas of geometry is the measurement of shapes. Often in this course it will be important to find the areas and perimeters of shapes. How these measurements change as a shape is enlarged or reduced in size is especially interesting. Today your team will apply algebraic skills as you **investigate** the areas and perimeters of similar shapes.

1-19. CARPETMART

Your friend Alonzo has come to your team for help. His family owns a rug manufacturing company, which is famous for its unique and versatile designs. One of their most popular designs is shown at right. Each rug design has an "original" size as well as enlargements that are exactly the same shape.

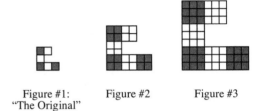

Figure #1:   Figure #2   Figure #3
"The Original"

Alonzo is excited because his family found out that the king of a far-away land is going to order an extremely large rug for one of his immense banquet halls. Unfortunately, the king is fickle and won't decide which rug he will order until the very last minute. The day before the banquet, the king will tell Alonzo which rug he wants and how big it will need to be. The king's palace is huge, so the rug will be VERY big!

Since the rugs are different sizes, and since each rug requires wool for the interior and fringe to wrap around the outside, Alonzo will need to quickly find the area and perimeter of each rug in order to obtain the correct quantities of wool and fringe.

Your Task: Your teacher will assign your team one of the rug designs to **investigate** (labeled (a) through (f) below). The "original" rug is shown in Figure 1, while Figures 2 and 3 are the next enlarged rugs of the series. With your team, create a table, graph, and rule for both the area and perimeter of your rug design. Then decide which representation will best help Alonzo find the area and perimeter for *any* figure number.

Problem continues on next page →

1-19.   *Problem continued from previous page.*

Be ready to share your analysis with the rest of the class. Your work must include the following:

- Diagrams for the rugs of the next two sizes (Figures 4 and 5) following the pattern shown in Figures 1, 2, and 3.

- A description of Figure 20. What will it look like? What are its area and perimeter?

- A table, graph, and rule representing the perimeter of your rug design.

- A table, graph, and rule representing the area of your rug design.

Rug Designs:

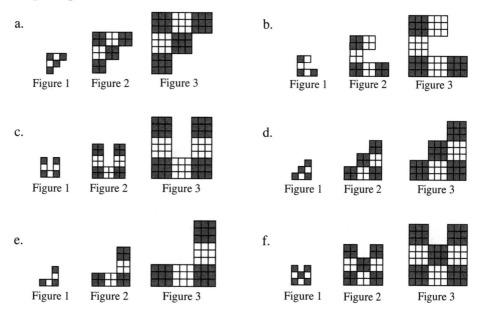

a.
Figure 1   Figure 2   Figure 3

b.
Figure 1   Figure 2   Figure 3

c.
Figure 1   Figure 2   Figure 3

d.
Figure 1   Figure 2   Figure 3

e.
Figure 1   Figure 2   Figure 3

f.
Figure 1   Figure 2   Figure 3

Further Guidance

1-20.   To start problem 1-20, first analyze the pattern your team has been assigned on graph paper, draw diagrams of Figures 4 and 5 for your rug design. Remember to shade Figures 4 and 5 the same way Figures 1 through 3 are shaded.

1-21.   Describe Figure 20 of your design. Give as much information as you can. What will it look like? How will the squares be arranged? How will it be shaded?

1-22.   A table can help you learn more about how the perimeter changes as the rugs get bigger.

a.   Organize your perimeter data in a table like the one shown below.

Figure number	1	2	3	4	5	20
Perimeter (in units)						

b.   Graph the perimeter data for Figures 1 through 5. (You do not need to include Figure 20.) What shape is the graph?

c.   How does the perimeter grow? **Examine** your table and graph and describe how the perimeter changes as the rugs get bigger.

d.   Generalize the patterns you have found by writing an algebraic rule (equation) that will find the perimeter of any size rug in your design. That is, what is the perimeter of Figure n? Show how you got your answer.

1-23.   Now analyze how the area changes with a table and graph.

a.   Make a new table, like the one below, to organize information about the area of each rug in your design.

Figure number	1	2	3	4	5	20
Area (in square units)						

b.   On a new set of axes, graph the area data for Figures 1 through 5. (You do not need to include Figure 20.) What shape is the graph?

c.   How does the area grow? Does it grow the same way as the perimeter? **Examine** your table and graph and describe how the area changes as the rugs get bigger.

d.   Write a rule that will find the area of Figure n. How did you find your rule? Be ready to share your **strategy** with the class.

———— *Further Guidance section ends here.* ————

1-24.   The King has arrived! He demands a Rug #100, which is Figure 100 in your design. What will its perimeter be? Its area? **Justify** your answer.

Geometry Connections

METHODS AND MEANINGS

The Perimeter and Area of a Figure

The **perimeter** of a two-dimensional figure is the distance around its exterior (outside) on a flat surface. It is the total length of the boundary that encloses the interior (inside) region. See the example at right.

Perimeter $= 5 + 8 + 4 + 6 = 23$ units

The **area** indicates the number of square units needed to fill up a region on a flat surface. For a rectangle, the area is computed by multiplying its length and width. The rectangle at right has a length of 5 units and a width of 3 units, so the area of the rectangle is 15 square units.

Area $= 5 \cdot 3 = 15$ square units

1-25.   Read the Math Notes box for this lesson, which describes how to find the area and perimeter of a shape. Then **examine** the rectangle at right. If the perimeter of this shape is 120 cm, which equation below represents this fact? Once you have selected the appropriate equation, solve for x.

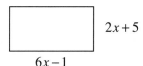

a. $2x + 5 + 6x - 1 = 120$

b. $4(6x - 1) = 120$

c. $2(6x - 1) + 2(2x + 5) = 120$

d. $(2x + 5)(6x - 1) = 120$

1-26.   Delilah drew 3 points on her paper. When she connects these points, must they form a triangle? Why or why not? Draw an example on your paper to support your **reasoning**.

1-27.   Copy the table below onto your paper. Complete it and write a rule relating x and y.

x	3	−1	0	2	−5	−2	1
y	0			−1			−2

1-28. Rebecca placed a transparent grid of square units over each of the shapes she was measuring below. Using her grid, determine the area of each shape.

a.

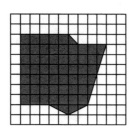

b.

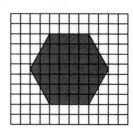

1-29. Evaluate each expression below if $a = -2$ and $b = 3$.

a. $3a^2 - 5b + 8$ b. $\frac{2}{3}b - 5a$ c. $\frac{a+2b}{4} + 4a$

1.1.4 Are you convinced?

Logical Arguments

"I don't have my homework today because…" Is your teacher going to be convinced? Will it make a difference whether you say that the dog ate your homework or whether you bring in a note from the doctor? Imagine your friend says, "I know that shape is a square because it has four right angles." Did your friend tell you enough to convince you?

Many jobs depend on your ability to convince other people that your ideas are correct. For instance, a defense lawyer must be able to form logical arguments to persuade the jury or judge that his or her client is innocent. Today you are going to use a Way of Thinking called **reasoning and justifying** to focus on what makes a statement convincing.

1-30.   TRIAL OF THE CENTURY

The musical group Apple Core has
accused your math teacher, Mr.
Bosky, of stealing its newest pop CD,
"Rotten Gala." According to the
police, someone stole the CD from
the BigCD Store last Saturday at
some time between 6:00 PM and 7:00
PM. Because your class is so well
known for only reaching conclusions
when sufficient evidence is presented,
the judge has made you the jury! You
are responsible for determining
whether or not there is enough
evidence to convict Mr. Bosky.

Carefully listen to the evidence that is presented. As each statement is read, decide:

- *Does the statement convince you? Why or why not?*

- *What could be changed or added to the statement to make it more convincing?*

Testimony

Mr. Bosky: *"But I don't like that CD! I wouldn't take it even if you paid me."*

Mr. Bosky: *"I don't have the CD. Search me."*

Mr. Bosky: *"I was at home having dinner Saturday."*

Casey: *"There were several of us having dinner with Mr. Bosky at his house. He made us a wonderful lasagna."*

Mrs. Thomas: *"All of us at dinner with Mr. Bosky left his house at 6:10 PM."*

Police Officer Yates: *"Driving as quickly as I could, it took me 30 minutes to go from Mr. Bosky's house to the BigCD store."*

Coach Teller: *"Mr. Bosky made a wonderful goal right at the beginning of our soccer game, which started at 7:00 PM. You can check the score in the local paper."*

Police Officer Yates: *"I also drove from the BigCD store to the field where the soccer game was. It would take him at least 40 minutes to get there."*

1-31.   THE FAMILY FORTUNE

You are at home when the phone rings. It is a good friend of yours who says, "Hey, your last name is Marston. Any chance you have a grandmother named Molly Marston who was REALLY wealthy? Check out today's paper." You glance at the front page:

> ### Family Fortune Unclaimed
>
> City officials are amazed that the county's largest family fortune may go unclaimed. Molly "Ol' Granny" Marston died earlier this week and it appears that she was survived by no living relatives. According to her last will and testament, "Upon my death, my entire fortune is to be divided among my children and grandchildren." Family members have until noon tomorrow to come forward with a written statement giving evidence that they are related to Ms. Marston or the money will be turned over to the city.

You're amazed – Molly is your grandmother, so your friend is right! However, you may not be able to collect your inheritance unless you can convince city officials that you are a relative. You rush into your attic where you keep a trunk full of family memorabilia.

a.   You find several items that you think might be important in an old trunk in the attic. With your team, decide which of the items listed below will help prove that Ol' Molly was your grandmother.

> **Family Portrait** — a photo showing three young children. On the back you see the date 1968.

> **Newspaper Clipping** — an article from 1972 titled "Triplets Make Music History." The first sentence catches your eye: "Jake, Judy, and Jeremiah Marston, all eight years old, were the first triplets ever to perform a six-handed piano piece at Carnegie Hall."

> **Jake Marston's Birth Certificate** — showing that Jake was born in 1964, and identifying his parents as Phillip and Molly Marston.

> **Your Learner's Permit** — signed by your father, Jeremiah Marston.

> **Wilbert Marston's Passport** — issued when Wilbert was fifteen.

b.   Your team will now write a statement that will convince the city official (played by your faithful teacher!) that Ol' Molly was your grandmother. Be sure to support any claims that you make with appropriate evidence. Sometimes it pays to be convincing!

METHODS AND MEANINGS

Solving Linear Equations

In Algebra, you learned how to solve a linear equation. This course will help you apply your algebra skills to solve geometric problems. Review how to solve equations by reading the example below.

- **Simplify**. Combine like terms on each side of the equation whenever possible.

$$3x - 2 + 4 = x - 6 \quad \text{Combine like terms}$$
$$3x + 2 = x - 6$$

$$-x = -x \quad \text{Subtract } x \text{ on both sides}$$

- **Keep equations balanced.** The equal sign in an equation tells you that the expressions on the left and right are balanced. Anything done to the equation must keep that balance.

$$2x + 2 = -6$$
$$-2 = -2 \quad \text{Subtract 2 on both sides}$$
$$\frac{2x}{2} = \frac{-8}{2} \quad \text{Divide both sides by 2}$$
$$x = -4$$

- **Move your x-terms to one side of the equation.** Isolate all variables on one side of the equation and the constants on the other.

- **Undo operations.** Use the fact that addition is the opposite of subtraction and that multiplication is the opposite of division to solve for x. For example, in the equation $2x = -8$, since the 2 and the x are multiplied, then dividing both sides by 2 will get x alone.

1-32.   One goal of this course will be to review and enhance your algebra skills. Read the Math Notes box for this lesson. Then solve for x in each equation below, show all steps leading to your solution, and check your answer.

a. $34x - 18 = 10x - 9$

b. $4x - 5 = 4x + 10$

c. $3(x - 5) + 2(3x + 1) = 45$

d. $-2(x + 4) + 6 = -3$

Chapter 1: Shapes and Transformations

1-33.   The day before Gerardo returned from a two-week trip, he wondered if he left his plants inside his apartment or outside on his deck. He knows these facts:

- If his plants are indoors, he must water them at least once a week or they will die.

- If he leaves his plants outdoors and it rains, then he does not have to water them. Otherwise, he must water them at least once a week or they will die.

- It has not rained in his town for 2 weeks.

When Gerardo returns, will his plants be dead? Explain your **reasoning**.

1-34.   For each of the equations below, solve for *y* in terms of *x*.

a. $2x - 3y = 12$ b. $5x + 2y = 7$

1-35.   Find the area of the rectangle at right. Be sure to include units in your answer.

8 mm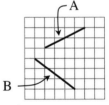
13 mm

1-36.   The **slope** of a line is a measure of its steepness and indicates whether it goes up or down from left to right. For example, the slope of the line segment A at right is $\frac{1}{2}$, while the slope of the line segment B is $-\frac{3}{4}$.

For each line segment below, find the slope. You may want to copy each line segment on graph paper in order to draw slope triangles.

a.   b.

c.   d.

1.1.5 What shapes can you find?

Building a Kaleidoscope

Today you will learn about angles and shapes as you study how a kaleidoscope works.

1-37.   BUILDING A KALEIDOSCOPE

How does a kaleidoscope create the complicated, colorful images you see when you look inside? A hinged mirror and a piece of colored paper can demonstrate how a simple kaleidoscope creates its beautiful repeating designs.

Your Task: Place a hinged mirror on a piece of colored, unlined paper so that its sides extend beyond the edge of the paper as shown at right. Explore what shapes you see when you look directly at the mirror, and how those shapes change when you change the angle of the mirror. Discuss the questions below with your team. Be ready to share your responses with the rest of the class.

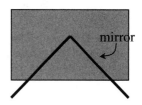

Discussion Points

What is this problem about? What is it asking you to do?

What happens when you change the angle (opening) formed by the sides of the mirror?

How can you describe the shapes you see in the mirror?

1-38.   To complete your exploration, answer these questions together as a team.

a.   What happens to the shape you see as the angle formed by the mirror gets bigger (wider)? What happens as the angle gets smaller?

b.   What is the smallest number of sides the shape you see in the mirror can have? What is the largest?

c.   With your team, find a way to form a **regular hexagon** (a shape with six equal sides and equal angles).

d.   How might you describe to another team how you set the mirrors to form a hexagon? What types of information would be useful to have?

1-39.   A good way to describe an angle is by measuring how *wide* or *spread apart* the angle is. For this course, you should think of the measurement of an angle as representing the amount of rotation that occurs when you separate the two sides of the mirror from a closed position. The largest angle you can represent with a hinged mirror is 360°. This is formed when you open a mirror all the way so that the backs of the mirror touch. This is a called a **circular angle** and is represented by the diagram at right.

360°

   a.   Other angles may be familiar to you. For example, an angle that forms a perfect "L" or a quarter turn is a 90° angle, called a **right angle** (shown at right). You can see that four of these angles would form a circular angle.

90°

   What if the two mirrors are opened to form a straight line? What measure would that angle have? Draw this angle and label its degrees. How is this angle related to a circular angle?

   b.   Based on the examples above, estimate the measures of these angles shown below. Then confirm your answer using a **protractor**, a tool that measures angles.

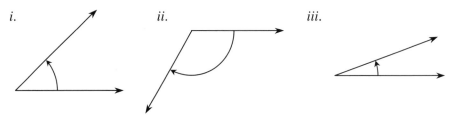

i.   *ii.*   *iii.*

1-40.   Now use your understanding of angle measurement to create some specific shapes using your hinged mirror. Be sure that both mirrors have the same length on the paper, as shown in the diagram at right.

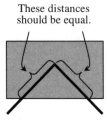

These distances should be equal.

   a.   Antonio says he can form an **equilateral triangle** (a triangle with three equal sides and three equal angles) using his hinged mirror. How did he do this? Once you can see the triangle in your mirror, place the protractor on top of the mirror. What is the measure of the angle formed by the sides of the mirror?

Problem continues on next page →

1-40.   *Problem continued from previous page.*

b.   Use your protractor to set your mirror so that the angle formed is 90°. Be sure that the sides of the mirror intersect the edge of the paper at equal lengths. What is this shape called? Draw and label a picture of the shape on your paper.

c.   Carmen's mirror shows the image at right, called a **regular pentagon**. She noticed that the five triangles in this design all meet at the hinge of her mirrors. She also noticed that the triangles must all the same size and shape, because they are reflections of the triangle formed by the mirrors and the paper.

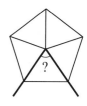

What must the sum of these five angles at the hinge be? And what is the angle formed by Carmen's mirrors? Test your conclusion with your mirror.

d.   Discuss with your team and predict how many sides a shape would have if the angle that the mirror forms measures 40°. Explain how you made your prediction. Then check your prediction using the mirror and a protractor. Describe the shape you see with as much detail as possible.

1-41.   Reflect on what you learned during today's activity.

a.   Based on this activity, what are some things that you think you will be studying in Geometry?

b.   This activity was based on the question, *"What shapes can be created using reflections?"* What ideas from this activity would you want to learn more about? Write a question that could prompt a different, but related, future **investigation**.

MᴇᴛHODS AND Mᴇᴀɴɪɴɢѕ

MATH NOTES

Types of Angles

When trying to describe shapes, it is convenient to classify types of angles. This course will use the following terms to refer to angles:

ACUTE: Any angle with measure **between** (but not including) 0° and 90°.

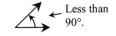

RIGHT: Any angle that measures 90˚.

OBTUSE: Any angle with measure **between** (but not including) 90° and 180°.

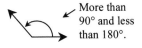

STRAIGHT: Straight angles have a measure of 180° and are formed when the sides of the angle form a straight line.

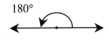

CIRCULAR: Any angle that measures 360°.

1-42.   Estimate the size of each angle below to the nearest 10°. A right angle is shown for reference so you should not need a protractor.

a.

b.

c.

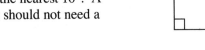

1-43.   Using graph paper, draw a pair of xy-axes and scale them for $-6 \le x \le 6$ (x-values between and including –6 and 6) and $-10 \le y \le 10$ (y-values between and including –10 and 10). Complete the table below, substituting the x-values (inputs) to find the corresponding y-values (outputs) for the rule $y = x + 2$. Plot and connect the resulting points.

x (input)	–4	–3	–2	–1	0	1	2	3	4
y (output)									

1-44.   Angela had a rectangular piece of paper and then cut a rectangle out of a corner as shown at right. Find the area and perimeter of the resulting shape. Assume all measurements are in centimeters.

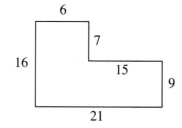

1-45.   For each equation below, solve for the given variable. If necessary, refer to the Math Notes box in Lesson 1.1.4 for guidance. Show the steps leading to your solution and check your answer.

a. $75 = 14y + 5$

b. $-7r + 13 = -71$

c. $3a + 11 = 7a - 13$

d. $2m + m - 8 = 7$

1-46.   On graph paper, draw four different rectangles that each have an area of 24 square units. Then find the perimeter of each one.

1.2.1 How do you see it?

•••

Spatial Visualization and Reflections

Were you surprised when you looked into the hinged mirror during the Kaleidoscope Investigation of Lesson 1.1.5? Reflection can create many beautiful and interesting shapes and can help you learn more about the characteristics of other shapes. However, one reason you may have been surprised is because it is sometimes difficult to predict what a reflection will be. This is where spatial visualization plays an important role. **Visualizing**, the act of "picturing" something in your mind, is often helpful when working with shapes. In order to be able to **investigate** and describe a geometric concept, it is first useful to **visualize** a shape or action.

Today you will be **visualizing** in a variety of ways and will develop the ability to find reflections. As you work today, keep the following focus questions in mind:

How do I see it?

How can I verify my answer?

How can I describe it?

1-47.   BUILDING BOXES

Which of the nets (diagrams) below would form a box with a lid if folded along the interior lines? Be prepared to defend your answer.

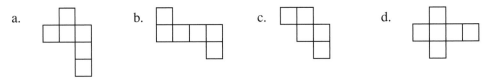

a.   b.   c.   d.

1-48.   Have you ever noticed what happens when you look in a mirror? Have you ever tried to read words while looking in a mirror? What happens? Discuss this with your team. Then re-write the following words as they would look if you held this book up to a mirror. Do you notice anything interesting?

a.   GEO   b.   STAR   c.   WOW

1-49.   When Kenji spun the flag shown at right very quickly about its
         pole, he noticed a three-dimensional shape emerge.

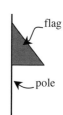
flag

pole

    a.   What shape did he see? Draw a picture of the three-
          dimensional shape on your paper and be prepared to
          defend your answer.

    b.   What would the flag need to look like so that a **sphere** (the
          shape of a basketball) is formed when the flag is rotated
          about its pole? Draw an example.

1-50.   REFLECTIONS

         The shapes created in the Kaleidoscope
         Investigation in Lesson 1.1.5 were the result
         of reflecting a triangle several times in a
         hinged mirror. However, other shapes can
         also be created by a reflection. For example,
         the diagram at right shows the result of
         reflecting a snowman across a line.

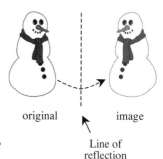
original   image

Line of
reflection

    a.   Why do you think the image is called a reflection?
          How is the image different from the original?

    b.   On the Lesson 1.2.1 Resource Page provided by your teacher, use your
          visualization skills to imagine the reflection of each shape across the given
          line of reflection. Then draw the reflection. Check your work by folding the
          paper along the line of reflection.

(1)   (2)   (3)

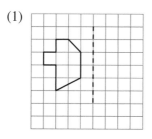   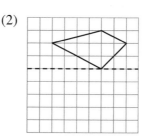   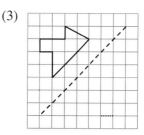

(4)   (5)   (6)

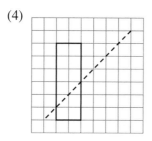   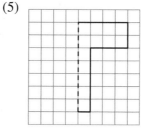   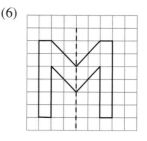

1-51.   Sometimes, a motion appears to be a reflection when it really isn't. How can you tell if a motion is a reflection? Consider each pair of objects below. Which diagrams represent reflections across the given lines of reflection? Study each situation carefully and be ready to explain your thinking.

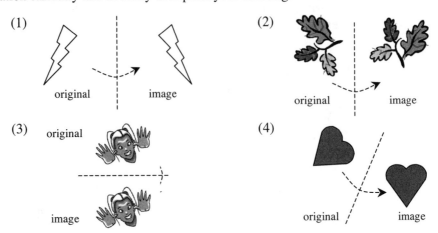

1-52.   CONNECTIONS WITH ALGEBRA

What other ways can you use reflections? Consider how to reflect a graph as you answer the questions below.

a.   On your Lesson 1.2.1 Resource Page, graph the parabola $y = x^2 + 3$ and the line $y = x$ for $x = -3, -2, -1, 0, 1, 2, 3$ on the same set of axes.

b.   Now reflect the parabola over the line $y = x$. What do you observe? What happens to the x- and y-values of the original parabola?

1-53.   LEARNING LOG

Throughout this course, you will be asked to reflect on your understanding of mathematical concepts in a Learning Log. Your Learning Log will contain explanations and examples to help you remember what you have learned throughout the course. It is important to write each entry of the Learning Log in your own words so that later you can use your Learning Log as a resource to refresh your memory. Your teacher will tell you where to write your Learning Log entries and how to structure or label them. Remember to label each entry with a title and a date so that it can be referred to later.

In this first Learning Log entry, describe what you learned today. For example, is it possible to reflect any shape? Is it possible to have a shape that, when reflected, doesn't change? How does reflection work? If it helps you to explain, sketch and label pictures to illustrate what you write. Title this entry "Reflections" and include today's date.

METHODS AND **M**EANINGS

MATH NOTES

Graphing an Equation

In Algebra, you learned how to graph an equation. During this course, you will apply your algebra skills to solve geometric problems. Review how to graph an equation by reading the example below.

- **Create a table of *x*-values.** Choose *x*-values that will show you any important regions of the graph of the equation. If you do not know ahead of time what the graph will look like, use the values $-4 \le x \le 4$ as shown at right.

$y = -x - 2$

x	−4	−3	−2	−1	0	1	2	3	4
y	2	1	0	−1	−2	−3	−4	−5	−6

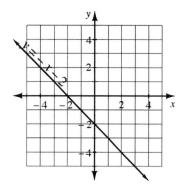

- **Use the equation to find *y*-values.** Substitute each value of *x* into your equation and find the corresponding *y*-value. For example, for $y = -x - 2$, when $x = -3$, $y = -(-3) - 2 = 1$.

- **Graph the points using the coordinates from your table onto a set of $x \rightarrow y$ axes.** Connect the points and, if appropriate, use arrows to indicate that the graph of the equation continues in each direction.

- **Complete the graph.** Be sure your axes are scaled and labeled. Also label your graph with its equation as shown in the example above.

1-54.   Review how to graph by reading the Math Notes box for this lesson. Then graph each line below on the same set of axes.

a. $y = 3x - 3$ b. $y = -\frac{2}{3}x + 3$ c. $y = -4x + 5$

1-55.   Estimate the size of each angle below to the nearest 10°. A
         right angle is shown at right for reference so you do not need a
         protractor.

a.   b.   c.

1-56.   The distance along a straight road is measured as shown in the diagram below. If the
         distance between towns A and C is 67 miles, find the distance between towns A
         and B.

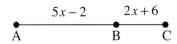

1-57.   For each equation below, solve for x. Show all work. The answers are provided so
         that you can check them. If you are having trouble with any solutions and cannot
         find your errors, you may need to see your teacher for extra help (you can ask your
         team as well). [**Solutions: a: 3.75, b: 3, c: 0, d: 3, e: $\approx$ 372.25, f: –3.4**]

a. $5x - 2x + x = 15$ b. $3x - 2 - x = 7 - x$

c. $3(x - 1) = 2x - 3 + 3x$ d. $3(2 - x) = 5(2x - 7) + 2$

e. $\frac{26}{57} = \frac{849}{5x}$ f. $\frac{4x+1}{3} = \frac{x-5}{2}$

1-58.   The three-dimensional shape at right is called a **cylinder**.
         Its bottom and top bases are both circles, and its side is
         perpendicular to the bases. What would the shape of a flag
         need to be in order to generate a cylinder when it rotates
         about its pole? (You may want to refer to problem 1-49 to
         review how flags work.)

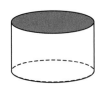

1.2.2 What if it is reflected more than once?

Rigid Transformations: Rotations and Translations

In Lesson 1.2.1, you learned how to change a shape by reflecting it across a line, like the ice cream cones shown at right. Today you will learn more about reflections and learn about two new types of transformations: rotations and translations.

original   image

1-59.   As Amanda was finding reflections, she wondered, "What if I reflect a shape twice over parallel lines?" **Investigate** her question as you answer the questions below.

a.   On the Lesson 1.2.2 Resource Page provided by your teacher, find △*ABC* and lines *n* and *p* (shown below). What happens when △*ABC* is reflected across line *n* to form △*A'B'C'* and then △*A'B'C'* is reflected across line *p* to form △*A"B"C"*? First **visualize** the reflections and then test your idea of the result by **drawing** both reflections.

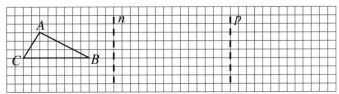

b.   **Examine** your result from part (a). Compare the original triangle △*ABC* with the final result, △*A"B"C"*. What single motion would change △*ABC* to △*A"B"C"*?

c.   Amanda analyzed her results from part (a). "It looks like I could have just slid △*ABC* over!" Sliding a shape from its original position to a new position is called **translating**. For example, the ice cream cone at right has been **translated**. Notice that the image of the ice cream cone has the same *orientation* as the original (that is, it is not turned or flipped). What words can you use to describe a translation?

original   image

d.   The words **transformation** and **translation** sound alike and can easily be confused. Discuss in your team what these words mean and how they are related to each other.

1-60.   After answering Amanda's question in problem 1-59, her teammate asks, "What if the lines of reflection are not parallel? Is the result still a translation?" Find △*EFG* and lines *v* and *w* on the Lesson 1.2.2 Resource Page.

a.   First **visualize** the result when △*EFG* is reflected over *v* to form △*E'F'G'*, and then △*E'F'G'* is reflected over *w* to form △*E"F"G"*. Then **draw** the resulting reflections on the resource page. Is the final image a translation of the original triangle? If not, describe the result.

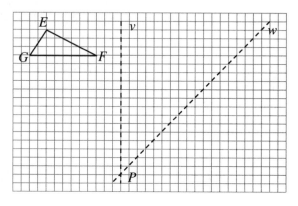

b.   Amanda noticed that when the reflecting lines are not parallel, the original shape is **rotated**, or turned, to get the new image. For example, the diagram at right shows the result when an ice cream cone is rotated about a point.

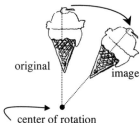

original   image

center of rotation

In part (a), the center of rotation is at point *P*, the point of intersection of the lines of reflection. Use a piece of tracing paper to test that △*E"F"G"* can be obtained by rotating △*EFG* about point *P*. To do this, first trace △*EFG* and point *P* on the tracing paper. While the tracing paper and resource page are aligned, apply pressure on *P* so that the tracing paper can rotate about this point. Then turn your tracing paper until △*EFG* rests atop △*E"F"G"*.

c.   The rotation of △*EFG* in part (a) is an example of a 90° clockwise rotation. The term "clockwise" refers to a rotation that follows the direction of the hands of a clock, namely ↻. A rotation in the opposite direction (↺) is called "counter-clockwise."

On your resource page, rotate the "block L" 90° counter-clockwise (↺) about point *Q*.

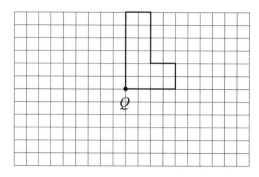

Geometry Connections

1-61. NOTATION FOR TRANSFORMATIONS

In the diagram at right, the original square
ABCD on the left was *translated* to the
image square on the right. (We label a shape
by labeling each corner.) The image location
is different from the original, so we use
different letters to label its **vertices**
(corners).

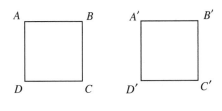

To keep track of how the vertices correspond, we call the image *A′B′C′D′*. The ′
symbol is read as "prime," so the shape on the right is called, "*A* prime *B* prime *C*
prime *D* prime." *A′* is the image of *A*, *B′* is the image of *B*, etc. This notation tells
you which vertices correspond.

a. The diagram at right shows a different
    transformation of *ABCD*. Look
    carefully at the correspondence
    between the vertices. Can you rotate
    or reflect the original square to make
    the letters correspond as shown? If
    you can reflect, where would the line
    of reflection be? If you can rotate,
    where would the point of rotation be?

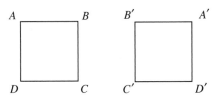

b. The diagram at right shows a rotation.
    Copy the diagram (both squares and the
    point) and label the corners of the image
    square on the right. If you have trouble,
    ask your teacher for tracing paper.
    Describe the rotation.

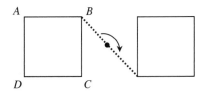

1-62. CONNECTIONS WITH ALGEBRA

Examine how translations and rotations can affect graphs. On graph paper, graph
the equation $y = \frac{1}{2}x - 3$.

a. Draw the result if $y = \frac{1}{2}x - 3$ is translated up 4 units. You may want to use
    tracing paper. Write the equation of the result. What is the relationship of
    $y = \frac{1}{2}x - 3$ and its image?

b. Now use tracing paper to rotate $y = \frac{1}{2}x - 3$ 90° clockwise (↻) about the point
    (0, 0). Write the equation of the result. What is the relationship of $y = \frac{1}{2}x - 3$
    and this new image?

MATH NOTES

Rigid Transformations

A transformation that preserves the size, shape, and orientation of a figure while *sliding* it to a new location is called a **translation.**

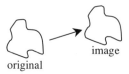

original image

A transformation that preserves the size and shape of a figure across a line to form a mirror image is called a **reflection**. The mirror line is a **line of reflection**. One way to find a reflection is to *flip* a figure over the line of reflection.

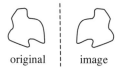

original image

A transformation that preserves the size and shape while *turning* an entire figure about a fixed point is called a **rotation**. Figures can be rotated either clockwise (↻) or counterclockwise (↺).

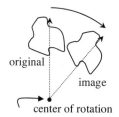

original image center of rotation

When labeling a transformation, the new figure (image) is often labeled with **prime notation**. For example, if $\triangle ABC$ is reflected across the vertical dashed line, its image can be labeled $\triangle A'B'C'$ to show exactly how the new points correspond to the points in the original shape. We also say that $\triangle ABC$ is **mapped** to $\triangle A'B'C'$.

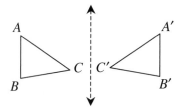

1-63. The diagram at right shows a flat surface containing a line and a circle with no points in common. Can you **visualize** moving the line and/or circle so that they intersect at exactly one point? Two points? Three points? Explain each answer and illustrate each with an example when possible.

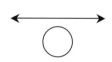

Geometry Connections

1-64.   Decide which transformation was used on each pair of shapes below. Some may
         have undergone more than one transformation.

a. 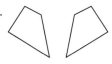   b. 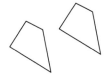   c.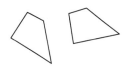

d.    e.    f.

1-65.   The **perimeter** (the sum of the side lengths) of the
         triangle at right is 52 units. Write and solve an
         equation based on the information in the diagram.
         Use your solution for x to find the measures of each
         side of the triangle. Be sure to confirm that your
         answer is correct.

1-66.   Bertie placed a transparent grid made up of unit squares over each of the shapes she
         was measuring below. Using her grid, approximate the area of each region.

a. 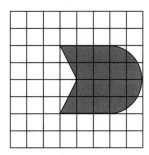   b.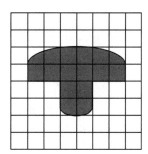

1-67.   For each equation below, find y if $x = -3$.

a. $y = -\frac{1}{3}x - 5$ b. $y = 2x^2 - 3x - 2$ c. $2x - 5y = 4$

1.2.3 How can I move it?

• •

Using Transformations

In Lesson 1.1.1, your class made a quilt using designs based on a geometric shape. Similarly, throughout American history, quilters have created quilts that use transformations to create intricate geometric designs. For example, the quilt at right is an example of a design based on rotation and reflection, while the quilt at left contains translation, rotation, and reflection.

Photo courtesy of the artist.

Sue Sales, *Hearts.*

Photo courtesy of the artist.

Sue Sales, *Balance From Within.*

Today, you will work with your team to discover ways to find the image of a shape after it is rotated, translated, or reflected.

1-68. Here are some situations that occur in everyday life. Each one involves one or more of the basic transformations: reflection, rotation or translation. State the transformation(s) involved in each case.

a. You look in a mirror as you comb your hair.

b. While repairing your bicycle, you turn it upside down and spin the front tire to make sure it isn't rubbing against the frame.

c. You move a small statue from one end of a shelf to the other.

d. You flip your scrumptious buckwheat pancakes as you cook them on the griddle.

e. The bus tire spins as the bus moves down the road.

f. You examine the footprints made in the sand as you walked on the beach.

1-69.   ROTATIONS ON A GRID

Consider what you know about rotation, a motion that turns a shape about a point. Does it make any difference if a rotation is clockwise (↻) versus counterclockwise (↺)? If so, when does it matter? Are there any circumstances when it does not matter? And are there any situations when the rotated image lies exactly on the original shape?

Investigate these questions as you rotate the shapes below about the given point on the Lesson 1.2.3 Resource Page. Use tracing paper if needed. Be prepared to share your answers to the questions posed above.

a.   180° ↺   b.   180° ↻   c.   90° ↻   d.   90° ↺

e.   270° ↺   f.   360° ↻   g.   180°   h.   90°

1-70.   TRANSLATIONS ON A GRID

The formal name for a slide is a translation. (Remember that translation and transformation are different words.) $\triangle A'B'C'$ at right is the result of translating $\triangle ABC$.

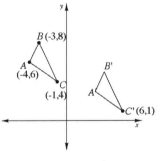

a.   Describe the translation. That is, how many units to the right and how many units down does the translation move the triangle?

b.   On graph paper, plot $\triangle EFG$ if $E(4, 5)$, $F(1, 7)$, and $G(2, 0)$. Find the coordinates of $\triangle E'F'G'$ if $\triangle E'F'G'$ is translated the same way as $\triangle ABC$ was in part (a).

c. $\triangle X'Y'Z'$ is the result of performing the same translation on $\triangle XYZ$. If $X'(2, -3)$, $Y'(4, -5)$, and $Z'(0, 1)$, name the coordinates of X, Y, and Z.

1-71.   Using tracing paper, reflect $\triangle ABC$ across line l at right to form $\triangle A'B'C'$. One way to do this is to trace the triangle and the line of reflection. Fold the tracing paper along the line of reflection and trace the triangle on the other side of the line of reflection. Then unfold the tracing paper and observe the original triangle and its reflection. Label the reflection $\triangle A'B'C'$.

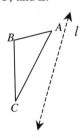

Problem continues on next page →

1-71. *Problem continued from previous page.*

Use your ruler to draw three dashed lines, each one joining one vertex of the original triangle with its image in the reflected triangle. Describe any geometric relationships you notice between each dashed line and the line of reflection.

1-72. How is the result from problem 1-71 useful? Consider reflecting a **line segment** (the portion of a line between two points) across a line that passes through one of its endpoints. For example, what would be the result when $\overline{AB}$ is reflected across $\overleftrightarrow{BC}$?

a. Copy $\overline{AB}$ and $\overleftrightarrow{BC}$ and draw $\overline{A'B}$, the reflection of $\overline{AB}$. When points A and A' are connected, what shape is formed?

b. Use what you know about reflection to make as many statements as you can about the shape from part (a). For example, are there any sides that must be the same length? Are there any angles that must be equal? Is there anything else special about this shape?

c. When two sides of a triangle have the same length, that triangle is called **isosceles**. In your Learning Log, describe all the facts you know about isosceles triangles. Be sure to include a diagram. Label this entry "Isosceles Triangles" and include today's date.

Ⓜ ETHODS AND MEANINGS

More on Reflections

When a figure is reflected across a line of reflection, such as the quadrilateral at right, it appears that the figure is "flipped" over the line.

However, there are other interesting relationships between the original figure and its image.

For instance, the line segment connecting each image point with its corresponding point on the original figure is perpendicular to (meaning that it forms a right angle with) the line of reflection.

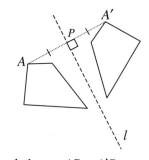

In addition, the line of reflection bisects (cuts in half) the line segment connecting each image point with its corresponding point on the original figure. For example, for reflection of the quadrilateral above, $AP = A'P$.

1-73.   Plot the following points on another sheet of graph paper and connect them in the order given. Then connect points A and D.

$$A(5, 3), B(5, -6), C(-4, -6), \text{ and } D(-4, 3)$$

a.   What is the resulting figure?

b.   Find the area of shape *ABCD*.

c.   If *ABCD* is rotated 90° clockwise (↻) about the origin to form $A'B'C'D'$, what are the coordinates of the vertices of $A'B'C'D'$?

1-74.   Solve for the variable in each equation below. Show the steps leading to your answer.

a. $8x - 22 = -60$ b. $\frac{1}{2}x - 37 = -84$

c. $\frac{3x}{4} = \frac{6}{7}$ d. $9a + 15 = 10a - 7$

1-75.   Graph the rule $y = -\frac{3}{2}x + 6$ on graph paper. Label the points where the line intersects the *x*- and *y*-axes.

1-76.   On graph paper, graph the line through the point (0, –2) with slope $\frac{4}{3}$.

a.   Write the equation of the line.

b.   Translate the graph of the line up 4 and to the right 3 units. What is the result? Write the equation for the resulting line.

c.   Now translate the original graph down 5 units. What is the result? Write the equation for the resulting line.

1-77.   Evaluate the expression $\frac{1}{4}k^5 - 3k^3 + k^2 - k$ for $k = 2$.

1.2.4 What shapes can I create with triangles?

Using Transformations to Create Shapes

In Lesson 1.2.3, you practiced reflecting, rotating and translating figures. Since these are rigid transformations, the image always had the same size and shape as the original. In this lesson, you will combine the image with the original to make new, larger shapes from four basic "building-block" shapes.

As you create new shapes, consider what information the transformation gives you about the resulting new shape. By the end of this lesson, you will have generated most of the shapes that will be the focus of this course.

1-78. THE SHAPE FACTORY

The Shape Factory, an innovative new company, has decided to branch out to include new shapes. As Product Developers, your team is responsible for finding exciting new shapes to offer your customers. The current company catalog is shown at right.

Since your boss is concerned about production costs, you want to avoid buying new machines and instead want to reprogram your current machines.

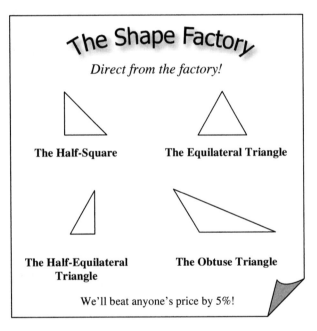

The factory machines not only make all the shapes shown in the catalog, but they also are able to rotate or reflect a shape. For example, if the half-equilateral triangle is rotated 180° about the **midpoint** (the point in the middle) of its longest side, as shown at right, the result is a rectangle.

Your Task: Your boss has given your team until the end of this lesson to find as many new shapes as you can. Your team's reputation, which has suffered recently from a series of blunders, could really benefit by an impressive new line of shapes formed by a triangle and its transformations. For each triangle in the catalog, determine which new shapes are created when it is rotated or reflected so that the image shares a side with the original triangle. Be sure to make as many new shapes as possible. Use tracing paper or any other reflection tool to help.

1-79. Since there are so many possibilities to test, it is useful to start by considering the shapes that can be generated just from one triangle. For example:

    a. Test what happens when the half-square is reflected across each side. For each result (original plus image), draw a diagram and describe the shape that you get. If you know a name for the result, state it.

    b. The point in the middle of each side is called its middle point, or **midpoint** for short. Try rotating the half-square 180° about the midpoint of each side to make a new shape. For each result, draw a diagram. If you know its name, write it near your new shape.

    c. Repeat parts (a) and (b) with each of the other triangles offered by the Shape Factory.

―――――――― *Further Guidance* ――――――――
section ends here.

1-80. EXTENSIONS

What other shapes can be created by reflection and rotation? Explore this as you answer the questions below. You can **investigate** these questions in any order. Remember that the resulting shape includes the original shape and all of its images. Remember to record and name each result.

• What if you reflect an equilateral triangle twice, once across one side and another time across a different side?

• What if an equilateral triangle is repeatedly rotated about one vertex so that each resulting triangle shares one side with another until new triangles are no longer possible? Describe the resulting shape.

• What if you rotate a trapezoid 180° around the midpoint of one if its non-parallel sides?

1-81. BUILDING A CATALOG

Your boss now needs you to create a catalog page that includes your shapes. Each entry should include a diagram, a name, and a description of the shape. List any special features of the shape, such as if any sides are the same length or if any angles must be equal. Use color and arrows to highlight these attributes in the diagram.

METHODS AND MEANINGS

Polygons

A **polygon** is defined as a two-dimensional closed figure made up of straight line segments connected end-to-end. These segments may not cross (intersect) at any other points.

At right are some examples of polygons.

Shape A at right is an example of a **regular polygon** because its sides are all the same length and its angles have equal measure.

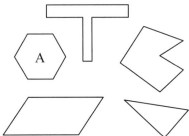

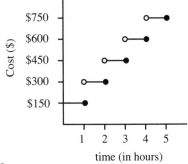

=== Review & Preview ===

1-82.   According to the graph at right, how much money would it cost to speak to an attorney for 2 hours and 25 minutes?

1-83.   Lourdes has created the following challenge for you: She has given you three of the four points necessary to determine a rectangle on a graph. She wants you to find the points that "complete" each of the rectangles below.

    a.   (−1, 3), (−1, 2), (9, 2)

    b.   (3, 7), (5, 7), (5, −3)

    c.   (−5, −5), (1, 4), (4, 2)

    d.   (−52, 73), (96, 73), (96, 1483)

1-84.   Find the area of the rectangles formed in parts (a), (b), and (d) of problem 1-83.

Geometry Connections

1-85.   Copy the diagrams below on graph paper. Then find the result when each indicated transformation is performed.

a.   Reflect Figure A across line *l*.

b.   Rotate Figure B 90° clockwise (↻) about point P.

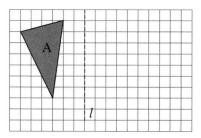

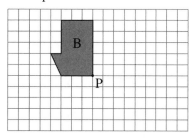

c.   Reflect Figure C across line *m*.

d.   Rotate Figure D 180° about point Q.

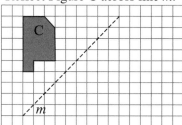

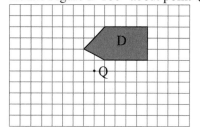

1-86.   At right is a diagram of a **regular hexagon** with center *C*. A polygon is regular if all sides are equal and all angles are equal. Copy this shape on your paper, then answer the questions below.

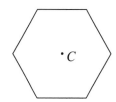

a.   Draw the result of rotating the hexagon about its center 180°. Explain what happened. When this happens, the shape has **rotation symmetry**.

b.   What is the result when the original hexagon is reflected across line *n*, as shown at right? A shape with this quality is said to have **reflection symmetry** and line *n* is a **line of symmetry** of the hexagon (not of the reflection).

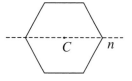

c.   Does a regular hexagon have any other lines of symmetry? That is, are there any other lines you could fold across so that both halves of the hexagon will match up? Find as many as you can.

1.2.5 What shapes have symmetry?

Symmetry

You have encountered symmetry several times in this chapter. For instance, the quilt your class created in Lesson 1.1.1 contained symmetry. The shapes you saw in the hinged mirrors during the kaleidoscope investigation were also symmetric. But so far, you have not developed a test for determining whether a polygon is symmetric. And since symmetry is related to transformations, how can we use transformations to describe this relationship? This lesson is designed to deepen your understanding of symmetry.

By the end of this lesson, you should be able to answer these target questions:

What is symmetry?

How can I determine whether or not a polygon has symmetry?

What types of symmetry can a shape have?

1-87.   REFLECTION SYMMETRY

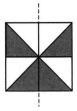

In problem 1-1, you created a quilt panel that had **reflection symmetry** because if the design were reflected across the line of symmetry, the image would be exactly the same as the original design. See an example of a quilt design that has reflection symmetry at right.

Obtain the Lesson 1.2.5 Resource Page. On it, **examine** the many shapes that will be our focus of study in this course. Which of these shapes have reflection symmetry? Consider this as you answer the questions below.

a.   For each shape on the resource page, draw all the possible lines of symmetry. If you are not sure if a shape has reflection symmetry, trace the shape onto tracing paper and try folding the paper so that both sides of the fold match.

b.   Which types of triangles have reflection symmetry?

c.   Which types of quadrilaterals (shapes with four sides) have reflection symmetry?

d.   Which shapes on your resource page have more than three lines of symmetry?

Geometry Connections

1-88.   ROTATION SYMMETRY

In problem 1-87, you learned that many shapes have reflection symmetry. These
shapes remain unaffected when they are reflected across a line of symmetry.
However, some shapes remain unchanged when rotated about a point.

a.   **Examine** the shape at right. Can this shape be
      rotated so that its image has the same position
      and orientation as the original shape? Trace
      this shape on tracing paper and test your
      conclusion. If it is possible, where is the point
      of rotation?

b.   Jessica claims that she can rotate <u>all</u>
      shapes in such a way that they will not
      change. How does she do it?

c.   Since all shapes can be rotated 360°
      without change, that is not a very
      special quality. However, the shape in
      part (a) above was special because it
      could be rotated less than 360° and
      still remain unchanged. A shape with
      this quality is said to have **rotation
      symmetry**.

      But what shapes have rotation symmetry? **Examine** the shapes on your Lesson
      1.2.5 Resource Page and identify those that have rotation symmetry.

d.   Which shapes on the resource page have 90° rotation symmetry? That is,
      which can be rotated about a point 90° and remain unchanged?

1-89.   TRANSLATION SYMMETRY

In problems 1-87 and 1-88, you identified shapes that have reflection and rotation
symmetry. What about translation symmetry? Is there an object that can be
translated so that its end result is exactly the same as the original object? If so, draw
an example and explain why it has **translation symmetry**.

1-90.   CONNECTIONS WITH ALGEBRA

During this lesson, you have focused on the types of symmetry that can exist in geometric objects. But what about shapes that are created on graphs? What types of graphs have symmetry?

a.   **Examine** the graphs below. Decide which have reflection symmetry, rotation symmetry, translation symmetry, or a combination of these.

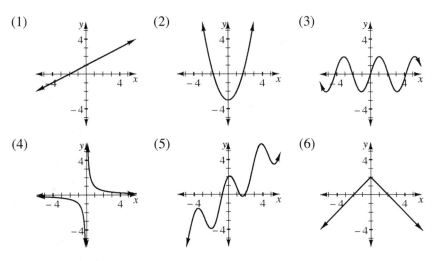

b.   If the y-axis is a line of symmetry of a graph, then its rule is referred to as **even**. Which of the graphs in part (a) have an even rule?

c.   If the graph has rotation symmetry about the origin (0, 0), its rule is called **odd**. Which of the graphs in part (a) have an odd rule?

1-91.   Reflect on what you have learned during this lesson. In your Learning Log, answer the questions posed at the beginning of this lesson, reprinted below. When helpful, give examples and draw a diagram.

What is symmetry?

How can I determine whether or not a polygon has symmetry?

What types of symmetry can a shape have?

MᴇTHODS AND Mᴇᴀɴɪɴɢꜱ

MATH NOTES

Slope of a Line and
Parallel and Perpendicular Slopes

During this course, you will use your algebra tools to learn more about shapes. One of your algebraic tools that can be used to learn about the relationship of lines is slope. Review what you know about slope below.

The **slope** of a line is the ratio of the change in y (Δy) to the change in x (Δx) between any two points on the line. It indicates both how steep the line is and its direction, upward or downward, left to right.

$$\text{slope} = \frac{\text{vertical change}}{\text{horizontal change}} = \frac{\Delta y}{\Delta x}$$

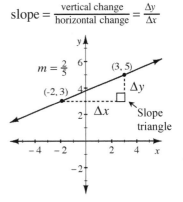

Lines that point upward from left to right have positive slope, while lines that point downward from left to right have negative slope. A horizontal line has zero slope, while a vertical line has undefined slope. The slope of a line is denoted by the letter m when using the $y = mx + b$ equation of a line.

One way to calculate the slope of a line is to pick two points on the line, draw a slope triangle (as shown in the example above), determine Δy and Δx, and then write the slope ratio. Be sure to verify that your slope correctly resulted in a negative or positive value based on its direction.

Parallel lines lie in the same plane (a flat surface) and never intersect. They have the same steepness, and therefore they grow at the same rate. Lines l and n at right are examples of parallel lines.

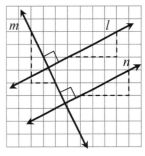

On the other hand, **perpendicular lines** are lines that intersect at a right angle. For example, lines m and n at right are perpendicular, as are lines m and l. Note that the small square drawn at the point of intersection indicates a right angle.

The **slopes of parallel lines** are the same. In general, the slope of a line parallel to a line with slope m is m.

The **slopes of perpendicular lines** are opposite reciprocals. For example, if one line has slope $\frac{4}{5}$, then any line perpendicular to it has slope $-\frac{5}{4}$. If a line has slope -3, then any line perpendicular to it has slope $\frac{1}{3}$. In general, the slope of a line perpendicular to a line with slope m is $-\frac{1}{m}$.

1-92.   On graph paper, graph each of the lines below on the same set of axes. What is the relationship between lines (a) and (b)? What about between (b) and (c)?

   a. $y = \frac{1}{3}x + 4$ b. $y = -3x + 4$ c. $y = -3x - 2$

1-93.   The length of a side of a square is $5x + 2$ units. If the perimeter is 48 units, complete the following.

   a.   Write an equation to represent this information.

   b.   Solve for x.

   c.   What is the area of the square?

1-94.   When the shapes below are reflected across the given line of reflection, the original shape and the image (reflection) create a new shape. For each reflection below, name the new shape that is created.

   a.   b.   c.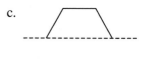

   d.   Use this method to create your own shape that has reflection symmetry. Add additional lines of symmetry. Note that the dashed lines of reflection in the figures above become lines of symmetry in the new shape.

1-95.   If a triangle has two equal sides, it is called **isosceles** (pronounced eye-SOS-a-lees). "Iso" means "same" and "sceles" is related to "scale." Decide whether each triangle formed by the points below is isosceles. Explain how you decided.

   a.   (6, 0), (0, 6), (6, 6)   b.   (–3, 7), (–5, 2), (–1, 2)

   c.   (4, 1), (2, 3), (9, 2)   d.   (1, 1), (5, –3), (1, –7)

1-96.   Copy the diagrams below on graph paper. Then find each result when each indicated transformation is performed.

a.   Reflect A across line *l*.

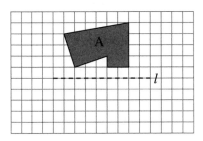

b.   Rotate B 90° counterclockwise (↺) about point P.

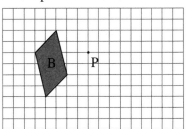

c.   Rotate C 180° about point Q.

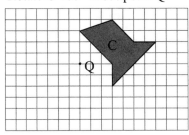

d.   Reflect D across line *m*.

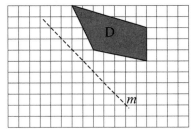

1.3.1 How can I classify this shape?

Attributes and Characteristics of Shapes

In Lesson 1.2.4, you generated a list of shapes formed by triangles and in Lesson 1.2.5, you studied the different types of symmetries that a shape can have. In Section 1.3, you will continue working with shapes to learn more about their attributes and characteristics. For example, which shapes have sides that are parallel? And which of our basic shapes are equilateral?

By the end of this lesson you should have a greater understanding about the attributes that make shapes alike and different. Throughout the rest of this course you will study these qualities that set shapes apart as well as learn how shapes are related.

1-97.   INTRODUCTION TO THE SHAPE BUCKET

Obtain a Shape Bucket and a Lesson 1.3.1B Resource Page from your teacher. The
Shape Bucket contains most of the basic geometric shapes you will study in this
course. Count the items and verify that you have all 16 shapes. Take the shapes out
and notice the differences between them. Are any alike? Are any strangely
different?

Once you have examined the shapes in your bucket, work as a team to build the
composite figures below (also shown on the resource page). Composite figures are
made by combining two or more shapes to make a new figure. On the Lesson 1.3.1B
Resource Page, show the shapes you used to build the composite shapes by filling in
their outlines within each composite shape.

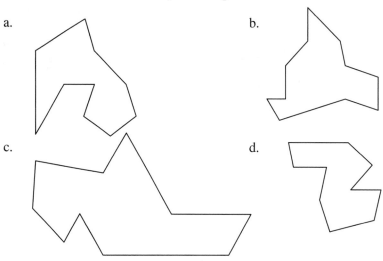

a.

b.

c.

d.

1-98.   VENN DIAGRAMS

Obtain a Venn diagram from your teacher.

a.   The **left** circle of
      the Venn
      diagram, Circle
      #1, will represent
      the attribute "has
      at least one pair
      of parallel sides"

**#1: Has at
least one pair
of parallel
sides**

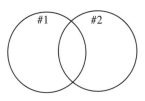

**#2: Has at
least two sides
of equal
length**

      and the **right** side, Circle #2, will represent the attribute "has at least two sides
      of equal length" as shown above. Sort through the shapes in the Shape Bucket
      and decide as a team where each shape belongs. Be sure to record your
      solution on paper. As you discuss this problem with your teammates, **justify**
      your statements with reasons such as, "I think this shape goes here **because**…"

Problem continues on next page →

Geometry Connections

1-98.   *Problem continued from previous page.*

b.   Next, reclassify the shapes for the new Venn diagram shown below. Describe each region in a sentence.

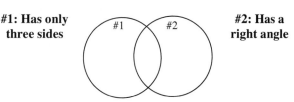

#1: Has only three sides   #1   #2   **#2: Has a right angle**

c.   Finally, reclassify the shapes for the new Venn diagram shown below. Describe each region in a sentence.

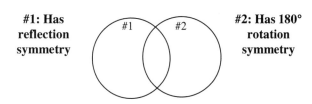

#1: Has reflection symmetry   #1   #2   **#2: Has 180° rotation symmetry**

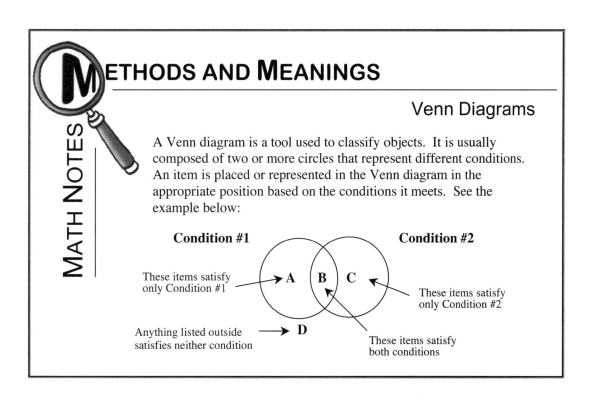

METHODS AND MEANINGS

MATH NOTES

Venn Diagrams

A Venn diagram is a tool used to classify objects. It is usually composed of two or more circles that represent different conditions. An item is placed or represented in the Venn diagram in the appropriate position based on the conditions it meets. See the example below:

Condition #1   **Condition #2**

These items satisfy only Condition #1 → **A** **B** **C** ← These items satisfy only Condition #2

Anything listed outside satisfies neither condition → **D**

These items satisfy both conditions

1-99. Copy the Venn diagram below on your paper. Then show where each person described should be represented in the diagram. If a portion of the Venn diagram remains empty, describe the qualities a person would need to belong there.

#1: Studies a lot for class 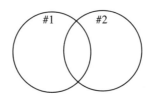 **#2: Has long hair**

a. Carol: *"I rarely study and enjoy braiding my long hair."*

b. Bob: *"I never do homework and have a crew cut."*

c. Pedro: *"I love joining after school study teams to prepare for tests and I like being bald!"*

1-100. Solve the equations below for x, if possible. Be sure to check your solution.

a. $\frac{3x-1}{4} = -\frac{5}{11}$

b. $(5-x)(2x+3)=0$

c. $6-5(2x-3)=4x+7$

d. $\frac{3x}{4}+2=4x-1$

1-101. **Examine** the figure graphed on the axes at right.

a. What happens when you rotate this figure about the origin 90°? 45°? 180°?

b. What other angle could the figure at right be rotated so that the shape does not appear to change?

c. What shape will stay the same no matter how many degrees it is rotated?

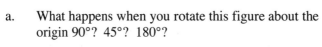

1-102. Copy $\triangle ABC$ at right on graph paper.

a. Rotate $\triangle ABC$ 90° counter-clockwise ($\circlearrowleft$) about the origin to create $\triangle A'B'C'$. Name the coordinates of C'.

b. Reflect $\triangle ABC$ across the vertical line $x=1$ to create $\triangle A''B''C''$.

c. Translate $\triangle ABC$ so that A''' is at (4, –5). Name the coordinates of B'''.

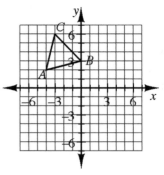

Geometry Connections

1-103. Copy the table below, complete it, and write a rule relating *x* and *y*.

x	−3	−2	−1	0	1	2	3	4
y	−7			2	5			14

$1.3.2$ How can I describe it?

More Characteristics of Shapes

In Lesson 1.3.1, you used shapes to build new, unique, composite shapes. You also started to analyze the attributes (qualities) of shapes. Today you will continue to look at their attributes as you learn new vocabulary.

1-104. Using your Venn diagram Resource Page from Lesson 1.3.1, categorize the shapes from the Shape Bucket in the Venn diagram as shown below. Record your results on paper.

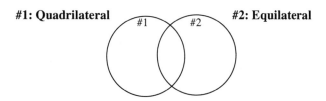

#1: Quadrilateral   #2: Equilateral

1-105. DESCRIBING A SHAPE

How can you describe a square? With your class, find a way to describe a square using its attributes (special qualities) so that anyone could draw a square based on your description. Be as complete as possible. You may not use the word "square" in the description.

Square

1-106. Each shape in the bucket is unique; that is, it differs from the others. You will be assigned a few shapes to describe **as completely as possible** for the class. But what is a complete description? As you work with your team to create a complete description, consider the questions below.

- What do you notice about your shape?

- What makes it different from other shapes?

- If you wanted to describe your shape to a friend on the telephone who cannot see it, what would you need to include in the description?

1-107.   GEOMETRY TOOLKIT

Obtain the Lesson 1.3.2A Resource Page entitled "Shapes Toolkit" from your
teacher.

a.   Label each shape that you have learned about so far with its geometrical name.

b.   In the space provided, describe the shape based on the descriptions generated
from problem 1-106. Leave space so that later observations can be added for
each shape. Note that the description for **rectangle** has been provided as an
example.

c.   On the diagram for each shape, mark sides that must have equal length or that
must be parallel. Also mark any angles that measure 90°. See the descriptions
below.

•   To show that two sides have the same length, use "tick marks" on the sides.
However, to show that one pair of equal sides may not be the same length as
the other pair of equal sides, you should use one tick mark on two of the
opposite, equal sides and two tick marks on the other two opposite, equal
sides, as shown below.

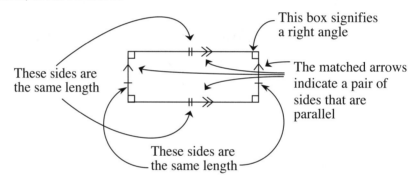

•   To show that the rectangle has two pairs of parallel sides, use one ">" mark
on one pair of parallel sides and two ">>" marks on the other two parallel
sides, as shown above.

•   Also mark any right angles by placing a small square at the right angle vertex
(the corner). See the example above.

d.   The Shapes Toolkit Resource Page is
the first page of a special information
organizer, called your Geometry
Toolkit, which you will be using for this
course. It is a reference tool that you
can use when you need to remember the
name or description of a shape. Find a
safe place in your Geometry binder to
keep your Toolkit.

54

1-108.   **Examine** the Venn diagram below.

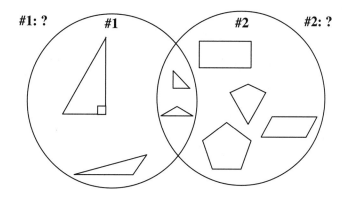

a.   What attribute does each circle represent? How can you tell?

b.   Where would the equilateral hexagon from your Shape Bucket go in this Venn? What about the trapezoid? Justify your **reasoning**.

c.   Create another shape that would belong outside both circles. Does your shape have a name that you have studied so far? If not, give it a new name.

1-109.   Elizabeth has a Venn diagram that she started at right. It turns out that the **only** shape in the Shape Bucket that could go in the intersection (where the two circles overlap) is a square! What are the possible attributes that her circles could represent? Discuss this with your team and be ready to share your ideas with the class.

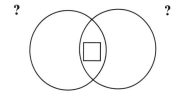

1-110.   If no sides of a triangle have the same length, the triangle is called **scalene** (pronounced SCALE-een). However, if the triangle has two sides that are the same length, the triangle is called **isosceles** (pronounced eye-SOS-a-lees). Use the markings in each diagram below to decide if △ABC is isosceles or scalene. Assume the diagrams are not drawn to scale.

a.

b.

c.   *ABDC* is a square

1-111. In Lesson 1.3.3, you will start to learn how to make predictions in mathematics using probability. Decide if the following events must happen, cannot happen, or possibly will happen.

    a.   You will drink water some time today.

    b.   An earthquake will strike in your region tomorrow.

    c.   You will read a book.

    d.   If a shape is a trapezoid, it will have a pair of parallel sides.

    e.   You will shave your head this month.

1-112. Without referring to your Shapes Toolkit, see if you can recall the names of each of the shapes below. Then check your answers with definitions from your Shapes Toolkit. How did you do?

a.

b.

c.

d.

e.

f.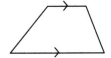

1-113. Copy the Venn diagram at right onto your paper. Then carefully place each capitalized letter of the alphabet below into your Venn diagram based on its type of symmetry.

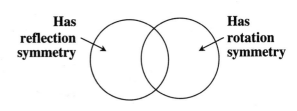

Has reflection symmetry   **Has rotation symmetry**

A, B, C, D, E, F, G, H, I, J, K, L, M, N, O, P, Q, R, S, T, U, V, W, X, Y, Z

1-114. **Multiple Choice:** Which equation below correctly represents the relationship of the sides given in the diagram at right?

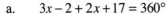

    a. $3x - 2 + 2x + 17 = 360°$

    b. $3x - 2 + 2x + 17 = 180°$

    c. $3x - 2 + 2x + 17 = 90°$

    d. $3x - 2 = 2x + 17$

1.3.3 What are the chances?

Introduction to Probability

You have learned a great deal about the shapes in the Shape Bucket and with that knowledge can draw conclusions and make predictions. For example, if you are told that a shape is not a triangle, you could predict that it might be a quadrilateral.

One mathematical tool that can be used to predict how likely it is that something will happen is **probability**, the formal word for chance. There are many examples of the uses of probability in our lives. One such example is the weather. If a forecast predicts that there is a 95% likelihood of rain, you would probably bring a raincoat and umbrella with you for that day. If you buy a lottery ticket, the small print on the back of your ticket usually tells you the probability of randomly selecting winning numbers, which, unfortunately, is very low.

Your task today is to learn the principles of probability so that you can make predictions.

1-115.   WHAT'S THE CHANCE OF THAT HAPPENING?

In your teams, discuss the following list of events and consider the likelihood of each event occurring:

1. Your school's basketball team having a winning season this year

2. The sky being cloudy tomorrow

3. Your math teacher assigning homework tonight

4. The sun setting this evening

5. Meeting a famous person on the way home

6. Snow falling in Arizona in July

a.   Copy the probability line shown below on your paper. Where should these events lie? Place each event listed above at the appropriate location on the probability line. Do any events have no chance of happening? These should be placed on the left end of the number line. If you are sure the event will happen, place it on the right end.

NO CHANCE OF HAPPENING	MIGHT HAPPEN	CERTAIN TO HAPPEN

Problem continues on next page →

1-115.   *Problem continued from previous page.*

   b.   With your team, make up three new situations and place them on the
         probability line. One event should be certain to happen, one event should have
         no chance of happening, and one should possibly occur.

   c.   **Probability** can be written as a ratio, as a decimal, or as a percent. When
         using ratios or decimals, probability is expressed using numbers from 0 to 1.
         An event that has no chance of happening is said to have a probability of 0. An
         event that is certain to happen is said to have a probability of 1. Events that
         "might happen" will have values somewhere between 0 and 1. See the Math
         Notes box at the end of this lesson for more information about probability.

         Which events from part (a) have a probability of 0? A probability of 1? Which
         event has closest to a 25% chance? 50% chance?

1-116.   Probability is used to make predictions. For example, if you were to reach into the
          Shape Bucket and randomly pull out a shape, you could use probability to predict the
          chances of the shape having a right angle. Since there are 16 total shapes and 4 that
          have right angles, the probability is:

$$P(\text{right angle}) = \frac{4 \text{ shapes with right angles}}{16 \text{ total shapes}} = \frac{4}{16} = \frac{1}{4} = 0.25 = 25\%$$

          The example above shows all forms of writing probability: $\frac{4}{16}$ (read "4 out of 16") is
          the probability as a ratio, 0.25 is its decimal form, and 25% is its equivalent percent.

          Find the probabilities of randomly selecting the following shapes from a Shape
          Bucket that contains all 16 basic shapes.

   a.   P(quadrilateral)   b.   P(shape with an obtuse angle)

   c.   P(equilateral triangle)   d.   P(shape with parallel sides)

1-117.   What else can you predict about the shape randomly pulled from the Shape Bucket?

   a.   In your teams, describe at least two new probability conditions for the shapes
         in the bucket (like those in problem 1-116). One of the events should have a
         probability of 0 or a probability of 1. Write each P(condition) on a separate
         blank piece of paper and give it to your teacher.

   b.   Your teacher will select conditions generated by the class in part (a). For each
         condition, work with your team to determine the probability.

Geometry Connections

1-118.   Laura's team has a shape bucket that is missing several pieces. While talking to her teammates, Laura claims, "The probability of choosing a shape with four right angles is ¼ ."

Barbara adds, "Well…there are 2 shapes that have four right angles."

Montry responds, "I think you are both right."

What does Montry mean? Explain how Laura and Barbara could both be correct.

1-119.   What else can probability be used to predict? Analyze each of the situations below:

a.   What is the probability of randomly drawing a face card (king, queen or jack) from a deck of 52 cards, when you draw one card?

b.   Eduardo has in his pocket $1 in pennies, $1 in nickels, and $1 in dimes. If he randomly pulls out just one coin, what is the probability that he will pull out a dime?

c.   P(rolling an 8) with one regular die if you roll the die just once.

d.   P(dart hitting a shaded region) if the dart is randomly thrown and hits the target at right.

target

1-120.   In your Learning Log, write an entry describing what you learned today about probability. Be sure to show an example. Title this entry "Probability" and include today's date.

METHODS AND MEANINGS

Ratio and Probability

MATH NOTES

A comparison of two quantities is called a **ratio**. A ratio can be written as:

$$a:b \quad or \quad \tfrac{a}{b} \quad or \quad "a \, to \, b"$$

Each ratio has a numeric value that can be expressed as a fraction or decimal. For example, a normal deck of 52 playing cards has 13 cards with hearts. To compare the number of cards with hearts to the total number of playing cards, you could write:

$$\frac{\text{Number of Hearts}}{\text{Total Number of Cards}} = \frac{13}{52} = \frac{1}{4} = 0.25 \, .$$

Probability is a measure of the likelihood that an event will occur at random. It is expressed using numbers with values that range from 0 to 1, or from 0% to 100%. For example, an event that has no chance of happening is said to have a probability of 0 or 0%. An event that is certain to happen is said to have a probability of 1 or 100%. Events that "might happen" have values somewhere between 0 and 1 or between 0% and 100%.

The probability of an event happening is written as the ratio of the number of ways that the desired outcome can occur to the total number of possible outcomes (assuming that each possible outcome is equally likely).

$$P(\text{event}) = \frac{\text{Number of Desired Outcomes}}{\text{Total Possible Outcomes}}$$

For example, on a standard die, P(5) means the probability of rolling a 5. To calculate the probability, first determine how many possible outcomes exist. Since a die has six different numbered sides, the number of possible outcomes is 6. Of the six sides, only one of the sides has a 5 on it. Since the die has an equal chance of landing on any of its six sides, the probability is written:

$$P(5) = \frac{1 \text{ side with the number five}}{6 \text{ total sides}} = \frac{1}{6} \text{ or } 0.1\overline{6} \text{ or approximately } 16.7\%$$

1-121.   **Examine** the shapes listed on your Shapes Toolkit. Name a shape that meets the following conditions: It has a pair of congruent sides, it has one right angle, and it has three vertices.

1-122.   Augustin is in line to choose a new locker at school. The locker coordinator has each student reach into a bin and pull out a locker number. There is one locker at the school that all the kids dread! This locker, # 831, is supposed to be haunted, and anyone who has used it has had strange things happen to him or her! When it is Augustin's turn to reach into the bin and select a locker number, he is very nervous. He knows that there are 535 lockers left and locker # 831 is one of them. What is the probability that Augustin reaches in and pulls out the dreaded locker # 831? Should he be worried? Explain.

1-123.   The four triangles at right are placed in a bag. If you reach into the bag without looking and pull out one triangle at random, what is the probability that:

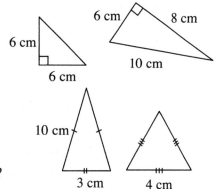

a.   the triangle is scalene?

b.   the triangle is isosceles?

c.   at least one side of the triangle is 6 cm?

1-124.   Rosalinda examined the angles at right and wrote the equation below.

$$(2x+1°)+(x-10°)=90°$$

a.   Does her equation make sense? If so, explain why her equation must be true. If it is not correct, determine what is incorrect and write the equation.

b.   If you have not already done so, solve her equation, clearly showing all your steps. What are the measures of the two angles?

c.   Verify that your answer is correct.

1-125.   Copy the table below on your paper and complete it for the rule $y = x^2 + 2x - 3$. Then graph and connect the points on graph paper. Name the x-intercepts.

x	−4	−3	−2	−1	0	1	2
y							

Chapter 1 Closure   What have I learned?

Reflection and Synthesis

The activities below offer you a chance to reflect on what you have learned in this chapter. As you work, look for concepts that you feel very comfortable with, ideas that you would like to learn more about, and topics you need more help with. Look for connections between ideas as well as connections with material you learned previously.

①   TEAM BRAINSTORM

With your team, brainstorm a list for each of the following two subjects. Be as detailed as you can. How long can you make your list? Challenge yourselves. Be prepared to share your team's ideas with the class.

Topics: What have you studied in this chapter? What ideas and words were important in what you learned? Remember to be as detailed as you can.

Connections: How are the topics, ideas, and words that you learned in previous courses connected to the new ideas in this chapter? Again, make your list as long as you can.

②   MAKING CONNECTIONS

The following is a list of the vocabulary used in this chapter. Make sure that you are familiar with all of these words and know what they mean. Refer to the glossary or index for any words that you do not yet understand.

angle	area	conjecture
equilateral	graph	image
line segment	measurement	parallel
perimeter	perpendicular	polygon
probability	protractor	random
ratio	reflection	regular hexagon
right angle	rotation	slope
solve	straight angle	symmetry
transformation	translation	triangle
Venn diagram	vertex (vertices)	

Problem continues on next page →

②   *Problem continued from previous page.*

Make a concept map showing all of the connections you can find among the key words and ideas listed above. To show a connection between two words, draw a line between them and explain the connection, as shown in the example below. A word can be connected to any other word as long as there is a justified connection. For each key word or idea, provide a sketch of an example.

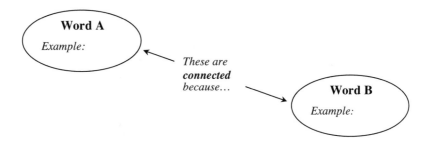

Your teacher may provide you with vocabulary cards to help you get started. If you use the cards to plan your concept map, be sure either to re-draw your concept map on your paper or to glue the vocabulary cards to a poster with all of the connections explained for others to see and understand.

While you are making your map, your team may think of related words or ideas that are not listed above. Be sure to include these ideas on your concept map.

③   SUMMARIZING MY UNDERSTANDING

This section gives you an opportunity to show what you know about certain math topics or ideas. Your teacher will direct you how to do this. Your teacher may give you a "GO" page to work on (or you can download this from www.cpm.org). "GO" stands for "Graphic Organizer," a tool you can use to organize your thoughts and communicate your ideas clearly.

④   WHAT HAVE I LEARNED?

This section will help you recognize those types of problems you feel comfortable with and those you need more help with. This section will appear at the end of every chapter to help you check your understanding. Even if your teacher does not assign this section, it is a good idea to try these problems and find out for yourself what you know and what you need to work on.

Solve each problem as completely as you can. The table at the end of this closure section has answers to these problems. It also tells you where you can find additional help and practice on similar problems.

CL 1-126. Trace the figures in parts (a) and (b) onto your paper and perform the indicated transformations. Copy the figure from part (c) onto graph paper and perform the indicated transformation. Label each image with prime notation ($A \rightarrow A'$).

a. Rotate *EFGHI* 90° clockwise ↻ about point *Z*

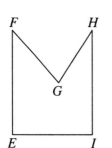

b. Reflect *JKLMN* over line *t*

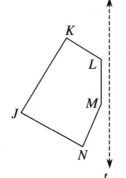

c. Translate *ABCD* down 5 units and right 3 units

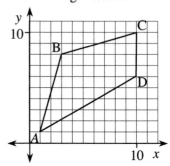

CL 1-127. Assume that all angles in the diagram at right are right angles and that all the measurements are in centimeters. Find the perimeter of the figure.

CL 1-128. Estimate the measures of the angles below. Are there any that you know for sure?

a.

b.

c.

d.

CL 1-129. **Examine** the angles in problem CL 1-128. If these four angles are placed in a bag, what is the probability of randomly selecting:

a. An acute angle

b. An angle greater than 60°

c. A 90° angle

d. An angle less than or equal to 180°

CL 1-130. **Examine** the shapes at right.

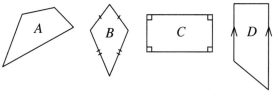

a. Describe what you know about each shape based on the information provided in the diagram.

b. Name the shape.

c. Decide where each shape would be placed in the Venn diagram at right.

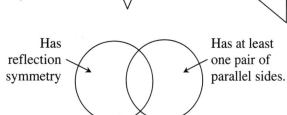

CL 1-131. Solve each equation below. Check your solution.

a. $3x - 12 + 10 = 8 - 2x$

b. $\frac{x}{7} = \frac{3}{2}$

c. $5 - (x + 7) + 4x = 7(x - 1)$

d. $x^2 + 11 = 36$

CL 1-132. Graph and connect the points in the table below. Then graph the equation in part (b) on the same set of axes. Also, find the equation for the data in the table.

a.

x	−4	−3	−2	−1	0	1	2	3	4	5	6
y	−5	−3	−1	1	3	5	7	9	11	13	15

b. $y = x^2 + x - 2$

CL 1-133. $\triangle ABC$ at right is equilateral. Use what you know about an equilateral triangle to write and solve an equation for x. Then find the perimeter of $\triangle ABC$.

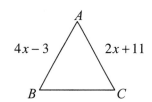

CL 1-134. Check your answers using the table at the end of this section. Which problems do you feel confident about? Which problems were hard? Have you worked on problems like these in math classes you have taken before? Use the table to make a list of topics you need help on and a list of topics you need to practice more.

HOW AM I THINKING?

This course focuses on five different **Ways of Thinking**: investigating, examining, reasoning and justifying, visualizing, and choosing a strategy/tool. These are some of the ways in which you think while trying to make sense of a concept or attempting to solve a problem (even outside of math class). During this chapter, you have probably used each Way of Thinking multiple times without even realizing it!

This closure activity will focus on one of these Ways of Thinking: **Investigating**. Read the description of this Way of Thinking at right.

Think about the **investigating** that you have done in this chapter. When have you tried something that you weren't specifically asked to do? What did you do when testing your idea did not work as you expected? You may want to flip through the chapter to refresh your memory about the problems that you have worked on. Discuss any of the **investigations** you have made with the rest of the class.

Once your discussion is complete, think about the way you think as you answer the following problems.

Investigating

Investigating is when you create your own questions and come up with something to do to answer those questions. You are often investigating when you ask questions for which you have not learned a formula or process. Common thoughts you might have when you are investigating are: *"I wonder what would happen if I . . . "* or, *"I might find the answer to my question if I . . . "* or, *"That didn't quite answer my question; now I'll try . . . "*

a.   In the Shape Factory, you **investigated** what shapes you could create with four basic triangles. But **what if** you had started with quadrilaterals instead? **Investigate** this question as you find all shapes that could be made by rotating or reflecting the four quadrilaterals below. Remember that once rotated or reflected, the image should share a side with the original shape. Use tracing paper to help generate the new shapes. If you know the name of the new figure, state it.

PARALLELOGRAM   **GENERAL TRAPEZOID**   **SQUARE**   **SPECIAL TRAPEZOID**

Problem continues on next page →

Geometry Connections

Problem continued from previous page.

b. Now write three "What if …?" questions that would extend your **investigation** in part (a).

c. Now choose <u>one</u> question from part (b) to **investigate**. Decide how you will **investigate** your question. For example, will you need tracing paper? Do you need to make a table and record information?

After you have answered your question, write the results of your **investigation** clearly so that someone else can understand what question you selected, how you **investigated** your question, and any conclusions you made.

Answers and Support for Closure Activity #4
What Have I Learned?

Problem	Solutions	Need Help?	More Practice
CL 1-126.	a. b.	Lesson 1.2.2, Lessons 1.2.2 and 1.2.3 Math Notes boxes, and Problem 1-50	Problems 1-50, 1-51, 1-61, 1-64, 1-68, 1-69, 1-70, 1-71, 1-73, 1-85, 1-86, 1-94, 1-96, 1-102

a.

E′ — F′

G′

I′ — H′

•Z

b.

K′

L′

M′

J′

N′

t

Problem	Solutions	Need Help?	More Practice
CL 1-126 continued	c.	Lesson 1.2.2, Lessons 1.2.2 and 1.2.3 Math Notes boxes, and Problem 1-50	Problems 1-50, 1-51, 1-61, 1-64, 1-68, 1-69, 1-70, 1-71, 1-73, 1-85, 1-86, 1-94, 1-96, 1-102
CL 1-127.	Perimeter = 30 centimeters	Lesson 1.1.3 Math Notes box	Problems 1-25, 1-44, 1-46, 1-65, 1-93
CL 1-128.	a. 90° b. ≈ 100° c. 180° d. ≈ 30°	Lesson 1.1.5 Math Notes box Problem 1-39	Problems 1-39, 1-42, 1-55
CL 1-129.	a. $\frac{1}{4}$ b. $\frac{3}{4}$ c. $\frac{1}{4}$ d. 1	Lesson 1.3.3 Math Notes box Problem 1-115	Problems 1-116, 1-118, 1-119, 1-122, 1-123
CL 1-130.	a. A: four sides make it a generic quadrilateral; B: Two pairs of equal sides make it a kite; C: Four right angles and a pair of parallel sides make it a rectangle; D: A quadrilateral with two parallel sides is a trapezoid. b.	Lesson 1.3.1 Math Notes box, Shapes Toolkit, catalog from problem 1-81, problem 1-107	Problems 1-3, 1-15, 1-86, 1-87, 1-94, 1-98, 1-99, 1-101, 1-104, 1-105, 1-108, 1-109, 1-110, 1-112, 1-113, 1-121, 1-123

Problem	Solutions	Need Help?	More Practice
CL 1-131.	a. $x = 2$ b. $x = \frac{21}{2}$ c. $x = \frac{5}{4}$ d. $x + = 5$	Lesson 1.1.4 Math Notes box	Problems 1-17, 1-25, 1-32, 1-34, 1-45, 1-57, 1-74, 1-100, 1-124
CL 1-132.	a. $y = 2x + 3$ 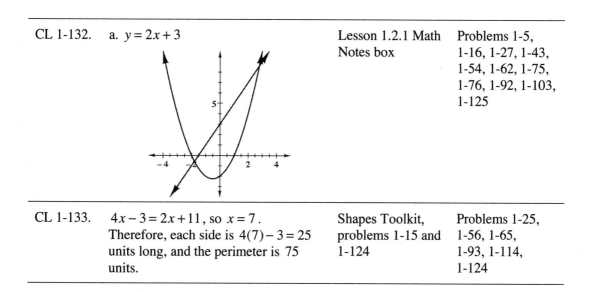	Lesson 1.2.1 Math Notes box	Problems 1-5, 1-16, 1-27, 1-43, 1-54, 1-62, 1-75, 1-76, 1-92, 1-103, 1-125
CL 1-133.	$4x - 3 = 2x + 11$, so $x = 7$. Therefore, each side is $4(7) - 3 = 25$ units long, and the perimeter is 75 units.	Shapes Toolkit, problems 1-15 and 1-124	Problems 1-25, 1-56, 1-65, 1-93, 1-114, 1-124

CHAPTER 2   Angles and Measurement

In Chapter 1, you studied many common geometric shapes and learned ways to describe a shape using its attributes. In this chapter, you will further **investigate** how to describe a complex shape by developing ways to accurately determine its angles, area, and perimeter. You will also use transformations from Chapter 1 to uncover special relationships between angles within a shape.

Throughout this chapter you will be asked to solve problems, such as those involving area or angles, in more than one way. This will require you to "see" shapes in multiple ways and to gain a broader understanding of problem solving.

Guiding Questions

Think about these questions throughout this chapter:

What is the relationship?

Can I generalize the process?

Is it always true?

What information do I need?

Is there another way?

In this chapter, you will learn:

➢ the relationships between pairs of angles formed by transversals and the angles in a triangle.

➢ how to find the area and perimeter of triangles, parallelograms, and trapezoids.

➢ the relationship among the three side lengths of a right triangle (the Pythagorean Theorem).

➢ how to estimate the value of square roots.

➢ how to determine when the lengths of three segments can and cannot form a triangle.

Chapter Outline

Section 2.1   You will broaden your understanding of angle, begun in Chapter 1, to include relationships between angles, such as those formed by intersecting lines or those inside a triangle.

Section 2.2   After examining how units of measure work, you will develop methods to find the areas of triangles, parallelograms, and trapezoids as well as more complicated shapes.

Section 2.3   You will discover a relationship among the sides of a right triangle called the Pythagorean Theorem. This will allow you to find the perimeter of triangles, parallelograms, and trapezoids, and to find the distance between two points on a graph.

2.1.1 What's the relationship?

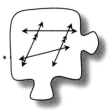

••

Complementary, Supplementary, and Vertical Angles

In Chapter 1, you compared shapes by looking at similarities between their parts. For example, two shapes might have sides of the same length or equal angles. In this chapter you will **examine** relationships between parts within a *single* shape or diagram. Today you will start by looking at angles to identify relationships in a diagram that make angle measures equal. As you **examine** angle relationships today, keep the following questions in mind to guide your discussion:

How can we name the angle?

What is the relationship?

How do we know?

2-1.   SOMEBODY'S WATCHING ME

Usually, to see yourself in a small mirror you have to be looking directly into it—if you move off to the side, you can't see your image any more. But Mr. Douglas knows a neat trick. He claims that if he makes a right angle with a hinged mirror, he can see himself in the mirror no matter from which direction he looks into it.

a.   By forming a right angle with a hinged mirror, test Mr. Douglas's trick for yourself. Look into the place where the sides of the mirror meet. Can you see yourself? What if you look in the mirror from a different angle?

b.   Does the trick work for *any* angle between the sides of the mirror? Change the angle between the sides of the mirror until you can no longer see your reflection where the sides meet.

c.   At right is a diagram of a student trying out the mirror trick. What appears to be true about the lines of sight? Can you explain why Mr. Douglas's trick works? Talk about this with your team and be ready to share your ideas with the class.

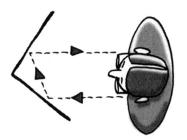

2-2. To completely understand how Mr. Douglas's reflection trick works, you need to learn more about the relationships between angles. But in order to clearly communicate relationships between angles, you will need a convenient way to refer to and name them. **Examine** the diagram of equilateral $\triangle ABC$ at right.

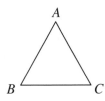

a. The "top" of this triangle is usually referred to as "angle A," written $\angle A$. Point A is called the **vertex** of this angle. The *measure* of $\angle A$ (the number of degrees in angle A) is written $m\angle A$. Since $\triangle ABC$ is equilateral, write an equation showing the relationship between its angles.

b. Audrey rotated $\triangle ABC$ around point A to form $\triangle AB'C'$. She told her teammate Maria, "I think the two angles at A are equal." Maria did not know which angles she was referring to. How many angles can you find at A? Are there more than three?

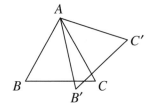

c. Maria asked Audrey to be more specific. She explained, "One of my angles is $\angle BAB'$." At the same time, she marked her two angles with the same marking at right to indicate that they have the same measure.

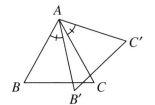

Name her other angle. Be sure to use three letters so there is no confusion about which angle you mean.

2-3. ANGLE RELATIONSHIPS

When you know two angles have a certain **relationship**, learning something about one of them tells you something about the other. Certain angle relationships come up often enough in geometry that we give them special names.

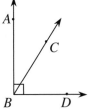

a. Two angles whose measures have a sum of 90° are called **complementary angles**. Since $\angle ABD$ is a right angle in the diagram at right, angles $\angle ABC$ and $\angle CBD$ are complementary. If $m\angle CBD = 76°$, what is $m\angle ABC$? Show how you got your answer.

b. Another special angle is 180°. If the sum of the measures of two angles is 180°, they are called **supplementary angles**. In the diagram at right, $\angle LMN$ is a straight angle. If $m\angle LMP = 62°$, what is $m\angle PMN$?

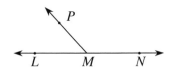

Problem continues on next page →

Geometry Connections

2-3.   *Problem continued from previous page.*

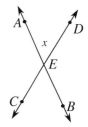

c.   Now consider the diagram at right, which shows $\overleftrightarrow{AB}$ and $\overrightarrow{CD}$ intersecting at *E*. If $x = 23°$, find $m\angle AEC$, $m\angle DEB$, and $m\angle CEB$. Show all work.

d.   When angles have equal measure, they are referred to as **congruent**. Which angle is congruent to $\angle AED$?

e.   When two lines intersect, the angles that lie on opposite sides of the intersection point are called **vertical angles**. For example, in the diagram above, $\angle AED$ and $\angle CEB$ are vertical angles. Find another pair of vertical angles in the diagram.

2-4.   Travis noticed that the vertical angles in parts (c) and (d) of problem 2-3 are congruent and wondered if pairs of vertical angles always have the same measure.

a.   Return to the diagram above and find $m\angle CEB$ if $x = 54°$. Show all work.

b.   Based on your observations, write a **conjecture** (a statement based on an educated guess that is unproven). Start with, "*Vertical angles are …*"

2-5.   In the problems below, you will use geometric relationships to find angle measures. Start by finding a special relationship between some of the sides or angles, and use that relationship to write an equation. Solve the equation for the variable, then use that variable value to answer the original question.

a.   Find $m\angle MNP$.

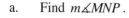

b.   Find $m\angle FGH$.

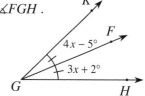

c.   Find $m\angle DBC$.

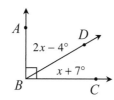

d.   Find $m\angle LPQ$ and $m\angle LPN$.

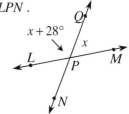

2-6.   When Jao answered part (b) of problem 2-4, he wrote the conjecture: *"Vertical angles are equal."* (Remember that a **conjecture** is an educated guess that has not yet been proven.)

   a.   **Examine** the diagram at right. Express the measures of every angle in the diagram in terms of *x*.

   b.   Do you think your vertical angle conjecture holds for *any* pair of vertical angles? Be prepared to convince the rest of the class.

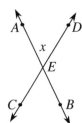

2-7.   Describe each of the angle relationships you learned about today in an entry in your Learning Log. Include a diagram, a description of the angles, and what you know about the relationship. For example, are the angles always equal? Do they have a special sum? Label this entry "Angle Relationships" and include today's date.

METHODS AND MEANINGS

MATH NOTES

Angle Relationships

If two angles have measures that add up to 90°, they are called **complementary angles**. For example, in the diagram at right, ∢*ABC* and ∢*CBD* are complementary because together they form a right angle.

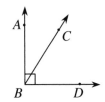

If two angles have measures that add up to 180°, they are called **supplementary angles**. For example, in the diagram at right, ∢*EFG* and ∢*GFH* are supplementary because together they form a straight angle.

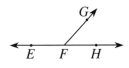

Two angles do not have to share a vertex to be complementary or supplementary. The first pair of angles at right are supplementary; the second pair of angles are complementary.

   **Supplementary**   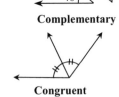   **Complementary**

When two angles have equal measure, they are called **congruent**. Their equality can be shown with matching markings, as shown in the diagram at right.

Congruent

Geometry Connections

2-8.   Find the area of each rectangle below:

a.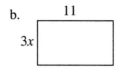

11 cm

3 cm

b.

11

3x

c.

3x   −4

11x

−2

2-9.   Mei puts the shapes at right into a bucket and asks Brian to pick one out.

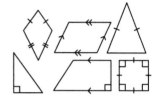

a.   What is the probability that he pulls out a quadrilateral with parallel sides?

b.   What is the probability that he pulls out a shape with rotation symmetry?

2-10.   Camille loves guessing games. She is going to tell you a fact about her shape to see if you can guess what it is.

a.   "My triangle has only one line of symmetry. What is it?"

b.   "My triangle has three lines of symmetry. What is it?"

c.   "My quadrilateral has no lines of symmetry but it does have rotation symmetry. What is it?"

2-11.   Jerry has an idea. Since he knows that an isosceles trapezoid has reflection symmetry, he **reasons**, "That means that it must have two pairs of congruent angles." He marks the congruent parts on his diagram at right.

ISOSCELES TRAPEZOID

Copy the shapes below onto your paper. Similarly mark which angles must be equal due to reflection symmetry.

a.

KITE

b.

ISOSCELES TRIANGLE

c.

REGULAR HEXAGON

d.

RHOMBUS

2-12.   Larry saw Javon's incomplete Venn diagram at right, and he wants to finish it. However, he does not know the condition that each circle represents. Find a possible label for each circle, and place two more shapes into the diagram.

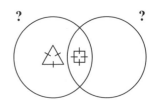

2.1.2   What's the relationship?

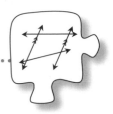

Angles Formed by Transversals

In Lesson 2.1.1, you **examined** vertical angles and found that vertical angles are always equal. Today you will look at another special relationship that guarantees angles are equal.

2-13.   **Examine** the diagrams below. For each pair of angles marked on the diagram, quickly decide what relationship their measures have. Your responses should be limited to one of three relationships: congruent (equal measures), complementary (have a sum of 90°), and supplementary (have a sum of 180°).

a.
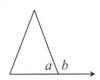

b.

c.
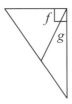

d.

2-14.   Marcos was walking home after school thinking about special angle relationships when he happened to notice a pattern of parallelogram tiles on the wall of a building. Marcos saw lots of special angle relationships in this pattern, so he decided to copy the pattern into his notebook.

The beginning of Marcos's diagram is shown at right and provided on the Lesson 2.1.2 Resource Page. This type of pattern is sometimes called a **tiling**. In a tiling, a shape is repeated without gaps or overlaps to fill an entire page. In this case, the shape being tiled is a parallelogram.

Problem continues on next page →

Geometry Connections

2-14.   *Problem continued from previous page.*

a.   Consider the angles inside a single parallelogram. Are any angles congruent? On your resource page, use color to show which angles must have equal measure. If two angles are not equal, make sure they are shaded with different colors.

b.   Since each parallelogram is a translation of another, what can be stated about the angles in the rest of Marcos' tiling? Use a dynamic geometry tool, transparencies on an overhead, or tracing paper to find determine which angles must be congruent. Color all angles that must be equal the same color.

c.   What about relationships between lines? Can you identify any lines that must be parallel? Mark all the lines on your diagram with the same number of arrows to show which lines are parallel.

2-15.   Julia wants to learn more about the angles in Marcos's diagram and has decided to focus on just a part of his tiling. An enlarged view of that section is shown in the image below right, with some points and angles labeled.

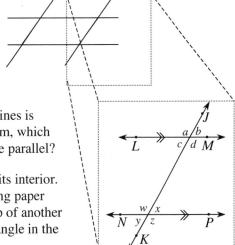

a.   A line that crosses two or more other lines is called a **transversal**. In Julia's diagram, which line is the transversal? Which lines are parallel?

b.   Trace $\measuredangle x$ on tracing paper and shade its interior. Then translate $\measuredangle x$ by sliding the tracing paper along the transversal until it lies on top of another angle and matches it exactly. Which angle in the diagram corresponds with x?

c.   What is the relationship between the measures of angles x and b? Must one be greater than the other, or must they be equal? Explain how you know.

d.   In this diagram, $\measuredangle x$ and $\measuredangle c$ are called **corresponding angles** because they are in the same position at two different intersections of the transversal. The corresponding angles in this diagram are equal because they were formed by translating a parallelogram. Name all the other pairs of equal corresponding angles you can find in Julia's diagram.

e.   Suppose $b = 60°$. Use what you know about vertical, supplementary, and corresponding angle relationships to find the measures of all the other angles in Julia's diagram.

2-16.   Frank wonders whether corresponding angles are *always* equal. For parts (a) through (d) below, decide whether you have enough information to find the measures of *x* and *y*. If you do, find the angle measures and state the relationship. Use tracing paper to help you find corresponding angles.

a.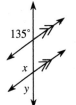

b.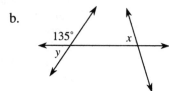

c.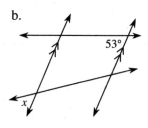

d.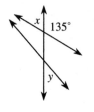

e.   Answer Frank's question: Are corresponding angles always equal?

f.   Conjectures are often written in the form, "*If...*, *then...*". A statement in if-then form is called a **conditional statement**. Make a conjecture about corresponding angles by completing this conditional statement: "If ..., then corresponding angles are equal."

2-17.   For each diagram below, find the value of *x* if possible. If it is not possible, explain how you know. State the relationships you use. Be prepared to **justify** every measurement you find to other members of your team.

a.

b.

c.
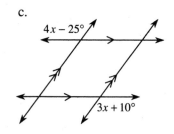

ETHODS AND MEANINGS

Naming Parts of Shapes

Part of geometry is the study of parts of shapes, such as points, line segments, and angles. To avoid confusion, standard notation is used to name these parts.

A point is named using a single capital letter. For example, the vertices (corners) of the triangle at right are named *A*, *B*, and *C*.

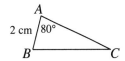

If a shape is transformed, the image shape is often named using **prime notation**. The image of point *A* is labeled *A′* (read as "A prime"), the image of *B* is labeled *B′* (read as "B prime"), etc. At right, $\triangle A'B'C'$ is the image of $\triangle ABC$.

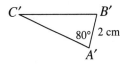

The side of a polygon is a **line segment**. A line segment is named by naming its endpoints and placing a bar above them. For example, one side of the first triangle above is named $\overline{AB}$. When referring to the length of a segment, the bar is omitted. In $\triangle ABC$ above, $AB = 2$ cm.

A **line**, which differs from a segment in that it extends infinitely in either direction, is named by naming two points on the line and placing a bar with arrows above them. For example, the line below is named $\overleftrightarrow{DE}$. When naming a segment or line, the order of the letters is unimportant. The line below could also be named $\overleftrightarrow{ED}$.

An angle can be named by putting an angle symbol in front of the name of the angle's vertex. For example, the angle measuring 80° in $\triangle ABC$ above is named $\angle A$. Sometimes using a single letter makes it unclear which angle is being referenced. For example, in the diagram at right, it is unclear which angle is referred to by $\angle G$. When this happens, the angle is named with three letters. For example, the angle measuring 10° is called $\angle HGI$ or $\angle IGH$. Note that the name of the vertex must be the second letter in the name; the order of the other two letters is unimportant.

To refer to an angle's measure, an *m* is placed in front of the angle's name. For example, $m\angle HGI = 10°$ means "the measure of $\angle HGI$ is 10°."

2-18. **Examine** the diagrams below. What is the geometric relationship between the labeled angles? What is the relationship of their measures? Then, use the relationship to write an equation and solve for x.

a.

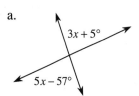

b.

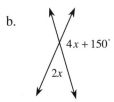

2-19. In problem 2-11, you determined that because an isosceles triangle has reflection symmetry, then it must have two angles that are congruent.

a. How can you tell which angles are congruent? For example, in the diagram at right, which angles must have equal measure? Name the angles and explain how you know.

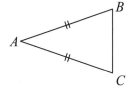

b. **Examine** the diagram for part (a). If you know that $m\angle B + m\angle C = 124°$, then what is the measure of $\angle B$? Explain how you know.

c. Use this idea to find the value of x in the diagram at right. Be sure to show all work.

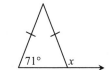

2-20. On graph paper, draw the quadrilateral with vertices $(-1, 3)$, $(4, 3)$, $(-1, -2)$, and $(4, -2)$.

a. What kind of quadrilateral is this?

b. Translate the quadrilateral 3 units to the left and 2 units up. What are the new coordinates of the vertices?

2-21. Find the equation for the line that passes through $(-1, -2)$ and $(4, 3)$. Is the point $(3, 1)$ on this line? Be sure to **justify** your answer.

Geometry Connections

2-22.   Juan decided to test what would happen if he rotated an angle.

a.   He copied the angle at right on tracing paper and rotated it 180°
about its vertex. What type of angle pair did he create? What
is the relationship of these angles?

b.   Juan then rotated the same angle 180° through a
different point (see the diagram at right). On your paper,
draw Juan's angle and the rotated image. Describe the
overall shape formed by the two angles.

2.1.3 What's the relationship?

More Angles Formed by Transversals

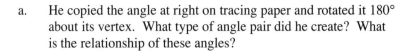

In Lesson 2.1.2, you looked at corresponding angles formed when a transversal intersects two
parallel lines. Today you will **investigate** other special angle relationships formed in this
situation.

2-23.   Suppose ∡a in the diagram at right measures 48°.

a.   Use what you know about vertical, corresponding, and
supplementary angle relationships to find the measure
of ∡b.

b.   Julia is still having trouble
seeing the angle relationships clearly in this
diagram. Her teammate, Althea explains,
"When I translate one of the angles along the
transversal, I notice its image and the other
given angle are a pair of vertical angles. That
way, I know that a and b must be congruent."

Use Althea's method and tracing paper to
determine if the following angle pairs are
congruent or supplementary. Be sure to state
whether the pair of angles created after the translation is a vertical pair or forms
a straight angle. Be ready to **justify** your answer for the class.

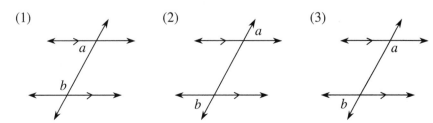

(1)   (2)   (3)

2-24.   In problem 2-23, Althea showed that the shaded angles in
         the diagram are congruent. However, these angles also
         have a name for their geometric relationship (their relative
         positions on the diagram). These angles are called
         alternate interior angles. They are called "alternate"
         because they are on opposite sides of the transversal, and
         "interior" because they are both inside the parallel lines.

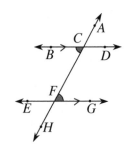

   a.   On tracing paper, trace and shade in ∡*CFG*. How
         can you transform ∡*CFG* so that it lands on
         ∡*BCF* ? Be sure your team agrees.

   b.   Find another pair of alternate interior angles in this diagram.

   c.   Think about the relationship between the measures of alternate interior angles.
         If the lines are parallel, are they always congruent? Are they always
         supplementary? Complete the conjecture, "*If lines are parallel, then alternate
         interior angles are…*".

   d.   Instead of writing conditional statements, Roxie likes to write **arrow diagrams**
         to express her conjectures. She expresses the conjecture from part (b) as

              Lines are parallel ➜ *alternate interior angles are congruent.*

         This arrow diagram says the same thing as the conditional statement you wrote
         in part (c). How is it different from your conditional statement? What does the
         arrow mean?

2-25.   The shaded angles in the diagram at right have another special
         angle relationship. They are called **same-side interior** angles.

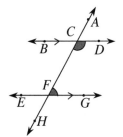

   a.   Why do you think they have this name?

   b.   What is the relationship between the angle measures of
         same-side interior angles? Are they always congruent?
         Supplementary? Talk about this with your team.

         Then write a conjecture about the relationship of the angle measures. Your
         conjecture can be in the form of a conditional statement or an arrow diagram.
         If you write a conditional statement, it should begin, "*If lines are parallel, then
         same-side interior angles are…*".

2-26.   THE REFLECTION OF LIGHT

You know enough about angle relationships now to start analyzing how light
bounces off mirrors. **Examine** the two diagrams below. Diagram A shows a beam
of light emitted from a light source at A. In Diagram B, someone has placed a
mirror across the light beam. The light beam hits the mirror and is reflected from its
original path.

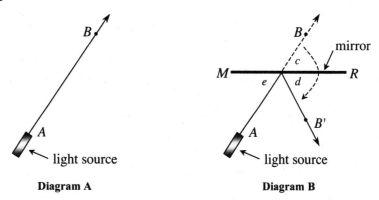

Diagram A   Diagram B

a.   What is the relationship between angles *c* and *d*? Why?

b.   What is the relationship between angles *c* and *e*? How do you know?

c.   What is the relationship between angles *e* and *d*? How do you know?

d.   Write a conjecture about the angle at which light hits a mirror and the angle at
      which it bounces off the mirror.

2-27.   A CD player works by bouncing a laser off the surface of the CD, which acts like a
         mirror. An emitter sends out the light, which bounces off the CD and then is
         detected by a sensor. The diagram below shows a CD held parallel to the surface of
         the CD player, on which an emitter and a sensor are mounted.

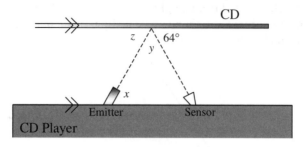

   a.   The laser is supposed to bounce off the CD at a 64° angle as shown in the
         diagram above. For the laser to go directly into the sensor, at what angle does
         the emitter need to send the laser beam? In other words, what does the
         measure of angle x have to be? **Justify** your conclusion.

   b.   The diagram above shows two parts of the laser beam: the one coming out of
         the emitter and the one that has bounced off the CD. What is the angle ($\measuredangle y$)
         between these beams? How do you know?

2-28.   ANGLE RELATIONSHIPS TOOLKIT

         Obtain a Lesson 2.1.3 Resource Page ("Angle
         Relationships Toolkit") from your teacher. This will be a
         continuation of the Geometry Toolkit you started in
         Chapter 1. Think about the new angle relationships you
         have studied so far in Chapter 2. Then, in the space
         provided, add a diagram and a description of the
         relationship for each special angle relationship you know.
         Be sure to specify any relationship between the measures
         of the angle (such as whether or not they are always
         congruent). In later lessons, you will continue to add relationships to this toolkit, so
         be sure to keep this resource page in a safe place. At this point, your toolkit should
         include:

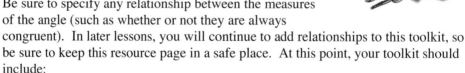

   •   Vertical angles   •   Straight angles

   •   Corresponding angles   •   Alternate interior angles

   •   Same-side interior angles

METHODS AND **M**EANINGS

MATH NOTES

Systems of Linear Equations

In a previous course, you learned that a **system of linear equations** is a set of two or more linear equations that are given together, such as the example at right. In a system, each variable represents the same quantity in both equations. For example, y represents the same quantity in *both* equations at right.

$$y = 2x$$
$$y = -3x + 5$$

To represent a system of equations graphically, you can simply graph each equation on the same set of axes. The graph may or may not have a **point of intersection**, as shown circled at right.

Sometimes two lines have *no* points of intersection. This happens when the two lines are parallel. It is also possible for two lines to have an *infinite* number of intersections. This happens when the graphs of two lines lie on top of each other. Such lines are said to **coincide**.

The **Substitution Method** is a way to change two equations with two variables into one equation with one variable. It is convenient to use when only one equation is solved for a variable. For example, to solve the system at right:

$$x = -3y + 1$$
$$4x - 3y = -11$$

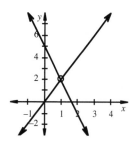

- Use substitution to rewrite the two equations as one. In other words, replace x with $(-3y + 1)$ to get $4(-3y + 1) - 3y = -11$. This equation can then be solved to find y. In this case, $y = 1$.

$$4(\quad) - 3y = -11$$
$$4(-3y + 1) - 3y = -11$$

- To find the point of intersection, substitute to find the other value.

$$-12y + 4 - 3y = -11$$
$$-15y + 4 = -11$$

- Substitute $y = 1$ into $x = -3y + 1$ and write the answer for x and y as an ordered pair.

$$-15y = -15$$
$$y = 1$$

- To test the solution, substitute $x = -2$ and $y = 1$ into $4x - 3y = -11$ to verify that it makes the equation true. Since $4(-2) - 3(1) = -11$, the solution $(-2, 1)$ must be correct.

$$x = -3(1) + 1 = -2$$
$$(-2, 1)$$

2-29.   The set of equations at right is an example of a **system of equations**. Read the Math Notes box for this lesson on how to solve systems of equations. Then answer the questions below.

$$y = -x + 1$$
$$y = 2x + 7$$

   a.   Graph the system on graph paper. Then write its solution (the point of intersection) in (x, y) form.

   b.   Now solve the system using the Substitution Method. Did your solution match your result from part (a)? If not, check your work carefully and look for any mistakes in your algebraic process or on your graph.

2-30.   On graph paper, graph the rectangle with vertices at $(2, 1)$, $(2, 5)$, $(7, 1)$, and $(7, 5)$.

   a.   What is the area of this rectangle?

   b.   Shirley was given the following points and asked to find the area, but her graph paper is not big enough. Find the area of Shirley's rectangle, and explain to her how she can find the area without graphing the points.

   Shirley's points: $(352, 150)$, $(352, 175)$, $(456, 150)$, and $(456, 175)$

2-31.   Looking at the diagram below, John says that $m\angle BCF = m\angle EFH$.

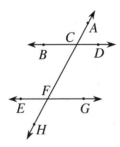

Note: This stoplight icon will appear periodically throughout the text. Problems with this icon display common errors that can be made. Be sure not to make the same mistakes yourself!

   a.   Do you agree with John? Why or why not?

   b.   Jim says, "You can't be sure those angles are equal. An important piece of information is missing from the diagram!" What is Jim talking about?

2-32. Use your knowledge of angle relationships to solve for x in the diagrams below. **Justify** your solutions by naming the geometric relationship.

a.

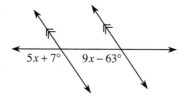

$5x + 7°$ $9x - 63°$

b.

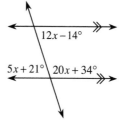

$12x - 14°$

$5x + 21°$ $20x + 34°$

2-33. On graph paper, draw line segment $\overline{AB}$ if $A(6, 2)$ and $B(3, 5)$.

a. Reflect $\overline{AB}$ across the line $x = 3$ and connect points B and B'. What shape is created by this reflection? Be as specific as possible.

b. What polygon is created when $\overline{AB}$ is reflected across the line $y = -x + 6$ and all endpoints are connected to form a polygon?

2.1.4 How can I use it?

· ·

Angles in a Triangle

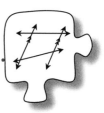

So far in this chapter, you have **investigated** the angle relationships created when two lines intersect, forming vertical angles. You have also **investigated** the relationships created when a transversal intersects two parallel lines. Today you will study the angle relationships that result when three non-parallel lines intersect, forming a triangle.

2-34. Marcos decided to study the angle relationships in triangles by making another tiling. Find his pattern, shown at right, on the Lesson 2.1.4 Resource Page.

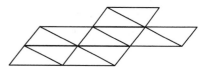

a. Copy one of Marcos's triangles onto tracing paper. Use a colored pen or pencil to shade one of the triangle's angles on the tracing paper. Then use the same color to shade every angle on the resource page that is equal to the shaded angle.

b. Repeat this process for the other two angles of the triangle, using a different color for each angle in the triangle. When you are done, every angle in your tiling should be shaded with one of the three colors.

Problem continues on next page →

2-34.   *Problem continued from previous page.*

    c.   Now **examine** your colored tiling. What relationship can you find between the three different-colored angles? You may want to focus on the angles that form a straight angle. What does this tell you about the angles in a triangle? Write a conjecture in the form of a conditional statement or an arrow diagram. If you write a conditional statement, it should begin, *"If a polygon is a triangle, then the measure of its angles…"*.

    d.   How can you convince yourself that your conjecture is true for all triangles? If a dynamic geometry tool is available, use it to convince yourself that your conjecture is always true. If technology is not available, use the diagram above and your Angle Relationships Toolkit to write a convincing argument that your conjecture is true.

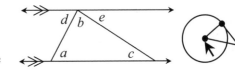

        Then add this angle relationship to your Angle Relationships Toolkit from Lesson 2.1.3. This will be referred to as the **Triangle Angle Sum Theorem.** (A theorem is a statement that has been proven.)

2-35.   Use your conjecture from problem 2-34 about the angles in a triangle to find *x* in each diagram below. Show all work.

    a.

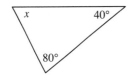

    b.

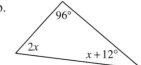

2-36.   What can the Triangle Angle Sum Theorem help you learn about special triangles?

    a.   Find the measure of each angle in an equilateral triangle. **Justify** your conclusion.

    b.   Consider the isosceles right triangle (also sometimes referred to as a "half-square") at right. Find the measures of all the angles in a half-square.

    c.   What if you only know one angle of an isosceles triangle? For example, if $m\angle A = 34°$, what are the measures of the other two angles?

Geometry Connections

2-37.   TEAM **REASONING** CHALLENGE

How much can you figure out about the
figure at right using your knowledge of
angle relationships? Work with your team
to find the measures of all the labeled angles
in the diagram at right. **Justify** your
solutions with the name of the angle
relationship you used. Carefully record your
work as you go and be prepared to share
your **reasoning** with the rest of the class.

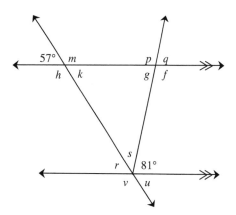

ETHODS AND MEANINGS

More Angle Pair Relationships

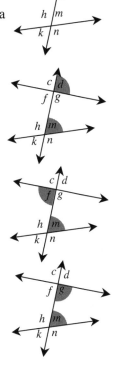

Vertical angles are the two opposite (that is, non-
adjacent) angles formed by two intersecting lines, such as
angles ∡c and ∡g in the diagram at right. ∡c by itself is not
a vertical angle, nor is ∡g, although ∡c and ∡g together are a
pair of vertical angles. Vertical angles always have equal
measure.

Corresponding angles lie in the same position but at
different points of intersection of the transversal. For
example, in the diagram at right, ∡m and ∡d form a pair of
corresponding angles, since both of them are to the right of
the transversal and above the intersecting line.
Corresponding angles are congruent when the lines
intersected by the transversal are parallel.

∡f and ∡m are **alternate interior angles** because one is
to the left of the transversal, one is to the right, and both are
between (inside) the pair of lines. Alternate interior angles
are congruent when the lines intersected by the transversal
are parallel.

∡g and ∡m are **same-side interior angles** because both
are on the same side of the transversal and both are
between the pair of lines. Same-side interior angles are
supplementary when the lines intersected by the transversal
are parallel.

2-38.   Find all missing angles in the diagrams below.

a.
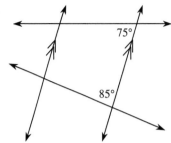

b.
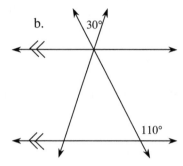

2-39.   Robert believes the lines graphed at right are perpendicular, but Mario is not convinced. Find the slope of each line, and explain how you know whether or not the lines are perpendicular.

2-40.   The diagram at right represents only half of a shape that has the graph of $y = 1$ as a line of symmetry. Draw the completed shape on your paper, and label the coordinates of the missing vertices.

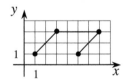

2-41.   Janine measured the sides of a rectangle and found that the sides were 12 inches and 24 inches. Howard measured the same rectangle and found that the sides were 1 foot and 2 feet. When their math teacher asked them for the area, Janine said 288, and Howard said 2. Why did they get two different numbers for the area of the same rectangle?

2-42.   Graph the system of equations at right on graph paper. Then state the solution to the system. If there is not a solution, explain why.

$$y = -\tfrac{2}{5}x + 1$$

$$y = -\tfrac{2}{5}x - 2$$

2.1.5 What's the relationship?

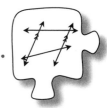

Applying Angle Relationships

During Section 2.1, you have been learning about various special angle relationships that are created by intersecting lines. Today you will **investigate** those relationships a bit further, then apply what you know to explain how Mr. Douglas's mirror trick (from problem 2-1) works. As you work in your teams today, keep the following questions in mind to guide your discussion:

What's the relationship?

Are the angles equal? Are they supplementary?

How can I be sure?

2-43.   Use your knowledge of angle relationships to answer
the questions below.

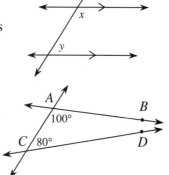

a.   In the diagram at right, what is the sum of angles
x and *y*? How do you know?

b.   While looking at the diagram at right,
Rianna exclaimed, "I think something is
wrong with this diagram." What do you
think she is referring to? Be prepared to
share your thinking with the class.

2-44.   Maria is still not convinced that the lines in
part (b) of problem 2-43 *must* be parallel.
She decides to assume that they are not
parallel and draws the diagram at right.

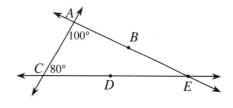

a.   Why must lines $\overleftrightarrow{AB}$ and $\overleftrightarrow{CD}$
intersect in Maria's diagram?

b.   What is $m\angle BED$? Discuss this question with your team and explain what it
tells you about $\overleftrightarrow{AB}$ and $\overleftrightarrow{CD}$.

c.   **Examine** the diagram at right. In this
diagram, must $\overleftrightarrow{FG}$ and $\overleftrightarrow{HI}$ be parallel?
Explain how you know.

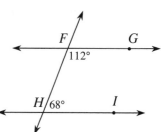

d.   Write a conjecture based on your conclusion to
this problem.

2-45.   Use your conjecture from problem 2-44 to explain why lines must be parallel in the diagrams below.

a.

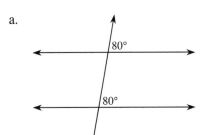

b.

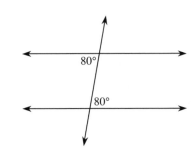

c.   Looking back at the diagrams in parts (a) and (b), write two new conjectures that begin, "*If corresponding angles are equal, …*" and "*If the measures of alternate interior angles are equal, …*".

2-46.   SOMEBODY'S WATCHING ME, Part Two

Remember Mr. Douglas' trick from problem 2-1? You now know enough about angle and line relationships to analyze why a hinged mirror set so the angle between the mirrors is 90° will reflect your image back to you from any angle. Since your reflection is actually light that travels from your face to the mirror, you will need to study the path of the light. Remember that a mirror reflects light, and that the angle the light hits the mirror will equal the angle it bounces off the mirror.

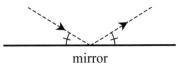

mirror

Your task: Explain why the mirror bounces your image back to you from any angle. Include in your analysis:

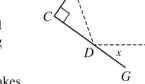

- Use angle relationships to find the measures of all the angles in the figure. (Each team member should choose a different x value and calculate all of the other angle measures using his or her selected value of x.)

- What do you know about the paths the light takes as it leaves you and as it returns to you? That is, what is the relationship between $\overrightarrow{BA}$ and $\overrightarrow{DE}$?

- Does Mr. Douglas' trick work if the angle between the mirrors is not 90°?

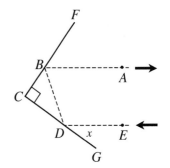

2-47. Since you are trying to show that the trick works for *any* angle at which the light could hit the mirror, each team will work with a different angle measure for *x* in this problem. Your teacher will tell you what angle *x* your team should use in the diagram at right.

    a. Using angle relationships and what you know about how light bounces off mirrors, find the measure of every other angle in the diagram.

    b. What is the relationship between $\angle ABD$ and $\angle EDB$? What does this tell you about the relationship between $\overline{BA}$ and $\overline{DE}$?

    c. What if $m\angle C = 89°$? Does the trick still work?

2-48. Explain why the 90° hinged mirror always sends your image back to you, no matter which angle you look into it from.

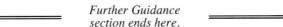

Further Guidance section ends here.

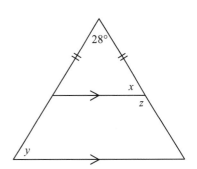

2-49. Use what you have learned in Section 2.1 to find the measures of *x*, *y*, and *z* at right. **Justify** each conclusion with the name of a geometric relationship from your Angle Relationships Toolkit.

2-50. EXTENSION

Hold a 90° hinged mirror at arm's length and find your own image. Now close your right eye. Which eye closes in the mirror? Look back at the diagram from problem 2-46. Can you explain why this eye is the one that closes?

Proof by Contradiction

The kind of argument you used in Lesson 2.1.5 to **justify** "If same-side interior angles are supplementary, then lines are parallel" is sometimes called a **proof by contradiction**. In a proof by contradiction, you prove a claim by thinking about what the consequences would be if it were false. If the claim's being false would lead to an impossibility, this shows that the claim must be true.

For example, suppose you know Mary's brother is seven years younger than Mary. Can you argue that Mary is at least five years old? A proof by contradiction of this claim would go:

> *Suppose* Mary is less than five years old.

> *Then* her brother's age is negative!

> *But* this is impossible, so Mary must be at least five years old.

To show that lines $\overleftrightarrow{AB}$ and $\overleftrightarrow{CD}$ must be parallel in the diagram at right, you used a proof by contradiction. You argued:

Suppose $\overleftrightarrow{AB}$ and $\overleftrightarrow{CD}$ intersect at some point E.

Then the angles in $\triangle AEC$ add up to more than 180°.

But this is impossible, so $\overleftrightarrow{AB}$ and $\overleftrightarrow{CD}$ must be parallel.

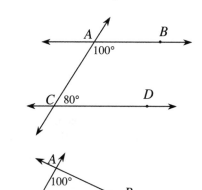

2-51. Solve for x in the diagram at right.

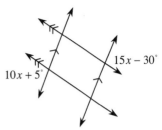

2-52. For each diagram below, set up an equation and solve for x.

a. Perimeter = 76 units

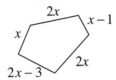

b.

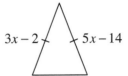

c.

d.

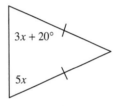

2-53. Solve each system of equation below. Then verify that your solution makes each equation true. You may want to refer to the Math Notes box in Lesson 2.1.3.

a. $y = 5x - 2$
$y = 2x + 10$

b. $x = -2y - 1$
$2x + y = -20$

2-54. Graph the line $y = \frac{3}{4}x$ on graph paper.

a. Draw a slope triangle.

b. Rotate your slope triangle 90° around the origin to get a new slope triangle. What is the new slope?

c. Find the equation of a line perpendicular to $y = \frac{4}{3}x$.

2-55. Mario has 6 shapes in a bucket. He tells you that the probability of pulling an isosceles triangle out of the bucket is $\frac{1}{3}$. How many isosceles triangles are in his bucket?

2.2.1 How can I measure an object?

Units of Measure

How tall are you? How large is the United States? How much water does your bathtub at home hold? All of these questions ask about the size of objects around you. How can we answer these questions more specifically than saying "big" or "small"? Today you will be **investigating** ways to answer these and other questions like them.

2-56.   Your teacher will describe with words a figure he or she has drawn. Your job is to try to draw the **exact same figure** on your paper so that if you placed your drawing on top of your teacher's, the figures would match perfectly. Redraw your figure as many times as necessary.

2-57.   Length often provides a direct way to answer the question, "How big?" In this activity, your teacher will give your team rulers with a unit of length to use to measure distances. For instance, if you have an object that has the same length as three of your units placed next to one another, then the object's length is "3 units."

    a.   Common units you may have used before are inches or meters. However, your unit does not match any of the familiar units. Give your unit of measure a unique name. Then continue marking and labeling units on your ruler as accurately as possible.

    b.   The **dimensions** of a figure are its measurements of length. For example, the measurements in each figure below describe the relative size of each object.

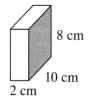

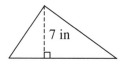

Dimensions can describe how long, wide, or tall the object is by measuring the lengths of the edges.

Or, dimensions can be found by measuring a length that is not a side.

Find the dimensions of the shape on the Lesson 2.2.1A Resource Page using your unit of length. That is, how wide is the shape? How tall? Compare your results with those of other teams. What happened?

    c.   The local baseball club is planning to make a mural that is the same general shape as the one you measured in part (b) (but fortunately a much larger version!). The club plans to frame the mural with neon tubes. Approximately how many units of neon tube will they need to do this?

Geometry Connections

2-58.   To paint the mural, the wall must first be covered with a coat of primer. How much of the wall will need to be painted with the primer? Remember that the measurement of the region inside a shape is called the **area** of the shape.

   a.   Just as you were able to use your unit of length to measure distance, you need a unit of area to measure a surface. Use your team's ruler to make unit squares that are 1 unit long on each side. The area that your square unit covers is called "one square unit" and can be abbreviated as 1 sq. unit or 1 un^2. What would you call this unit of measure, given the name you chose in problem 2-57?

   b.   What is the approximate area of the mural? That is, how many of your square units fit within your shape?

2-59.   Use your unit of measure to make a rectangle that has dimensions 3 units by 5 units.

   a.   What is the area of your rectangle? (That is, how many unit squares are there in your rectangle?)

   b.   When you answered part (a), did you count the squares? Did your team use a shortcut? If so, why does the shortcut work?

   c.   Compare your rectangle to rectangles that other teams made. What is the same about the rectangles and what is different?

2-60.   If you found out that your gym teacher is going to make you run 31,680 inches next period, how useful is this information? Or, if you knew that you were 0.003977 miles long at birth, do you have any idea how long that is?

   An important part of measurement is choosing an appropriate unit of measurement. With your team, suggest a unit of measurement that can best measure (and describe) each of these situations.

   a.   The distance you travel from home to get to school.

   b.   The surface of the school's soccer field.

   c.   The width of a strand of hair.

   d.   The length of your nose.

   e.   The amount of room in your locker.

METHODS AND MEANINGS

Triangle Angle Sum Theorem

The **Triangle Angle Sum Theorem** states that the measures of the angles in a triangle add up to 180°. For example, in $\triangle ABC$ at right:

$$m\angle A + m\angle B + m\angle C = 180°$$

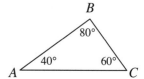

The Triangle Angle Sum Theorem can be verified by using a tiling of the given triangle (shaded at right). Because the tiling produces parallel lines, the alternate interior angles must be congruent. As seen in the diagram at right, the three angles of a triangle form a straight angle. Therefore, the sum of the angles of a triangle must be 180°.

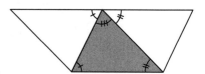

Review & Preview

2-61.   **Examine** the shapes in your Shape Toolkit. Then name at least three shapes in the Shape Toolkit that share the following qualities.

- They have only one line of symmetry.

- They have fewer than four sides.

2-62.   **Examine** the diagram at right. Then use the information provided in the diagram to find the measures of angles a, b, c, and d. For each angle, name the relationship from your Angle Relationships Toolkit that helped **justify** your conclusion. For example, did you use vertical angles? If not, what type of angle did you use?

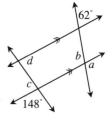

Geometry Connections

2-63.   **Examine** the triangle at right.

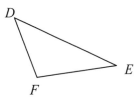

a.   If $m\angle D = 48°$ and $m\angle F = 117°$, then what is $m\angle E$?

b.   Solve for x if $m\angle D = 4x + 2°$, $m\angle F = 7x - 8°$, and $m\angle E = 4x + 6°$. Then find $m\angle D$.

c.   If $m\angle D = m\angle F = m\angle E$, what type of triangle is $\triangle FED$?

2-64.   Plot $\triangle ABC$ on graph paper if $A(6, 3)$, $B(2, 1)$, and $C(5, 7)$.

a. $\triangle ABC$ is rotated about the origin 180° to become $\triangle A'B'C'$. Name the coordinates of A', B', and C'.

b.   This time $\triangle ABC$ is rotated 180° about point C to form $\triangle A''B''C''$. Name the coordinates of B''.

c.   If $\triangle ABC$ is rotated 90° clockwise (↻) about the origin to form $\triangle A'''B'''C'''$, what are the coordinates of point A'''?

2-65.   **Examine** the graph at right.

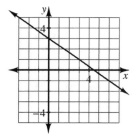

a.   Find the equation of the line.

b.   Is the line $y = \frac{3}{2}x + 1$ perpendicular to this line? How do you know?

c.   On graph paper, graph $\overleftrightarrow{AB}$ if $A(-2, 4)$ and $B(4, 7)$. Then find the equation of $\overleftrightarrow{AB}$.

d.   Find an equation of $\overleftrightarrow{AC}$ if $\overleftrightarrow{AC} \perp \overleftrightarrow{AB}$ from part (c).

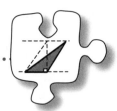

2.2.2 How can I find the area?

Areas of Triangles and Composite Shapes

How much grass would it take to cover a football field? How much paint would it take to cover a stop sign? How many sequins does it take to cover a dress? Finding the area of different types of shapes enables us to answer many questions. However, different people will see a shape differently. Therefore, during this lesson, be especially careful to look for different **strategies** that can be used to find area.

As you solve these problems, ask yourself the following focus questions:

What shapes do I see in the diagram?

Does this problem remind me of one I have seen before?

Is there another way to find the area?

2-66.   **STRATEGIES** TO MEASURE AREA

In Lesson 2.2.1 you used a grid to measure area. But what if a grid is not available? Or what if we want an exact measurement?

Examine the variety of shapes below. Work with your team to find the area of each one. If a shape has shading, then find the area of the shaded region. Be sure to listen to your teammates carefully and look for different **strategies**. Be prepared to share your team's method with the class.

a.
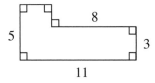

b.
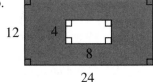

c.
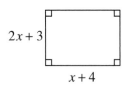

d.
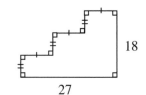

e.
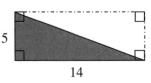

f.
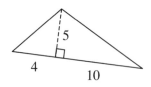

2-67.   Ismael claimed that he did not need to calculate the area for part (f) in problem 2-66 because it must be the same as the area for the triangle in part (e).

   a.   Is Ismael's claim correct? How do you know? Draw diagrams that show your thinking.

   b.   Do all triangles with the same base and height have the same area? Use your dynamic geometry tool to **investigate**. If no technology is available, obtain the Lesson 2.2.2 Resource Page and compare the areas of the given triangles.

   c.   Explain why the area of any triangle is half the area of a rectangle that has the same base and height. That is, show that the area of a triangle must be $\frac{1}{2}bh$.

2-68.   How do you know which dimensions to use when finding the area of a triangle?

   a.   Copy each triangle below onto your paper. Then find the area of each triangle. Draw any lines on the diagram that will help. Turning the triangles may help you discover a way to find their areas.

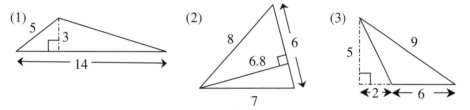

   b.   Look back at your work from part (a). Which numbers from each triangle did you use to find the area? For instance, in the center triangle, you probably used only the 6.8 and 6. Write an explanation and/or draw a diagram that would help another student understand how to **choose** which lengths to use when calculating the area.

   c.   Mario, Raquel, and Jocelyn are arguing about where the height is for the triangle at right. The three have written their names along the part they think should be the height. Determine which person is correct. Explain why the one you chose is correct and why the other two are incorrect.

2-69.   In a Learning Log entry, describe at least two different **strategies** that were used today to find the area of irregular shapes. For each method, be sure to include an example. Title this entry "Areas of Composite Figures" and include today's date.

METHODS AND MEANINGS

Multiplying Binomials

One method for multiplying binomials is to use a generic rectangle. That is, use each factor of the product as a dimension of a rectangle and find its area. If $(2x + 5)$ is the base of a rectangle and $(3x - 1)$ is the height, then the expression $(2x + 5)(3x - 1)$ is the area of the rectangle. See the example below.

	$2x$	$+5$	
-1	$-2x$	-5	-1
$3x$	$6x^2$	$15x$	$3x$
	$2x$	$+5$	

Multiply:
$$\begin{aligned}(2x+5)(3x-1) &= 6x^2 - 2x + 15x - 5 \\ &= 6x^2 + 13x - 5\end{aligned}$$

Review & Preview

2-70.   Review how to multiply binomials by reading the Math Notes box for this lesson. Then rewrite each of the expressions below by multiplying binomials and simplifying the resulting expression.

a. $(4x+1)(2x-7)$

b. $(5x-2)(2x+7)$

c. $(4x-3)(x-11)$

d. $(-3x+1)(2x-5)$

2-71.   The shaded triangle at right is surrounded by a rectangle. Find the area of the triangle.

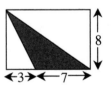

2-72.   For each diagram below, solve for x. Explain what relationship from your Angle Relationships Toolkit you used for each problem.

a.
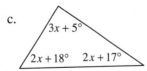

b.

c. $3x + 5°$ $2x + 18°$ $2x + 17°$

d. x $30°$

Geometry Connections

2-73.   Daniel and Mike were having an argument about where to place a square in the Venn diagram at right. Daniel wants to put the square in the intersection (the region where the two circles overlap).

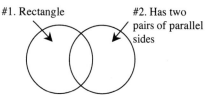
#1. Rectangle   #2. Has two pairs of parallel sides

Mike doesn't think that's right. *"I think it should go in the right region because it is a square, not a rectangle."*

"But a square IS a rectangle!" protests Daniel. Who is right? Explain your thinking.

2-74.   What is the probability of drawing each of the following cards from a standard playing deck? Remember that a standard deck of cards includes: 52 cards of four suits. Two suits are black: clubs and spades; two are red: hearts and diamonds. Each suit has 13 cards: 2 through 10, Ace, Jack, Queen, and King.

a.   P(Jack)   b.   P(spade)

c.   P(Jack of spades)   d.   P(not spade)

2.2.3 What's the area?

Areas of Parallelograms and Trapezoids

In Lesson 2.2.2, you used your knowledge of the area of a rectangle to develop a method to find the exact area of a triangle that works for all triangles. How can your understanding of the area of triangles and rectangles help with the study of other shapes? As you work today, ask yourself and your team members these focus questions:

What do you see?

What shapes make up the composite figure?

Is there another way?

2-75.   Find the areas of the figures below. Can you find more than one method for each shape?

a.

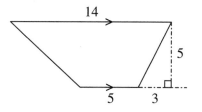

b.

2-76.   FINDING THE AREA OF A PARALLELOGRAM

One of the shape in your Shape Bucket is
shown at right. It is called a **parallelogram**: a
four-sided shape with two pairs of parallel
sides. How can you find the area of a
parallelogram? Consider this question as you
answer the questions below.

a.   Kenisha thinks that the rectangle and parallelogram below have the same area.
Her teammate Shaundra disagrees. Who is correct? **Justify** your conclusion.

Rectangle   **Parallelogram**

b.   In the parallelogram shown in part (a), the two lengths that you were given are
often called the **base** and **height**. Several more parallelograms are shown
below. In each case, find a related rectangle for which you know both the base
and height. Rotating your book might help. Use what you know about
rectangles to find the area of each parallelogram.

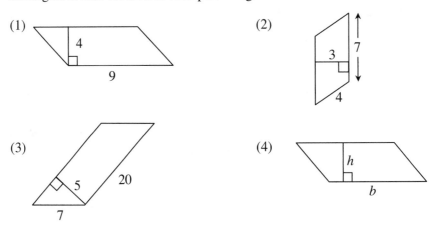

c.   Describe how to find the area of a parallelogram when given its base and
height.

d.   Does the angle at which the parallelogram slants matter? Does every
parallelogram have a related rectangle with equal area? Why or why not?
Explain how you know.

Geometry Connections

2-77.   Shaundra claims that the area of a parallelogram can be found by *only* using triangles.

a.   Do you agree? Trace the parallelogram at right onto your paper. Then divide it into two triangles. (Do you see more than one way to do this? If you do, ask some team members to divide the parallelogram one way, and the others a second way.)

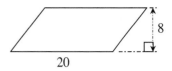

b.   Use what you know about calculating the area of a triangle to find the area of the parallelogram. It may help to trace each triangle separately onto tracing paper so that you can rotate them and label any lengths that you know.

c.   How does the answer to part (b) compare to the area you found in part (a) of problem 2-76?

2-78.   FINDING THE AREA OF A TRAPEZOID

Another shape you will study from the Shape Bucket is a **trapezoid**: a four-sided shape that has at least one pair of parallel sides. The sides that are parallel are called **bases**, as shown in the diagram at right. Answer the questions below with your team to develop a method to find the area of a trapezoid.

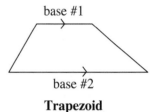

Trapezoid

a.   While playing with the shapes in her Shape Bucket, Shaundra noticed that two identical trapezoids can be arranged to form a parallelogram. Is she correct?

Trace the trapezoid shown at right onto a piece of tracing paper. Be sure to label its bases and height as shown in the diagram. Work with a team member to move and rearrange the trapezoid on each piece tracing paper so that they create a parallelogram.

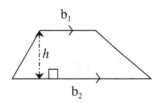

b.   Since you built a parallelogram from two trapezoids, you can use what you know about finding the area of a parallelogram to find the area of the trapezoid. If the bases of each trapezoid are b_1 and b_2 and the height of each is h, then find the area of the parallelogram. Then use this area to find the area of the original trapezoid.

Problem continues on next page →

2-78.   *Problem continued from previous page.*

   c.   Kenisha sees it differently. She sees two
         triangles inside the trapezoid. If she divides a
         trapezoid into two triangles, what area will she
         get? Again assume that the bases of the
         trapezoid are b_1 and b_2 and the height is h.

   d.   Are the area expressions you created in parts (b) and (c)
         equivalent? That is, will they calculate the same area? Use
         your algebra skills to demonstrate that they are equivalent.

2-79.   Calculate the exact areas of the shapes below.

   a.   b.   c.   d.

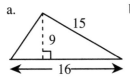   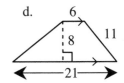

2-80.   EXTENSION

         Examine the diagram of the kite at right. Work with your team to
         find a way to show that its area must be half of the product of the
         diagonals. That is, if the length of the diagonals are a and b, provide
         a diagram or explanation of why the area of the kite must be $\frac{1}{2}ab$.

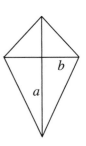

Ⓜ️ETHODS AND MEANINGS

Conditional Statements

MATH NOTES

A **conditional statement** is written in the form "**If …, then ….**" Here
are some examples of conditional statements:

> *If a shape is a rhombus, then it has four sides of equal length.*
>
> *If it is February 14th, then it is Valentine's Day.*
>
> *If a shape is a parallelogram, then its area is $A = bh$.*

2-81.   Solve for y in terms of x. That is, rewrite each equation so that it starts "$y = $".

    a. $6x + 5y = 20$ b. $4x - 8y = 16$

2-82.   Calculate the area of the shaded region at right. Use the appropriate units.

8"

17"

2-83.   How tall are you? How do you measure your height? Consider these questions as you answer the questions below.

    a.   Why do you stand up straight to measure your height?

    b.   Which diagram best represents how you would measure your height? Why?

           Diagram 1   **Diagram 2**

    c.   When you measure your height, do you measure up to your chin? Down to your knees? Explain.

2-84.   Read the Math Notes box for this lesson. Then rewrite each of the following statements as a conditional statement.

    a.   Mr. Spelling is always unhappy when it rains.

    b.   The sum of two even numbers is always even.

    c.   Marla has a piano lesson every Tuesday.

2-85.   Simplify the following expressions.

    a. $2x + 8 + 6x + 5$ b. $15 + 3(2x - 4) - 4x$

    c. $(x - 3)(3x + 4)$ d. $5x(2x + 7) + x(3x - 5)$

2.2.4 How can I find the height?

Heights and Area

In Lesson 2.2.2, you learned that triangles with the same base and height must have the same area. But what if multiple dimensions of a shape are labeled? How can you determine which dimension is the height?

2-86. Candice missed the lesson about finding the area of the triangle. Not knowing where to start, she drew a triangle and measured its sides, as shown at right. After drawing her triangle, Candice said, "Well, I've measured all of the sides. I must be ready to find the area!"

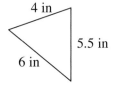

If you think she is correct, write a description of how to use the side lengths to find the area. If you think she needs to measure anything else, copy the figure on your paper and add a line segment to represent a measurement she needs.

2-87. HEIGHT LAB

What is the height of a triangle? Is it like walking up any side of the triangle? Or is it like standing at the highest point and looking straight down? Today your team will build triangles with string and consider different ways height can be seen for triangles of various shapes.

a. Use the materials given to you by your teacher to make a triangle like the one in the diagram below.

(1) Tie one end of the short string to the weight and the other to the end of a pencil (or pen).

(2) Tape a 15 cm section of the long string along the edge of a desk or table. Be sure to leave long ends of string hanging off each side.

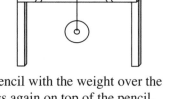

(3) Bring the loose ends of string up from the table and cross them as shown in the diagram. Then put the pencil with the weight over the crossing of the string. Cross the strings again on top of the pencil.

Problem continues on next page →

Geometry Connections

2-87.   *Problem continued from previous page.*

b.   Now, with your team, build and sketch triangles that meet the three conditions below. To organize your work, assign each team member one of the jobs described at right.

Student jobs:
• hold the pencil (or pen) with the weight.
• make sure that the weight hangs freely
• draw accurate sketches.
• hold the strings that make the sides of the triangles.

(1) The height of the triangle is inside the triangle.

(2) The height of the triangle is a side of the triangle.

(3) The height of the triangle is outside of the triangle.

c.   Now make sure that everyone in your team has sketches of the triangles that you made.

2-88.   How can I find the height of a triangle if it is not a right triangle?

a.   On the Lesson 2.2.4 Resource Page there are four triangles labeled (1) through (4). Assume you know the length of the side labeled "base." For each triangle, draw in the height that would enable you to find the area of the triangle. **Note:** You do not need to find the area.

b.   Find the triangle for part (b) at the bottom of the same resource page. For this triangle, draw all three possible heights. First **choose** one side to be the base and draw in the corresponding height. Then repeat the process of drawing in the height for the other two sides, one at a time.

c.   You drew in three pairs of bases and heights for the triangle in part (b). Using centimeters, measure the length of all three sides and all three heights. Find the area three times using all three pairs of bases and heights. Since the triangle remains the same size, your answers should match.

2-89.   AREA TOOLKIT

Over the past several days, you have explored how to find the areas of triangles, parallelograms, and trapezoids. Today you will start a new page of your Geometry Toolkit, called the Area Toolkit. Keep this toolkit in a safe place. You will want it for reference in class and when doing homework.

At this time, describe what you know about finding the areas of triangles, rectangles, parallelograms, and trapezoids. Be sure to include an example for each shape.

METHODS AND MEANINGS

Areas of a Triangle, Parallelogram, and Trapezoid

The area of a triangle is half the area of a rectangle with the same base and height. If the base of the triangle is length b and the height length h, then the area of the triangle is:

$$A = \tfrac{1}{2}bh \,.$$

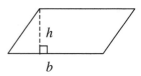

The area of a parallelogram is equal to a rectangle with the same base and height. If the base of the parallelogram is length b and the height length h, then the area of the parallelogram is:

$$A = bh \,.$$

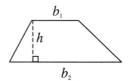

Finally, the area of a trapezoid is found by averaging the two bases and multiplying by the height. If the trapezoid has bases b_1 and b_2 and height h, then the area is:

$$A = \tfrac{1}{2}(b_1 + b_2)h \,.$$

Review & Preview

2-90.   Find the area of each figure below. Show all work. Remember to include units in your answer.

a.   a square:

7 cm

b.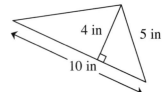

4 in   5 in

10 in

c.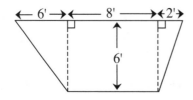

2-91.   Multiply the expressions below. Then simplify the result, if possible.

a. $3x(5x + 7)$

b. $(x + 2)(x + 3)$

c. $(3x + 5)(x - 2)$

d. $(2x + 1)(5x - 4)$

2-92.   Graph the following equations on the same set of axes. Label each line or curve with its equation. Where do the two curves intersect?

$$y = -x - 3 \qquad\qquad y = x^2 + 2x - 3$$

2-93.   On graph paper, plot quadrilateral $ABCD$ if $A(2, 7)$, $B(4, 8)$, $C(4, 2)$, and $D(2, 3)$.

    a.   What is the best name for this shape? **Justify** your conclusion.

    b.   Quadrilateral $A'B'C'D'$ is formed by rotating $ABCD$ 90° clockwise about the origin. Name the coordinates of the vertices.

    c.   Find the area of $ABCD$. Show all work.

2-94.   What is the probability of drawing each of the following cards from a standard playing deck? Refer to problem 2-74 if you need information about a deck of cards.

    a.   P(face card)

    b.   P(card printed with an even number)

    c.   P(red ace)

    d.   P(purple card)

2.3.1 How can I use a square?

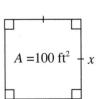

Squares and Square Roots

You now have the tools to find the area of many complex shapes and are also able to use transformations to create new shapes. In this lesson you will combine these skills to explore the area of a square and to develop a method for finding the length of the longest side of a right triangle.

2-95.   What do you notice about the rectangle at right?

    a.   Elyse does not know how to solve for x. Explain to her how to find the missing dimension.

    b.   What if the area of the shape above is instead 66 ft^2? What would x be in that case?

$A = 100$ ft^2 x

2-96.   While Alexandria was doodling on graph
        paper, she made the design at right. She
        started with the shaded right triangle. She
        then rotated it 90° clockwise and translated
        the result so that the right angle of the image
        was at *B*. She continued this pattern until she
        completed the diagram.

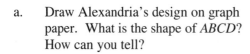
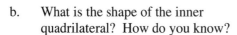
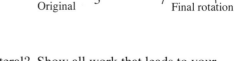

   a.   Draw Alexandria's design on graph
         paper. What is the shape of *ABCD*?
         How can you tell?

   b.   What is the shape of the inner
         quadrilateral? How do you know?

   c.   What is the area of the inner quadrilateral? Show all work that leads to your
         conclusion.

   d.   What's the length of the longest side of the shaded triangle?

2-97.   DONNA'S DILEMMA

        Donna needs help! She needs to find PQ (the length
        of $\overline{PQ}$) in △*PQR* shown at right.

        Alexandria realized that the
        triangle looked a lot like a piece
        of the square she drew earlier
        (shown in problem 2-96).
        Alexandria explained to Donna
        that if they could find the area
        of a square with side $\overline{PQ}$, then
        finding the length of $\overline{PQ}$ would
        be easy.

   a.   Draw △*PQR* on your paper.

   b.   Using the method from problem 2-96, help the students construct the square by
         rotating the triangle 90° and translating it. Then find the length of $\overline{PQ}$. Is the
         answer reasonable?

   c.   Find the perimeter of her triangle.

2-98.   Use Donna and Alexandria's method to find the length of side
        $\overline{AC}$ in the diagram at right. That is, draw a square on $\overline{AC}$
        using what you know about slope. Then use the area of the
        square to find $\overline{AC}$.

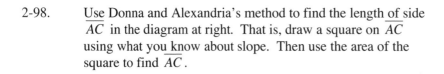

2-99. In a Learning Log entry, describe the process you developed today to find the length of the longest side of a right triangle when given the lengths of its legs. Label this entry "Finding the Length of the Hypotenuse" and include today's date.

METHODS AND MEANINGS

Square Root

The **square root** of a number x (written $\sqrt{x}$) is defined as the length of a side of a square with area x. For example, $\sqrt{16} = 4$. Therefore if the side of a square has a length of 4 units, then its area is 16 square units.

When the area of a square is not a "perfect square" such as 49, 25, or 9 un^2 then the square root is not an integer, and instead is an **irrational number.** Note that an irrational number cannot be written as a fraction of two integers. For example, $\sqrt{17} = 4.123\ldots$ is an irrational number because 17 is not a perfect square. The symbol "$\approx$" means "approximately equal to." Since 4.123 is an approximate value of $\sqrt{17}$, it can be stated that $\sqrt{17} \approx 4.123$.

The square root of a number can often be estimated by comparing the number under the square root with its closest perfect squares. For example, since $\sqrt{16} = 4$, then $\sqrt{17}$ must be a little bigger than 4. Therefore, a good estimation of $\sqrt{17}$ is 4.1. Likewise, if you want to estimate $\sqrt{97}$, you may want to start with $\sqrt{100} = 10$. Since 97 is smaller than 100, a good estimate of $\sqrt{97}$ would be 9.8 or 9.9.

2-100. Read the Math Notes box for this lesson.

a. What would be a reasonable estimate of $\sqrt{68}$? Explain your thinking. After you have made an estimate, check your estimation with a calculator.

b. Repeat this process to estimate the values below.

(1) $\sqrt{5}$ (2) $\sqrt{85}$ (3) $\sqrt{50}$ (4) $\sqrt{22}$

2-101.   Draw the triangle at right on graph paper. Then draw a square
          on $\overline{KM}$ and use it to find the length of $\overline{KM}$.

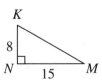

2-102.   **Examine** the rectangle at right.

          a.   What is the perimeter in terms of x? In other
                words, find the perimeter.

          b.   If the perimeter is 78 cm, find the dimensions of
                the rectangle. Show all your work.

          c.   Verify that the area of this rectangle is 360 sq. cm. Explain how you know
                this.

2-103.   **Examine** the arrow diagram below.

          Polygon is a parallelogram → *area of the polygon equals base times height.*

          a.   Write this conjecture as a conditional ("If, then") statement.

          b.   Write a similar conjecture about triangles, both as a conditional statement and
                as an arrow diagram.

2-104.   Berti is the Shape Factory's top
          employee. She has received awards
          every month for having the top sales
          figures so far for the year. If she stays
          on top, she'll receive a $5000 bonus
          for excellence. She currently has sold
          16,250 shapes and continues to sell
          340 per month.

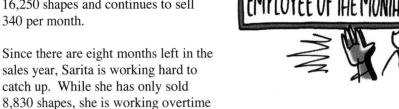

          Since there are eight months left in the
          sales year, Sarita is working hard to
          catch up. While she has only sold
          8,830 shapes, she is working overtime
          and on weekends so that she can sell 1,082 per month. Will Sarita catch up with
          Berti before the end of the sales year? If so, when?

2.3.2 Is the answer reasonable?

Triangle Inequality

You now have several tools for describing triangles (lengths, areas, and angle measures), but can **any** three line segments create a triangle? Or are there restrictions on the side lengths of a triangle? And how can you know that the length you found for the side of a triangle is accurate? Today you will **investigate** the relationship between the side lengths of a triangle.

2-105.   Roiri (pronounced "ROAR-ree") loves right
triangles and has provided the one at right to
analyze. He wants your team to find the length of
the **hypotenuse** ($\overline{AC}$) using the method from
problem 2-97.

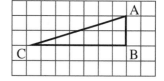

a.   Estimate the length of $\overline{AC}$.

b.   Draw Roiri's triangle on graph paper and then construct a square on the
hypotenuse. Use the area of the square to find the length of the hypotenuse.

c.   Was your result for the hypotenuse close to your estimate? Why or why not?

d.   For a different triangle $\triangle ABC$ where $AB = 3$ units and
$BC = 4$ units, Roiri found that $AC = 25$. Donna is not
sure that is possible. What do you think? **Visualize** this
triangle and explain if you think this triangle is possible or
not.

2-106.   PINK SLIP

Oh no! During your last shift at the Shape
Factory everything seemed to be going fine-
-until the machine that was producing
triangles made a huge CLUNK and then
stopped. Since your team was on duty, all
of you will be held responsible for the
machine's breakdown.

Luckily, your boss has informed you that if
you can figure out what happened and how
to make sure it will not happen again, you
will keep your job. The last order the
machine was processing was for a triangle
with sides 3 cm, 5 cm, and 10 cm.

a.   Use the manipulative (such as a dynamic geometry tool or pasta)
       provided by your teacher to **investigate** what happened today at
       the factory. Can a triangle be made with <u>any</u> three side lengths?
       If not, what condition(s) would make it impossible to build a
       triangle? Try building triangles with the side lengths listed
       below:

       (1)   3 cm, 5 cm, and 10 cm   (2)   4 cm, 9 cm, and 12 cm

       (3)   2 cm, 4 cm, and 5 cm   (4)   3 cm, 5 cm, and 8 cm

b.   For those triangles that could not be built, what happened? Why were they
       impossible?

c.   Use your dynamic geometry tool (or dry pasta) to **investigate** the restrictions
       on the three side lengths that can form a triangle. For example, if two sides of
       a triangle are 5 cm and 12 cm long, respectively, what is the longest side that
       could join these two sides to form a triangle? (Could the third side be 12 cm
       long? 19 cm long?) What is the shortest possible length that could be used to
       form a triangle? (Does 5 cm work? 9 cm?)

d.   Write a memo to your boss explaining what happened. If you can convince
       your boss that the machine's breakdown was not your fault **and** show the
       company how to fix the machine so that this does not happen again, you might
       earn a promotion!

2-107.   The values you found in parts (a) and (c) of problem 2-106 were the **minimum** and **maximum limits** for the length of the third side of any triangle with two sides of lengths 5 cm and 12 cm. The fact that there are restrictions on the side lengths that may be used to create a triangle is referred to as the **Triangle Inequality.**

Determine the minimum and maximum limits for each missing side length in the triangles below.

a.
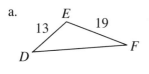

b.
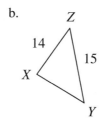

c.

2-108.   In a Learning Log entry, explain how you can tell if three sides will form a triangle or not. Draw diagrams to support your statements. Title this entry, "Triangle Inequality" and include today's date.

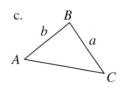

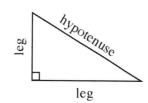

ETHODS AND MEANINGS

Right Triangle Vocabulary

All of the triangles that we have been working with in this section are right triangles, that is, triangles that contain a 90° angle. The two shortest sides of the right triangle (the sides that meet at the right angle) are called the **legs** of the triangle and the longest side (the side opposite the right angle) is called the **hypotenuse** of the triangle.

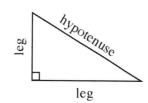

2-109.   Draw a right triangle with legs of length 6 and 8 units, respectively, onto graph paper. Construct a square on the hypotenuse and use the square's area to find the length of the hypotenuse.

2-110.   Write the equation of each line described below in slope-intercept form ($y = mx + b$).

   a. $m = \frac{6}{5}$ and $b = -3$ b. $m = -\frac{1}{4}$ and $b = 4.5$

   c. $m = \frac{1}{3}$ and the line passes   d. $m = 0$ and $b = 2$
         through the origin $(0, 0)$

2-111.   **Examine** the diagram at right. Based on the information in the diagram, which angles can you determine? Copy the diagram on your paper and find <u>only</u> those angles that you can **justify**.

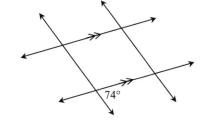

2-112.   Hannah's shape bucket contains an equilateral triangle, an isosceles right triangle, a regular hexagon, an isosceles trapezoid, a rhombus, a kite, a parallelogram and a rectangle. If she reaches in and selects a shape at random, what is the probability that that the shape will meet the criterion described below?

   a.   At least two sides congruent.

   b.   Two pairs of parallel sides.

   c.   At least one pair of parallel sides.

2-113.   On graph paper, plot $ABCD$ if $A(-1, 2)$, $B(0, 5)$, $C(2, 5)$, and $D(6, 2)$.

   a.   What type of shape is $ABCD$? **Justify** your answer.

   b.   If $ABCD$ is rotated 90° counterclockwise (↺) about the origin, name the coordinates of the image $A'B'C'D'$.

   c.   On your graph, reflect $ABCD$ across the y-axis to find $A''B''C''D''$. Name the coordinates of A''' and C'''.

   d.   Find the area of $ABCD$. Show all work.

2.3.3 Is there a shortcut?

· ·

The Pythagorean Theorem

During Lessons 2.3.1 and 2.3.2, you have learned a method to find the
length of a hypotenuse of a right triangle by finding the area of the square
built on the hypotenuse, as shown in the diagram at right. However, what
if the sides of the triangle make it difficult to draw (such as very large
numbers or decimal values)? Or what if you do not even know one of the
lengths of the legs?

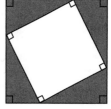

Today, you will work with your team to find the relationship between the legs and hypotenuse
of a right triangle. By the end of this lesson, you should be able to find the side of any right
triangle, when given the lengths of the other two sides.

2-114.   Roiri complained that while the method
          from problem 2-97 works, it seems like
          too much work! He remembers that
          rearranging a shape does not change its
          area and thinks he can find a shortcut
          by rearranging Donna's square. Obtain
          a Lesson 2.3.3 Resource Page for your
          team and cut out the shaded triangles.
          Note that the lengths of the sides of the
          triangles are a, b, and c units
          respectively.

a.   First, arrange the triangles to look like Donna's in the
      diagram at right. Draw this diagram on your paper.
      What is the area of the unshaded square?

b.   Roiri claims that moving the triangles within the outer
      square won't change the area of the unshaded square.
      Is Roiri correct? Why or why not?

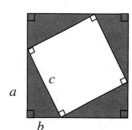

c.   Move the shaded triangles to match the diagram at
      right. In this configuration, what is the total area that
      is unshaded?

d.   Write an equation that relates the two ways that you
      found to represent the unshaded area in the figure.

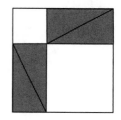

2-115.   The relationship you found in problem 2-114 (between the square of the lengths of the legs and the square of the length of the hypotenuse in a right triangle) is known as the **Pythagorean Theorem**. This relationship is a powerful tool because once you know the lengths of any two sides of a right triangle, you can find the length of the third side. But does this relationship always hold true?

a.   Use a dynamic geometry tool to test whether the square of the hypotenuse always equals the sum of the squares of the legs of a right triangle.

b.   Add an entry in your Learning Log for the Pythagorean Theorem, explaining what it is and how to use it. Be sure to include a diagram.

2-116.   Apply the Pythagorean Theorem to answer the questions below.

a.   For each triangle below, find the value of the variable.

(1)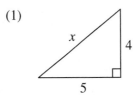

(2)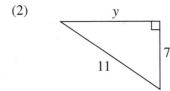

b.   **Examine** the rectangle shown at right. Find its perimeter and area.

c.   On graph paper, draw $\overline{AC}$ if $A(2, 6)$ and $B(5, -1)$. Then draw a slope triangle. Use the slope triangle to find the length of $\overline{AC}$.

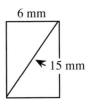

2-117.   The Garcia family took a day trip from Cowpoke Gulch. Their online directions told them to drive four miles north, six miles east, three miles north, and then one mile east to Big Horn Flat. Draw a diagram and calculate the direct distance (straight) from Cowpoke Gulch to Big Horn Flat.

METHODS AND **M**EANINGS

The Pythagorean Theorem

The **Pythagorean Theorem** states that in a right triangle,

$$(\text{length of leg \#1})^2 + (\text{length of leg \#2})^2 = (\text{length of hypotenuse})^2$$

The Pythagorean Theorem can be used to find a missing side length in a right triangle. See the example below.

$$5^2 + x^2 = 8^2$$
$$25 + x^2 = 64$$
$$x^2 = 39$$
$$x = \sqrt{39} \approx 6.24$$

In the example above, $\sqrt{39}$ is an example of an **exact** answer, while 6.24 is an **approximate** answer.

2-118.   One of the algebra topics you have reviewed during this chapter is solving systems of equations. Assess what you know about solving systems as you answer the questions below.

a.   Find the points of intersection of the lines below using any method. Write your solutions as a point (x, y).

(1) $y = -x + 8$ (2) $2x - y = 10$
        $y = x - 2$ $y = -4x + 2$

b.   Find the equation for each line on the graph at right. Remember, the general form of any line in the **slope-intercept form** is $y = mx + b$.

c.   Solve the system of equations you found in part (b) algebraically. Verify that your solution matches the one shown in the graph at right.

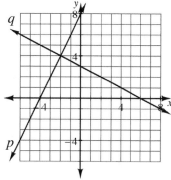

2-119.   Lines *p* and *q* graphed in problem 2-118 form a triangle with the *x*-axis.

   a.   How can you describe this triangle? In other words, what is the most appropriate name for this triangle? How do you know?

   b.   Find the area of the triangle.

   c.   What is the perimeter?

2-120.   Find the area of the trapezoid at right. What **strategies** did you use?

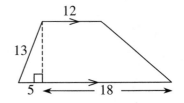

2-121.   Use the relationships in the diagrams below to solve for *x*, if possible. If it is not possible, state how you know. If it is possible, **justify** your solution by stating which geometric relationships you use.

   a.

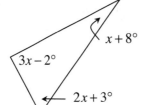

   b.

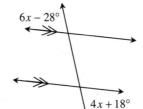

   c.

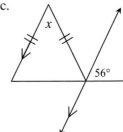

2-122.   Find the minimum and maximum limits for the length of a third side of a triangle if the other two sides are 8" and 13".

Chapter 2 Closure What have I learned?

Reflection and Synthesis

The activities below offer you a chance to reflect on
what you have learned in this chapter. As you
work, look for concepts that you feel very
comfortable with, ideas that you would like to learn
more about, and topics you need more help with.
Look for connections between ideas as well as
connections with material you learned previously.

①   TEAM BRAINSTORM

With your team, brainstorm a list for each of the following two subjects. Be as
detailed as you can. How long can you make your list? Challenge yourselves. Be
prepared to share your team's ideas with the class.

Topics: What have you studied in this chapter? What ideas and
words were important in what you learned? Remember to
be as detailed as you can.

Connections: How are the topics, ideas, and words that you learned in
previous courses connected to the new ideas in this
chapter? Again, make your list as long as you can.

②   MAKING CONNECTIONS

The following is a list of the vocabulary used in this chapter. The words that appear
in bold are new to this chapter. Make sure that you are familiar with all of these
words and know what they mean. Refer to the glossary or index for any words that
you do not yet understand.

alternate interior angles	area	**arrow diagram**
base	**complementary angles**	**conditional statement**
conjecture	**corresponding angles**	**dimension**
height	**hypotenuse**	leg
line	**parallelogram**	perimeter
proof by contradiction	**Pythagorean Theorem**	**rectangle**
right angle	**same-side interior angles**	**square**
square root	straight angle	**supplementary angles**
theorem	**transversal**	**trapezoid**
triangle inequality	**unit of measure**	**vertical angles**

Problem continues on next page →

②   *Problem continued from previous page.*

Make a concept map showing all of the connections you can find among the key words and ideas listed above. To show a connection between two words, draw a line between them and explain the connection, as shown in the example below. A word can be connected to any other word as long as there is a justified connection. For each key word or idea, provide a sketch of an example.

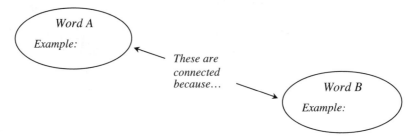

Your teacher may provide you with vocabulary cards to help you get started. If you use the cards to plan your concept map, be sure to either re-draw your concept map on your paper or glue the vocabulary cards to a poster with all of the connections explained for others to see and understand.

While you are making your map, your team may think of related words or ideas that are not listed above. Be sure to include these ideas on your concept map.

③   SUMMARIZING MY UNDERSTANDING

This section gives you an opportunity to show what you know about certain math topics or ideas. Your teacher will direct you how to do this.

④   WHAT HAVE I LEARNED?

This section will help you recognize those types of problems you feel comfortable with and those you need more help with. This section will appear at the end of every chapter to help you check your understanding. Even if your teacher does not assign this section, it is a good idea to try these problems and find out for yourself what you know and what you need to work on.

Solve each problem as completely as you can. The table at the end of this closure section has answers to these problems. It also tells you where you can find additional help and practice on similar problems.

CL 2-123. Sandra's music bag contains:

> 3 CDs of country music
>
> 6 CDs of rock and roll
>
> 4 CDs of rap music
>
> 5 cassette tapes of country music
>
> 1 cassette tape of rock and roll
>
> 3 cassette tapes of rap music

a. As Sandra drives, she often reaches into her bag and randomly pulls out some music (she does not want to take her eyes off the road). What is the probability that she will select some rap music?

b. Find P(CD), that is, the probability that she will randomly select a CD (of any type of music).

c. Find P(cassette of country music).

d. Find P(not rock and roll), the probability that the music she randomly selects is **not** rock and roll.

CL 2-124. Find the area of each figure.

a.

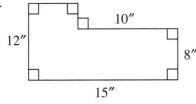

b.

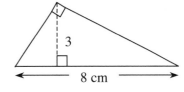

c.

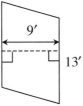

d.

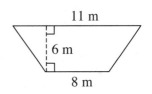

CL 2-125. Find the perimeter of each figure.

a.

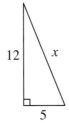

b.

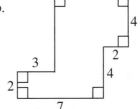

CL 2-126. Graph the segment that connects the points $A(-4,8)$ and $B(7,3)$.

    a.   What is the slope of $\overline{AB}$?   b.   How long is $\overline{AB}$?

CL 2-127. Identify the geometric angle relationship(s) in each diagram. Use what you know about those relationships to write an equation and solve for x.

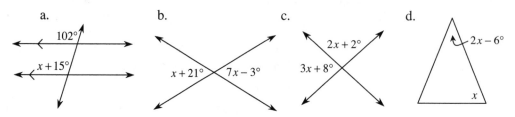

    a.   b.   c.   d.

CL 2-128. **Examine** the system of equations at right.

$$y = -2x + 6$$
$$y = \tfrac{1}{2}x - 9$$

    a.   Solve the system below <u>twice</u>: graphically and algebraically. Verify that your solutions from the different methods are the same.

    b.   What is the relationship between the two lines? How can you tell?

CL 2-129. Charlotte was transforming the hexagon *ABCDEF*.

    a.   What single transformation did she perform in Diagram #1?

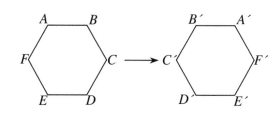

Diagram #1

    b.   What single transformation did she perform in Diagram #2?

    c.   What transformation didn't she do? Write directions for this type of transformation for hexagon *ABCDEF* and perform it.

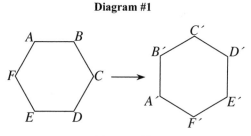

Diagram #2

CL 2-130. Explain what you are doing when you find the perimeter of a shape. How is that different than finding the area?

CL 2-131. Check your answers using the table at the end of this section. Which problems do you feel confident about? Which problems were hard? Have you worked on problems like these in math classes you have taken before? Use the table to make a list of topics you need help on and a list of topics you need to practice more.

Geometry Connections

⑤   HOW AM I THINKING?

This course focuses on five different **Ways of Thinking**: investigating, examining, reasoning and justifying, visualizing, and choosing a strategy/tool. These are some of the ways in which you think while trying to make sense of a concept or to solve a problem (even outside of math class). During this chapter, you have probably used each Way of Thinking multiple times without even realizing it!

This closure activity will focus on one of these Ways of Thinking: **examining**. Read the description of this Way of Thinking at right.

Think about the diagrams that you have looked at or drawn during this chapter. When have you had to understand information from a diagram or picture? What helps you to see what is in the diagram? What diagrams or parts of diagrams have you seen in a previous math class? You may want to flip through the chapter to refresh your memory about the problems that you have worked on. Discuss any of the methods you have developed to **examine** the problem in order to identify what is important or what information is conveyed in a diagram.

Once your discussion is complete, think about the way you think as you answer the following problems.

Examining

To **examine** means to look at a diagram and recognize its geometric properties. As you develop this way of thinking you will learn what specifically to look for in a diagram depending on the problem you are solving. Often, examining is what you do to answer to the question, "How does this diagram help me?" When you catch yourself thinking, "I can see from the diagram that...", you are examining.

a.   Sometimes, **examining** a shape means you have to disregard how it *looks* and concentrate on the information provided by the markings.

For example, **examine** the shape at right. This shape looks like a square, but is it? Make as many statements as you can state about this shape based on the markings. What shape is *ABCD*? Which statements are obvious from the diagram and which did you have to think about?

Problem continues on next page →

⑤ *Problem continued from previous page.*

b. At other times, examining a diagram suggests that you notice different parts that contribute to the entire diagram. For example, look at the diagram at right. Assuming that the diagram is drawn to scale, what shapes can you find in the diagram?

c. In part (a), you looked for attributes of a shape, while in part (b), you looked for shapes within a shape.

However, another aspect of examining a diagram is to look for relationships between parts of the diagram. For example, **examine** the diagram at right. What relationships do you see between the angles in the diagram? What relationships can you find between the lines and/or line segments? List as many relationships as you can based on the information in the diagram.

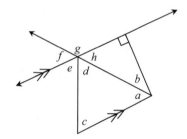

Answers and Support for Closure Activity #4
What Have I Learned?

Problem	Solutions	Need Help?	More Practice
CL 2-123.	a. $\frac{7}{22} \approx 0.32$ b. $\frac{13}{22} \approx 0.59$ c. $\frac{5}{22} \approx 0.23$ d. $\frac{15}{22} \approx 0.68$	Lesson 1.3.3 Math Notes box	Problems 2-9, 2-55, 2-74, 2-94, 2-112
CL 2-124.	a. 140 sq. in. b. 12 sq. cm. c. 117 sq. ft. d. 57 sq. m.	Lesson 2.2.4 Math Notes box	Problems 2-66, 2-68, 2-71, 2-75, 2-76, 2-78, 2-79, 2-82, 2-83, 2-90, 2-120
CL 2-125.	a. $x = 13$ units; $p = 30$ units b. $p = 34$ units	Lessons 1.1.3 and 2.3.3 Math Notes boxes	Problems 1-44, 1-65, 1-93, 2-52(a), 2-57, 2-97, 2-102, 2-116(b), 2-119

Problem	Solutions	Need Help?	More Practice
CL 2-126.	a. $m = -\frac{5}{11}$ b. length = $\sqrt{146} \approx 12.1$ units	Lessons 1.2.5 and 2.3.3 Math Notes boxes	Problems 1-36, 1-76, 2-116, 2-117
CL 2-127.	a. $x + 15° = 102°$; $x = 87°$; corresponding angles b. $7x - 3° = x + 21°$; $x = 4°$; vertical angles c. $2x + 2° + 3x + 8° = 180°$; $x = 34°$; supplementary angles d. $x + x + 2x - 6° = 180°$; $x = 46.5°$; sum of angles in a triangle is $180°$	Lessons 2.1.1 and 2.1.4 Math Notes boxes	Problems 2-13, 2-16, 2-17, 2-18, 2-19, 2-23, 2-31, 2-32, 2-35, 2-36, 2-37, 2-38, 2-49, 2-51, 2-52, 2-62, 2-63, 2-72, 2-111, 2-121
CL 2-128.	a. $(6, -6)$ b. They are perpendicular because the slopes are opposite reciprocals. 	Lessons 2.1.3, 1.2.1, and 1.2.5 Math Notes boxes	Problems 1-62, 1-92, 2-29, 2-39, 2-42, 2-53, 2-54(c), 2-65, 2-92, 2-118
CL 2-129.	a. Reflection (flip) b. Rotation (turn counterclockwise 90º) c. Translation (slide). Answers will vary, and example is provided below: 	Lessons 1.2.2 Math Notes box	Problems 1-50, 1-51, 1-59, 1-60, 1-61, 1-64, 1-69, 1-73, 1-85, 1-96, 2-11, 2-19, 2-20, 2-33, 2-64, 2-113
CL 2-130.	The perimeter of a shape is the length of that shape's boundary. While to find the area, consider everything *inside* the boundary.	Lesson 1.1.3 Math Notes box	Problems 1-65, 1-66, 1-74, 1-84, 2-57, 2-116

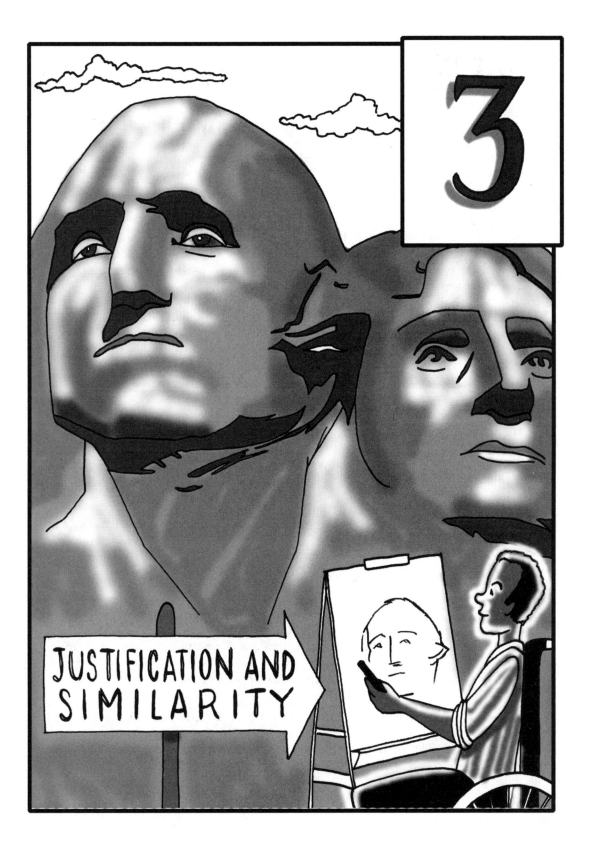

JUSTIFICATION AND SIMILARITY

CHAPTER 3
Justification and Similarity

Measuring, describing, and transforming: these are three major skills in geometry that you have been developing. In this chapter, you will focus on comparing; you will explore ways to determine if two figures have the same shape (called **similar**). You will also develop ways to use the information about one figure to learn more about another that has the same shape.

Making logical and convincing arguments that support specific ideas about the shapes you are studying is another important skill. In this chapter you will learn how you can document facts to support a conclusion in a flowchart.

In this chapter, you will learn:

➢ how to support a mathematical statement using flowcharts and conditional statements.

➢ about the special relationships between shapes that are similar or congruent.

➢ how to determine if triangles are similar or congruent.

Guiding Questions

Think about these questions throughout this chapter:

Is it still the same shape?

How did the shape grow or shrink?

What do I need to show for it to be similar?

How can I justify that?

What evidence can I state?

Chapter Outline

Section 3.1 Through an exploration with rubber bands, students will generate similar figures, which will launch a focus on similarity for Sections 3.1 and 3.2. During these lessons, students will determine what common qualities similar shapes have.

Section 3.2 As students discover the conditions that cause triangles to be similar or congruent, they will learn about using a flowchart to organize facts and support their conclusions.

3.1.1 What do these shapes have in common?

Similarity

In Section 1.3, you organized shapes into groups based on their size, angles, sides, and other characteristics. You identified shapes using their characteristics and investigated relationships between different kinds of shapes, so that now you can tell if two shapes are both parallelograms or trapezoids, for example. But what makes two figures look alike?

Today you will be introduced to a new transformation that enlarges a figure while maintaining its shape, called a **dilation**. After creating new enlarged shapes, you and your team will explore the interesting relationships that exist between figures that have the same shape.

In your teams, you should keep the following questions in mind as you work together today:

What do the shapes have in common?

How can you demonstrate that the shapes are the same?

What specifically is different about the shapes?

3-1. WARM-UP STRETCH

Before computers and copy machines existed, it sometimes took hours to enlarge documents or to shrink text on items such as jewelry. A pantograph device (like the one below) was often used to duplicate written documents and artistic drawings. You will now employ the same geometric principles by using rubber bands to draw enlarged copies of a design. Your teacher will show you how to do this.

3-2.   In problem 3-1, you created designs that were
        similar, meaning that they have the same shape.
        But how can you determine if two figures are
        similar? What do similar shapes have in common?
        To find out, your team will need to create similar
        shapes that you can measure and compare.

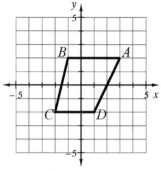

Diagram #1

   a.   Obtain a Lesson 3.1.1
        Resource Page from your
        teacher (or download it from
        www.cpm.org). On it, find
        the quadrilateral shown in
        Diagram #1 at right.

        Dilate (stretch) the
        quadrilateral from the origin
        by a factor of 2, 3, 4, or 5 to
        form $A'B'C'D'$. Each team
        member should pick a
        different enlargement factor.
        You may want to imagine that
        your rubber band chain is
        stretched from the origin so
        that the knot traces the
        perimeter of the original
        figure.

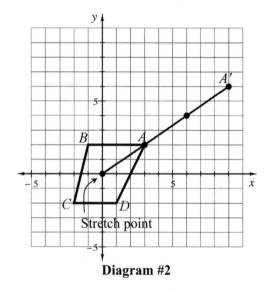

        For example, if your job is to
        stretch *ABCD* by a factor of 3,
        then A' would be located as
        shown in Diagram #2 at right.

Diagram #2

   b.   Carefully cut out your enlarged shape and compare it to your teammates'
        shapes. How are the four shapes different? How are they the same? As you
        investigate, make sure you record what qualities make the shapes different and
        what qualities make the shapes the same. Be ready to report your conclusions
        to the class.

Geometry Connections

3-3.   WHICH SHAPE IS THE EXCEPTION?

Sometimes shapes look the same and
sometimes they look very different.
What characteristics make figures alike
so that we can say that they are the same
shape? How are shapes that look the
same but are different sizes related to
each other? Understanding these
relationships will allow us to know if
shapes that appear to have the same
shape actually do have the same shape.

Your task: For each set of shapes below, three shapes are similar, and one is an
exception, which means that it is not like the others. Find the exception in each set
of shapes. Your teacher may give you tracing paper to help you in your
investigation.

Answer each of these questions for both sets of shapes below:

• Which shape appears to be the exception? What makes that shape different
  from the others?

• What do the other three shapes have in common?

• Are there commonalities in the angles? Are there differences?

• Are there commonalities in the sides? Are there differences?

a.

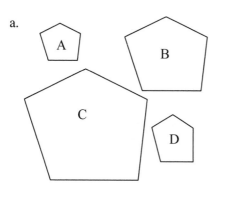

b.

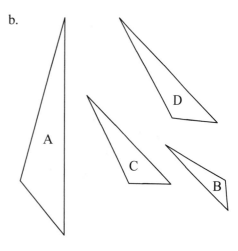

3-4.   Write an entry in your Learning Log about the characteristics that
        figures with different sizes need to have in order to maintain the same
        shape. Add your own diagrams to illustrate the description. Title this
        entry "Same Shape, Different Size" with today's date.

METHODS AND MEANINGS

Dilations

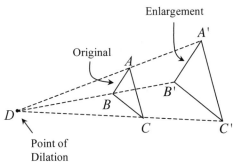

The transformations you studied in Chapter 1 (translations, rotations, and reflections) are called rigid transformations because they all maintain the size and shape of the original figure.

However, a **dilation** is a transformation that maintains the shape of a figure but multiplies its dimensions by a chosen factor. In a dilation, a shape is stretched proportionally from a particular point, called the **point of dilation** or **stretch point**. For example, in the diagram at right, $\triangle ABC$ is dilated to form $\triangle A'B'C'$. Notice that while a dilation changes the size and location of the original figure, it does not rotate or reflect the original.

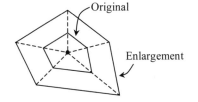

Note that if the point of dilation is located inside a shape, the enlargement encloses the original, as shown at right.

3-5.   Plot the rectangle *ABCD* formed with the points $A(-1, -2)$, $B(3, -2)$, $C(3, 1)$, and $D(-1, 1)$ onto graph paper. Use the method used in problem 3-2 to enlarge it from the origin by a factor of 2 (using two "rubber bands"). Label this new rectangle $A'B'C'D'$.

   a.   What are the dimensions of the enlarged rectangle, $A'B'C'D'$?

   b.   Find the area and the perimeter of $A'B'C'D'$.

   c.   Find AC (the length of $\overline{AC}$).

3-6.   Solve each equation below for *x*. Show all work and check your answer by substituting it back into the equation and verifying that it makes the equation true.

   a. $\frac{x}{3} = 6$ b. $\frac{5x+9}{2} = 12$ c. $\frac{x}{4} = \frac{9}{6}$ d. $\frac{5}{x} = \frac{20}{8}$

3-7.   **Examine** the triangle at right.

a.   Estimate the measure of each angle
      of the triangle at right.

b.   Given only its shape, what is the best name for this triangle?

3-8.   On graph paper, graph line $\overleftrightarrow{MU}$ if $M(-1, 1)$ and $U(4, 5)$.

a.   Find the slope of $\overleftrightarrow{MU}$.

b.   Find MU (the distance from M to U).

c.   Are there any similarities to the calculations used in parts (a) and (b)? Any differences?

3-9.   Rewrite the statements below into conditional ("If …, then …") form.

a.   All equilateral triangles have 120° rotation symmetry.

b.   A rectangle is a parallelogram.

c.   The area of a trapezoid is half the sum of the bases multiplied by the height.

3.1.2 How can I maintain the shape?

Proportional Growth and Ratios

So far you have studied several shapes that appear to be **similar** (exactly the same shape but not necessarily the same size). But how can we know for sure that two shapes are similar? Today you will focus on the relationship between the lengths of sides of similar figures by enlarging and reducing shapes and looking for patterns.

3-10.   Find your work from problem 3-5. The graph you
        created should resemble the graph at right.

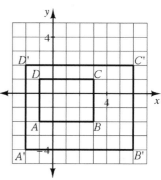

a.   In problem 3-5, you dilated (stretched) *ABCD*
      to create rectangle $A'B'C'D'$, which is similar
      to *ABCD*. Which side of $A'B'C'D'$
      corresponds to $\overline{CB}$? Which side corresponds to
      $\overline{AB}$?

b.   Compare the lengths of each pair of
      corresponding sides. What do you notice?
      How could you get the lengths of $A'B'C'D'$ from the dimensions of *ABCD*?

Problem continues on next page →

3-10.   *Problem continued from previous page.*

     c.   Could you get the dimensions of $A'B'C'D'$ by adding the same amount to each side of *ABCD*? Try this and explain what happened.

     d.   Monica enlarged *ABCD* to get a different rectangle $A''B''C''D''$ which is similar to *ABCD*. She knows that $\overline{A''B''}$ is 20 units long. How many times larger than *ABCD* is $A''B''C''D''$? (That is, how many "rubber bands" did she use?) And how long is $\overline{B''C''}$? Show how you know.

3-11.   In problem 3-10, you learned that you can create similar shapes by multiplying each side length by the same number. This number is called the **zoom factor**. You may have used a zoom factor before when using a copy machine. For example, if you set the zoom factor on a copier to 50%, the machine shrinks the image in half (that is, multiplies it by 0.5) but keeps the shape the same. In this course, the zoom factor will be used to describe the ratio of the new figure to the original figure.

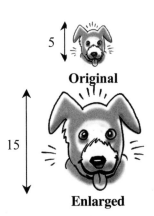

Original

Enlarged

What zoom factor was used to enlarge the puppy shown at right?

3-12.   Casey decided to enlarge her favorite letter: C, of course! Your team is going to help her out. Have each member of your team choose a different zoom factor below. Then on graph paper, enlarge (or reduce) the block "C" at right by your zoom factor.

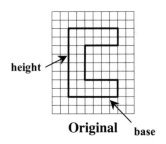

height

Original   base

     a.   3   b.   2

     c.   1   d. $\frac{1}{2}$

3-13.   Look at the different "C's" that were created in problem 3-12.

     a.   What happened when the zoom factor was 1?

     b.   When two shapes are the same shape <u>and</u> the same size (that is, the zoom factor is 1), they are called **congruent**. Compare the shapes below with tracing paper and determine which shapes are congruent.

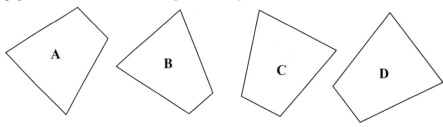

3-14.   EQUAL RATIOS OF SIMILARITY

Casey wants to learn more about her enlarged "C"s. Return to your work from problem 3-12.

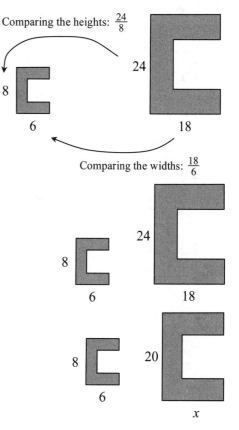

a.   Since the zoom factor multiplies each part of the original shape, then the **ratio** of the widths must equal the ratio of the lengths.

Casey decided to show these ratios in the diagram at right. Verify that her ratios are equal.

b.   When looking at Casey's work, her brother wrote the equation $\frac{8}{6} = \frac{24}{18}$. Are his ratios, in fact, equal? And how could he show his work on his diagram? Copy his diagram at right and add arrows to show what sides Casey's brother compared.

c.   She has decided to create an enlarged "C" for the door of her bedroom. To fit, it needs to be 20 units tall. If x is the width of this "C", write and solve an equation to find out how wide the "C" on Casey's door must be. Be ready to share your equation and solution with the class.

3-15.   Use your observations about ratios between similar figures to answer the following questions.

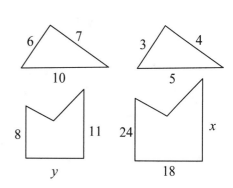

a.   Are the triangles above similar? How do you know?

b.   If the pentagons at right are similar, what are the values of x and y?

3-16.   In a new entry of your Learning Log, explain what you know about the ratio of similar figures. If you know the dimensions of one shape, how can you figure out the dimensions of another shape that is similar to it? Title this entry, "Common Ratios of Similar Figures" and include today's date.

Ratio of Similarity and Zoom Factor

The term "ratio" was introduced in Chapter 1 in the context of probability. But ratios are very important when comparing two similar figures. Review what you know about ratios below.

A comparison of two quantities (numbers or measures) is called a **ratio**. A ratio can be written as:

$$a{:}b \quad \text{or} \quad \frac{a}{b} \quad \text{or} \quad \text{"}a \text{ to } b\text{"}$$

Each ratio has a numeric value that can be expressed as a fraction or a decimal. For the two similar right triangles below, the ratio of the small triangle's hypotenuse to the large triangle's hypotenuse is $\frac{6}{10}$ or $\frac{3}{5}$. This means that for every three units of length in the small triangle's hypotenuse, there are five units of length in the large triangle's hypotenuse.

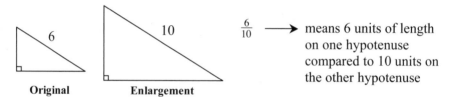

$\frac{6}{10} \longrightarrow$ means 6 units of length on one hypotenuse compared to 10 units on the other hypotenuse

Original   Enlargement

The ratio between any pair of corresponding sides in similar figures is called the **ratio of similarity**.

When a figure is enlarged or reduced, each side is multiplied (or divided) by the same number. While there are many names for this number, this text will refer to this number as the **zoom factor**. To help indicate if the figure was enlarged or reduced, the zoom factor is written as the ratio of the new figure to the original figure. For the two triangles above, the zoom factor is $\frac{10}{6}$ or $\frac{5}{3}$.

3-17.   Enlarge the shape at right on graph paper using a zoom factor of 4.

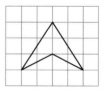

Geometry Connections

3-18.   The ratios Casey wrote from the table in part (a) of problem 3-14 are **common ratios between corresponding sides** of the two shapes. That is, they are ratios between the matching sides of two shapes.

   a.   Look at the two similar shapes below. Which sides correspond? Write common ratios with the names of sides and lengths, just like Bernhard did.

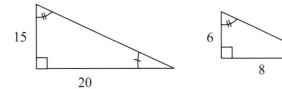

   b.   Find the hypotenuse of each triangle above. Is the ratio of the hypotenuses equal the ratios you found in part (a)?

3-19.   The temperature in San Antonio, Texas is currently 77°F and is increasing by 3° per hour. The current temperature in Bombay, India is 92°F and the temperature is dropping by 2° per hour. When will it be as hot in San Antonio as it is in Bombay? What will the temperature be?

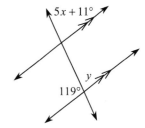

3-20.   **Examine** the relationships in the diagram at right. Then solve for x and y, if possible.

3-21.   Read the following statements and decide if, when combined, they present a *convincing* argument. You may need to refer back to your Shape Toolkit as you consider the following statements and decide if the conclusion is correct. Be sure to **justify** your reasoning.

   Fact #1: A square has four sides of equal length.

   Fact #2: A square is a rectangle because it has four right angles.

   Fact #3: A rhombus also has four sides of equal length

   Conclusion: Therefore, a rhombus is a rectangle.

3.1.3 How are the shapes related?

Using Ratios of Similarity

You have learned that when we enlarge or reduce a shape so that it remains similar (that is, it maintains the same shape), each of the side lengths have been multiplied by a common zoom factor. We can also set up ratios within shapes and make comparisons to other similar shapes. Today you will learn about what effect changing the size of an object has on its perimeter. You will also learn how ratios can help solve similarity problems when drawing the figures is impractical.

3-22. **Examine** the rectangles at right.

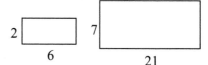

a. Use ratios to show that these shapes are similar (figures that have the same shape, but not necessarily the same size).

b. What other ratios could you use?

c. Linh claims that these shapes are not similar. When she compared the heights, she wrote $\frac{2}{7}$. Then she compared the bases and got $\frac{21}{6}$. Why is Linh having trouble? Explain completely.

3-23. Each pair of figures below is similar. Review what you have learned so far about similarity as you solve for x.

a.

b.

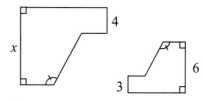

c.

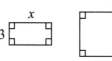

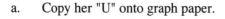

perimeter = 60

3-24. Casey's back at it! Now she wants you to enlarge the block "U" for her spirit flag.

a. Copy her "U" onto graph paper.

b. Now draw a larger "U" with a zoom factor of $\frac{3}{2} = 1.5$. What is the height of the new "U"?

c. Find the ratio of the perimeters. That is, find $\frac{\text{Perimeter New}}{\text{Perimeter Original}}$. What do you notice?

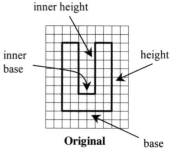

d. Casey enlarged "U" proportionally so that it has a height of 10. What was her zoom factor? What is the base of this new "U"? **Justify** your conclusion.

3-25.   After enlarging his "U" in problem
         3-24, Al has an idea. He drew a 60°
         angle, as shown in Diagram #1 at
         right. Then, he extended the sides of
         the angle so that they are twice as
         long, as shown in Diagram #2.
         "Therefore, the new angle must have
         measure 120°," he explained. Do
         you agree? Discuss this with your
         team and write a response to Al.

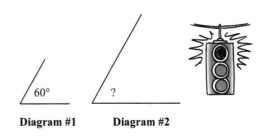

60°   ?

Diagram #1   **Diagram #2**

3-26.   Al noticed that the ratio of the perimeters of two similar
         figures is equal to the ratio of the side lengths. *"What about
         the area? Does it grow the same way?"* he wondered.

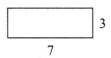

3

7

   a.   Find the area and perimeter of the rectangle above.

   b.   Test Al's question by enlarging the rectangle by a zoom factor of 2. Then find
         the new area and perimeter.

   c.   Answer Al's question: Does the perimeter double? Does the area double?
         Explain what happened.

Ⓜ️ETHODS AND MEANINGS

Proportional Equations

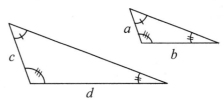

A **proportional equation** is one that compares two or more ratios.
Proportional equations can express comparisons between two similar objects
or compare two corresponding parts of an object.

For example, the following
equations can be written for the
similar triangles at right:

$$\frac{a}{c} = \frac{b}{d} \quad or \quad \frac{a}{b} = \frac{c}{d}.$$

MATH NOTES

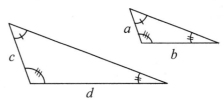

c

a

b

d

3-27. Rakisha is puzzled. She is working with the parallelogram drawn at right and wants to make it smaller instead of bigger.

a. What should she do if she wants the sides of her new shape to be *half as long* as the original sides? What zoom factor should she use? Find the dimensions of her new shape.

b. While drawing some other shapes, Rakisha ended up with a shape congruent to the original parallelogram. What is the common ratio between pairs of corresponding sides?

3-28. Enlarge the shape at right on graph paper using a zoom factor of 2. Then find the perimeter and area of both shapes. What do you notice when you compare the perimeters? The areas?

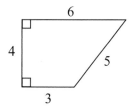

3-29. Solve each equation below. Show all work and check your answer.

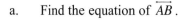

a. $\frac{14}{5} = \frac{x}{3}$ b. $\frac{10}{m} = \frac{5}{11}$ c. $\frac{t-2}{12} = \frac{7}{8}$ d. $\frac{x+1}{5} = \frac{x}{3}$

3-30. **Examine** the graph of line $\overrightarrow{AB}$ at right.

a. Find the equation of $\overrightarrow{AB}$.

b. Find the area and perimeter of $\triangle ABC$.

c. Write an equation of the line through A that is perpendicular to $\overrightarrow{AB}$.

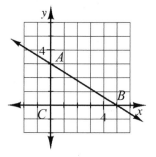

3-31. **Examine** the diagram at right. Name the geometric relationships of the angles below.

a. *d* and *e*   b. *e* and *h*

c. *a* and *e*   d. *c* and *d*

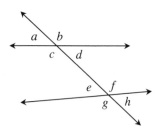

146   *Geometry Connections*

3.1.4 How can I use equivalent ratios?

Applications and Notation

Now that you have a good understanding of how to use ratios in similar figures to solve problems, how can you extend these ideas to situations outside the classroom? You will start by considering a situation for which we want to find the length of something that would be difficult to physically measure.

3-32.   GEORGE WASHINGTON'S NOSE

On her way to visit Horace Mann University, Casey stopped by Mount Rushmore in South Dakota. The park ranger gave a talk that described the history of the monument and provided some interesting facts. Casey could not believe that the carving of George Washington's face is 60 feet tall from his chin to the top of his head!

However, when a tourist asked about the length of Washington's nose, the ranger was stumped! Casey came to her rescue by measuring, calculating and getting an answer. How did Casey get her answer?

Your task: Figure out the length of George Washington's nose on the monument. Work with your study team to come up with a **strategy**. Show all measurements and calculations on your paper with clear labels so anyone could understand your work.

Discussion Points

What is this question asking you to find?

How can you use similarity to solve this problem?

Is there something in this room that you can use to compare to the monument?

What parts do you need to compare?

Do you have any math tools that can help you gather information?

3-33.   Solving problem 3-32, you may have written a proportional equation like the one below. When solving proportional situations, is very important that parts be labeled to help you follow your work.

$$\frac{Length\ of\ George's\ Nose}{Length\ of\ George's\ Head} = \frac{Length\ of\ Student's\ Nose}{Length\ of\ Student's\ Head}$$

Likewise, when working with geometric shapes such as the similar triangles below, it is easier to explain which sides you are comparing by using notation that everyone understands.

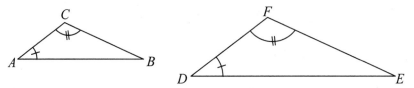

a.   One possible proportional equation for these triangles is $\frac{AC}{AB} = \frac{DF}{DE}$. Write at least three more proportional equations based on the similar triangles above.

b.   Jeb noticed that $m\angle A = m\angle D$ and $m\angle C = m\angle F$. But what about $m\angle B$ and $m\angle E$? Do these angles have the same measure? Or is there not enough information? **Justify** your conclusions.

3-34.   The two triangles at right are similar. Read the Math Notes box for this lesson to learn about how to write a statement to show that two shapes are similar.

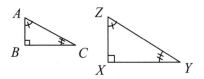

Then **examine** the two triangles below. Which of the following statements are correctly written and which are not? Note that more than one statement may be correct. Discuss your answers with your team.

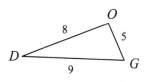
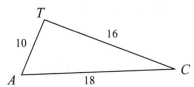

a. $\triangle DOG \sim \triangle CAT$ b. $\triangle DOG \sim \triangle CTA$

c. $\triangle OGD \sim \triangle ATC$ d. $\triangle DGO \sim \triangle CAT$

3-35.   Find the value of the variable in each pair of similar figures below. You may want to
         set up tables to help you write equations.

a.   *ABCD ~ JKLM*

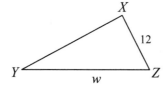

b.   △*NOP ~* △*XYZ*

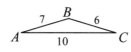

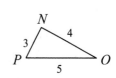

c.   △*GHI ~* △*PQR*

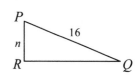

d.   △*ABC ~* △*XYZ*

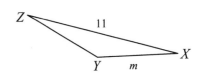

3-36.   Rochida was given the diagram at right and told that
         the two triangles are similar.

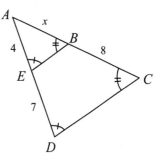

a.   Rochida knows that to be similar, all
      corresponding angles must be equal. Are all
      three sets of angles equal? How can you tell?

b.   Rochida decides to redraw the
      shape as two separate triangles,
      as shown at right. Write a
      proportional equation using the
      corresponding sides.

c.   Solve the equation for *x*. How
      long is $\overline{AB}$? How long is $\overline{AC}$?

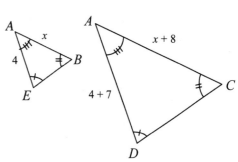

3-37.   Write a Learning Log entry describing the different ways you can
compare two similar objects or quantities with equivalent ratios.
Title this entry "Comparing With Ratios" and include today's date.

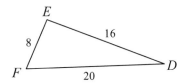

3-38.   Solve for the missing lengths in the sets of similar figures below. You may want to
set up tables to help you write equations.

a. $\triangle ABC \sim \triangle OPQ$

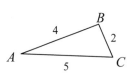

b.   *EFGHI ~ STUVW*

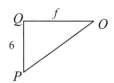
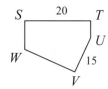

c.   *JKLM ~ WXYZ*

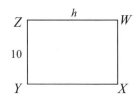

3-39.   In recent lessons, you have learned that similar triangles have equal corresponding angles. Is it possible to have equal corresponding angles when the triangles do not appear to match? What if you are not given all three angle measures? Consider the two cases below.

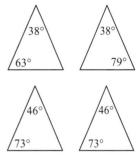

a.   Find the measure of the third angle in the first pair of triangles at right. Compare the two triangles. What do you notice?

b.   **Examine** the second pair of triangles at right. Without calculating, do you know that the unmarked angles must be equal? Why or why not?

3-40.   Sandy has a square, equilateral triangle, rhombus, and regular hexagon in her Shape Bucket, while Robert has a scalene triangle, kite, isosceles trapezoid, non-special quadrilateral, and obtuse isosceles triangle in his. Sandy will randomly select a shape from her Shape Bucket, while Robert will randomly select a shape from his.

a.   Who has a greater probability of selecting a quadrilateral? **Justify** your conclusion.

b.   Who has a greater probability of selecting an equilateral shape? **Justify** your conclusion.

c.   What is more likely to happen: Sandy selecting a shape with at least two sides that are parallel or Robert selecting a shape with at least two sides that are equal?

3-41.   Frank and Alice are penguins. At birth, Frank's beak was 1.95 inches long, while Alice's was 1.50 inches long. If Frank's beak grows by 0.25 inches per year and Alice's grows by 0.40 inches per year, how old will they be when their beaks are the same length?

3-42.   Plot *ABCDE* formed with the points $A(-3, -2)$, $B(5, -2)$, $C(5, 3)$, $D(1, 6)$, and $E(-3, 3)$ onto graph paper.

a.   Use the method from problem 3-2 to enlarge it from the origin by a factor of 2. Label this new shape $A'B'C'D'E'$.

b.   Find the area and the perimeter of both figures.

3.2.1 What information do I need?

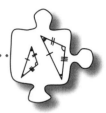

• •

Conditions for Triangle Similarity

Now that you know what similar shapes have in common, you are ready to turn to a related question: How much information do I need to know that two triangles are similar? As you work through today's lesson, remember that similar polygons have corresponding angles that are congruent and corresponding sides that are proportional.

3-43.   ARE THEY SIMILAR?

Erica thinks the triangles below might be
similar. However, she knows not to trust
the way figures look in a diagram, so she
asks for your help.

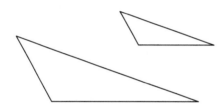

a.   If two shapes are similar, what must be true about their angles and sides?

b.   Obtain the Lesson 3.2.1 Resource Page from your teacher. Measure the angles
and sides of Erica's triangles and help her decide if the triangles are similar or
not.

3-44.   HOW MUCH IS ENOUGH?

Jovan is tired of measuring all the angles and sides to determine if two triangles are similar. "There must be an easier way," he thinks. "What if I know that all of the side lengths have a common ratio? Does that mean that the triangles are similar?"

a.   Before experimenting, make a prediction. Do you think that the triangles have to be similar if the pairs of corresponding sides share a common ratio?

b.   Experiment with Jovan's idea. To do this, use a dynamic geometry tool to test triangles with proportional side lengths. If you do not have access to a dynamic tool, cut straws or strands of linguini into the lengths below and create two triangles. Can you create two triangles that are not similar? **Investigate**, sketch your shapes, and write down your conclusion.

Triangle #1: side lengths 3, 5, and 6

Triangle #2: side lengths 6, 10, and 12

c.   Jovan then asks, "Is it enough to know that each pair of corresponding angles are congruent? Does that mean the triangles are similar?" Again use a dynamic geometry tool to test triangles with equal corresponding angles or use linguini with protractors to create three angles that form a triangle. Can you create two triangles with the same three angle measures that are not similar? **Investigate**, sketch your shapes, and write down your conclusion.

3-45.   Scott is looking at the set of shapes at right. He thinks that $\triangle EFG \sim \triangle HIJ$ but he is not sure that the shapes are drawn to scale.

a.   Are the corresponding angles equal? Convince Scott that these triangles are similar.

b.   How many pairs of angles need to be congruent to be sure that triangles are similar?

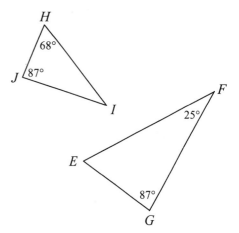

3-46.   Based on your conclusions from problems 3-44 and 3-45, decide if each pair of triangles below is similar. Explain your **reasoning**.

a.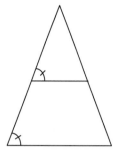

b.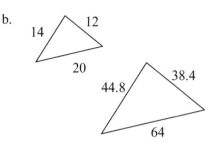

c.

d.

3-47.   Read the Math Notes box for this lesson, which introduces new names for the observations you made in problems 3-44 and 3-45. Then write a Learning Log entry about what you learned today. Be sure to address the question: *how much information do I need about a pair of triangles in order to be sure that they are similar?* Title this entry "AA ~ and SSS ~" and include today's date.

METHODS AND MEANINGS

Conditions for Triangle Similarity

For two shapes to be similar, corresponding angles must have equal measure and corresponding sides must be proportional.

However, if you are testing for similarity between two triangles, then it is sufficient to know that all three corresponding side lengths share a common ratio.

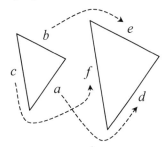

This guarantees similarity and is referred to as the **SSS Triangle Similarity Conjecture** (which can be abbreviated as "SSS Similarity" or "SSS ~" for short.)

SSS ~ : If $\frac{a}{d} = \frac{b}{e} = \frac{c}{f}$, then the triangles are similar.

Also, for two triangles it is sufficient to know that two pairs of corresponding angles have equal measures because then the third pair of angles must have equal measure.

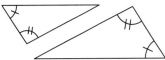

This is known as the **AA Triangle Similarity Conjecture** (which can be abbreviated as "AA Similarity" or "AA ~" for short).

AA ~ : If 2 pairs of angles have equal measure, then the triangles are similar.

Review & Preview

3-48.   Assume that all trees are green.

   a.   Does this statement mean that an oak tree must be green? Explain why or why not.

   b.   Does this statement mean that anything green must be a tree? Explain why or why not.

   c.   Are the statements "All trees are green" and "All green things are trees" saying the same thing? Explain why or why not.

3-49.   Decide if each pair of triangles below is similar. If the triangles are similar, **justify** your conclusion by stating the similarity conjecture you used. If the triangles are not similar, explain how you know.

a.

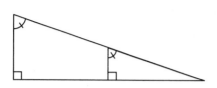

b.

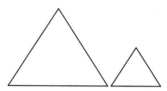

Equilateral Triangles

c.

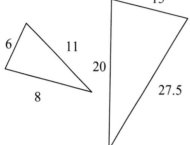

d.

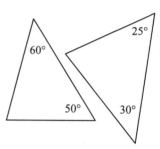

3-50.   Consider the following arrow diagram:

Lines are parallel ➜ *alternate interior angles are equal.*

a.   Write this arrow diagram as a conditional ("If…, then…") statement.

b.   Write a similar conditional statement about corresponding angles, then write it as an arrow diagram.

3-51.   Find the area and perimeter of the shape at right. Show all work.

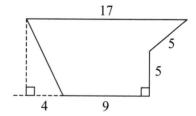

3-52.   Assume that each pair of figures below is similar. Write a similarity statement to illustrate which parts of each shape correspond. Remember: letter order is important!

a.   *ABCD ~ ?*

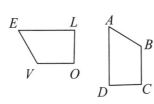

b.   *RIGHT ~ ?*

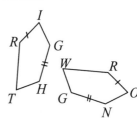

c.   △_____ ~△_____

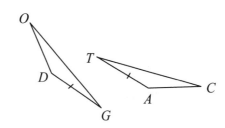

3.2.2   How can I organize my information?

Creating a Flowchart

In Lesson 3.2.1, you developed the AA ~ and SSS ~ conjectures to help confirm that triangles are similar. Today you will continue working with similarity and will learn how to use flowcharts to organize your **reasoning**.

3-53.   **Examine** the triangles at right.

a.   Are these triangles similar? Use full sentences to explain your **reasoning**.

b.   Julio decided to use a diagram (called a **flowchart**) to explain his reasoning.

Compare your explanation to Julio's flowchart. Did Julio use the same **reasoning** you used?

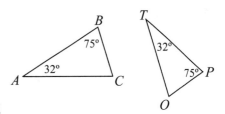

JULIO'S FLOWCHART

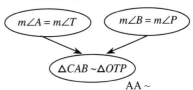

3-54. Besides showing your **reasoning**, a flowchart can be used to organize your work as you determine whether or not triangles are similar.

a. Are these triangles similar? Which triangle similarity conjecture (see the Math Notes box from Lesson 3.2.1) did you use?

b. What facts must you know to use the triangle similarity conjecture you chose? Julio started to list the facts in a flowchart at right. Copy them on your paper and complete the third oval.

c. Once you have the needed facts in place, you can conclude that you have similar triangles. Add to your flowchart by making an oval and filling in your conclusion.

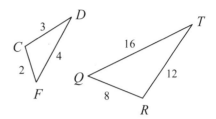

Facts:

Conclusion:

d. Finally, draw arrows to show the flow of the facts that lead to your conclusion and record the similarity conjecture you used, following Julio's example from problem 3-53.

3-55. Now **examine** the triangles at right.

a. Are these triangles similar? **Justify** your conclusion using a flowchart.

b. What is $m\angle C$? How do you know?

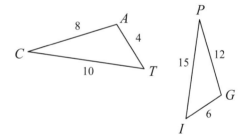

3-56. Lindsay was solving a math problem and drew the flowchart below:

a. Draw and label two triangles that could represent Lindsay's problem. What question did the problem ask her? How can you tell?

b. Lindsay's teammate was working on the same problem and made a mistake in his flowchart:

How is this flowchart different from Lindsay's? Why is this the wrong way to explain the **reasoning** in Lindsay's problem?

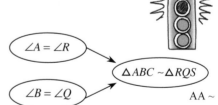

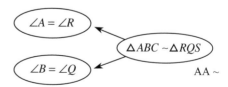

3-57.   Ramon is **examining**
the triangles at right.
He suspects they
may be similar by
SSS ~.

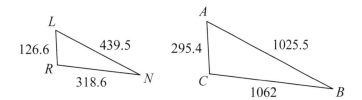

a.   Why is SSS ~ the best conjecture to test for these triangles?

b.   Set up ovals for the facts you need to know to show that the triangles are
similar. Complete any necessary calculations and fill in the ovals.

c.   Are the triangles similar? If so, complete your flowchart using an appropriate
similarity statement. If not, explain how you know.

3-58.   In your Learning Log, explain how to set up a flowchart. For
example, how do you know how many ovals you should use? How
do you know what to put inside the ovals? Provide an example.
Label this entry "Using Flowcharts" and include today's date.

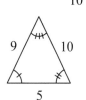

Ⓜ ETHODS AND MEANINGS

Congruent Shapes

If two figures have the same shape and are the
same size, they are **congruent**. Since the figures
must have the same shape, they must be similar.

Two figures are congruent if they meet both of the
following conditions:

- the two figures are similar, and
- their side lengths have a common ratio of 1.

MATH NOTES

3-59.   Solve for the missing lengths in the sets of similar figures below. You may want to set up tables to help you write equations.

a.   *ABCD ~ JKLM*

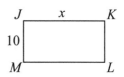
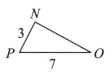

b.   △*NOP ~*△*XYZ*

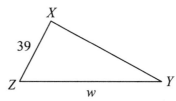
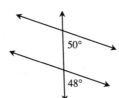
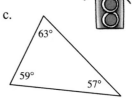

3-60.   **Examine** each diagram below. Which diagrams are possible? Which are impossible? **Justify** each conclusion.

a.
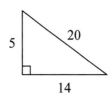

b.

c.

3-61.   Decide which transformations were used on each pair of shapes below. Note that there may have been more than one transformation.

a.

b.

c.
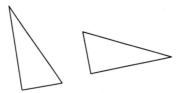

d.
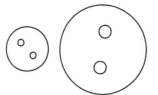

3-62.   Determine whether or not the **reasoning** in the flowchart at right is correct. If it is wrong, redo the flowchart to make it correct.

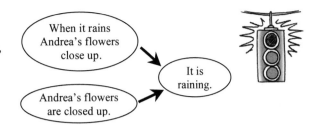

When it rains Andrea's flowers close up.

Andrea's flowers are closed up.

It is raining.

3-63.   Sketch each triangle if possible. If not possible, explain why not.

   a.   Right isosceles triangles

   b.   Right obtuse triangles

   c.   Scalene equilateral triangles

   d.   Acute scalene triangles

3.2.3   How can I use equivalent ratios?

Triangle Similarity and Congruence

By looking at side ratios and at angles, you are now able to determine whether two figures are similar. But how can you tell if two shapes are the same shape *and* the same size? In this lesson you will **examine** properties that guarantee that shapes are exact replicas of one another.

3-64.   Decide if each pair of triangles below is similar. Use a flowchart to organize your facts and conclusion for each pair of triangles.

   a.

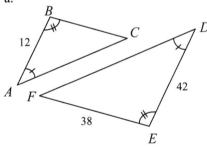

   b.

   c.

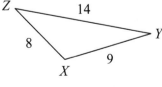

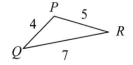

   d.

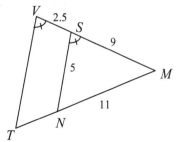

3-65. For the diagrams in problem 3-64, find the lengths of the sides listed below, if possible. If it is not possible, explain why not.

    a. $\overline{BC}$ b. $\overline{AC}$ c. $\overline{VT}$ d. $\overline{TN}$

3-66. **Examine** the triangles at right.

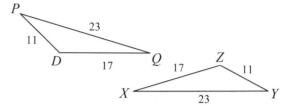

    a.   Are these triangles similar? How do you know? Use a flowchart to organize your explanation.

    b.   Kamraan says, "These triangles aren't just similar—they're congruent!" Is Kamraan correct? What special value in your flowchart indicates that the triangles are congruent?

    c.   Write a conjecture (in "If…, then…" form) that explains how you know when two shapes are congruent.

3-67. Flowcharts can also be used to represent real-life situations. For example, yesterday Joe found out that three people in his family (including Joe) wanted to see a movie, so he went to the theater and bought three tickets. Unfortunately, while he was gone, three more family members decided to go. When everyone arrived at the theater, Joe did not have enough tickets.

Joe sat down later that night and tried to create a flowchart to describe what had happened. Here are the three possibilities he came up with:

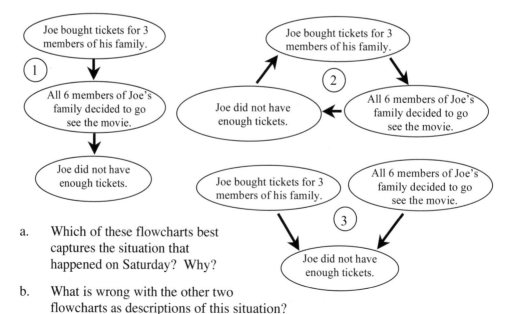

    a.   Which of these flowcharts best captures the situation that happened on Saturday? Why?

    b.   What is wrong with the other two flowcharts as descriptions of this situation?

                    Geometry Connections

MATH NOTES

METHODS AND **M**EANINGS

Solving a Quadratic Equation

In a previous course, you learned how to solve **quadratic equations** (equations that can be written in the form $ax^2 + bx + c = 0$). Review two methods for solving quadratic equations below.

Some quadratic equations can be solved by **factoring** and using the **Zero Product Property**. For example, because $x^2 - 3x - 10 = (x - 5)(x + 2)$, the quadratic equation $x^2 - 3x - 10 = 0$ can be rewritten as $(x - 5)(x + 2) = 0$. The Zero Product Property states that if $ab = 0$, then $a = 0$ or $b = 0$. So, if $(x - 5)(x + 2) = 0$, then $x - 5 = 0$ or $x + 2 = 0$. Therefore, $x = 5$ or $x = -2$.

Another method for solving quadratic equations is the **Quadratic Formula.** This method is particularly helpful for solving quadratic equations that are difficult or impossible to factor. Before using the Quadratic Formula, the quadratic equation you want to solve must be in standard form, that is, written as $ax^2 + bx + c = 0$.

In this form, a is the coefficient of the x^2 term, b is the coefficient of the x term, and c is the constant term. The Quadratic Formula states:

$$x = \frac{-b \pm \sqrt{b^2 - 4ac}}{2a}$$

This formula gives two possible answers due to the "$\pm$" symbol. This symbol (read as "plus or minus") is shorthand notation that tells us to calculate the formula twice: once using addition and once using subtraction. Therefore, every Quadratic Formula problem must be simplified twice to give:

$$x = \frac{-b + \sqrt{b^2 - 4ac}}{2a} \quad \text{or} \quad x = \frac{-b - \sqrt{b^2 - 4ac}}{2a}$$

To solve $x^2 - 3x - 10 = 0$ using the Quadratic Formula, substitute $a = 1$, $b = -3$, and $c = -10$ into the formula, as shown below.

$$x = \frac{-(-3) \pm \sqrt{(-3)^2 - 4(1)(-10)}}{2(1)} \rightarrow \frac{3 \pm \sqrt{49}}{2} \rightarrow \frac{3 \pm 7}{2} \rightarrow x = 5 \text{ or } x = -2$$

3-68.   On graph paper, graph the parabola $y = 2x^2 - 5x - 3$.

   a.   What are the roots (x-intercepts) of the parabola?

   b.   Read the Math Notes box for this lesson. Then solve the equation
         $2x^2 - 5x - 3 = 0$ algebraically. Did you solutions match your roots from part
         (a)?

3-69.   On graph paper, plot $ABCD$ if $A(0, 3)$, $B(2, 5)$, $C(6, 3)$, and $D(4, 1)$.

   a.   Rotate $ABCD$ 90° clockwise (↻) about the origin to form $A'B'C'D'$. Name
         the coordinates of B'.

   b.   Translate $A'B'C'D'$ up 8 units and left 7 units to form $A''B''C''D''$. Name the
         coordinates of C''.

   c.   After rotating $ABCD$ 180° to form $A'''B'''C'''D'''$, Arah noticed that
         $A'''B'''C'''D'''$ position and orientation was the same as $ABCD$. What was the
         point of rotation? How did you find it?

3-70.   Use the relationships in each diagram below to solve for x. **Justify** your solution by
         stating which geometry relationships you used.

   a.   b.   c.

   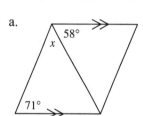   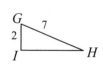   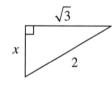
                                 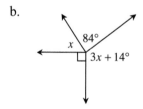

3-71.   Solve for the missing lengths in the sets of similar figures below. Show all work.

   a. $\triangle GHI \sim \triangle PQR$

   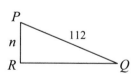

   b. $\triangle ABC \sim \triangle XYZ$

   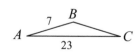   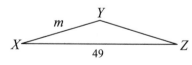

Geometry Connections

3-72.   Explain how you know that the shapes at
right are similar.

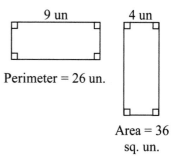

Perimeter = 26 un.

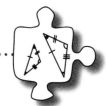

Area = 36
sq. un.

3.2.4 What information do I need?

....

More Conditions for Triangle Similarity

So far, you have worked with two methods for determining that triangles are similar: the AA ~
and the SSS ~ Conjectures. Are these the only ways to determine if two triangles are similar?
Today you will **investigate** similar triangles and complete your list of triangle similarity
conjectures.

Keep the following questions in mind as you work together today:

How much information do you need?

Are the triangles similar? How can you tell?

Can I find a triangle with this information that is not similar?

3-73.   Robel's team is using the SSS ~ Conjecture to show that two
triangles are similar. "This is too much work," Robel says. "When
we're using the AA ~ Conjecture, we only need to look at two
angles. Let's just calculate the ratios for *two* pairs of corresponding
sides to determine that triangles are similar."

Is SS ~ a valid similarity conjecture for triangles? That is, if two pairs of
corresponding side lengths share a common ratio, must the triangles be similar?

In this problem, you will **investigate** this question using a dynamic geometry tool or
another manipulative (such as linguini) provided by your teacher.

a.   Robel has a triangle with side lengths 4 cm and 5 cm. If your triangle has two
sides that share a common ratio with Robel's, does your triangle have to be
similar to his? Use a dynamic geometry tool or straws or linguini to
investigate this question.

Problem continues on next page →

3-73.   *Problem continued from previous page.*

b.   Kashi asks, "What if the angles between the two sides have the same measure? Would that be enough to know the triangles are similar?" Use the dynamic geometry tool or straws and linguini to answer Kashi's question.

c.   Kashi calls this the "SAS ~ Conjecture," placing the "A" between the two "S"s because the angle is *between* the two sides. He knows it works for Robel's triangle, but does it work on all other triangles? Test this method on a variety of triangles using the dynamic geometry tool or straws and linguini.

3-74.   What other triangle similarity conjectures involving sides and angles might there be? List the names of every other possible triangle similarity conjecture you can think of that involves sides and angles.

3-75.   Cori's team put "SSA ~" on their list of possible triangle similarity conjectures. Use a dynamic geometry tool or straws or linguini to **investigate** whether SSA ~ is a valid triangle similarity conjecture. If a triangle has two sides sharing a common ratio with Robel's, and has the same angle "outside" these sides as Robel's, must it be similar to Robel's triangle? If you determine SSA ~ is not a valid similarity conjecture, cross it off your list!

3-76.   Betsy's team came up with a similarity conjecture they call "AAS ~," but Betsy thinks they should cross it off their list. Betsy says, "This similarity conjecture has extra, unnecessary information. There's no point in having it on our list."

a.   What is Betsy talking about? Why does the AAS ~ method contain more information than you need?

b.   Go through your list of possible triangle similarity conjectures, crossing off all the invalid ones and all the ones that contain unnecessary information.

c.   How many valid triangle similarity conjectures (without extra information) are there? List them.

3-77.   Reflect on what you have learned today. In your Learning Log, write down the triangle similarity conjectures that help to determine if triangles are similar. You can write these conjectures as conditional statements (in "If…, then…" form) or as arrow diagrams. Title this entry "Triangle Similarity Conjectures" and include today's date.

METHODS AND MEANINGS

MATH NOTES

Writing a Flowchart

A **flowchart** helps to organize facts and indicate which facts lead to a conclusion. The bubbles contain facts, while the arrows point to a conclusion that can be made from a fact or multiple facts.

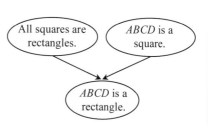

For example, in the flowchart at right, two independent (unconnected) facts are stated: *"All squares are rectangles"* and *"ABCD is a square."* However, these facts lead to the conclusion that *ABCD* must be a rectangle. Note that the arrows point toward the conclusion.

3-78. If possible, draw a triangle that has exactly the following number of lines of symmetry. Then name the kind of triangle drawn.

a. 0   b. 1   c. 2   d. 3

3-79. Do two lines always intersect? Consider this as you answer the questions below.

a. Write a system of linear equations that does not have a solution. Write each equation in your system in **slope-intercept form** ($y = mx + b$). Graph your system on graph paper and explain why it does not have a solution.

b. How can you tell algebraically that a system of linear equations has no solution? Solve your system of equations from part (a) algebraically and demonstrate how you know that the system has no solution.

3-80.   Determine which of the following pairs of triangles are similar. Explain your work.

a.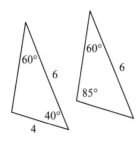

b.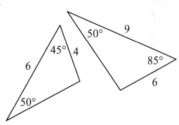

c.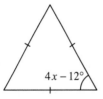

3-81.   The dashed line at right represents the line of symmetry of the shaded figure. Find the area and perimeter of the shaded region. Show all work.

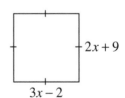

3-82.   **Examine** the diagrams below. For each one, write and solve an equation to find x. Show all work.

a.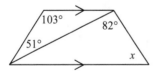

b.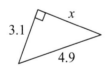

c.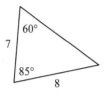

d.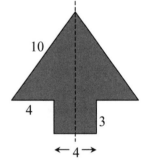

3.2.5 Are the triangles similar?

Determining Similarity

You now have a complete list of the three Triangles Similarity Conjectures (AA ~, SSS ~, and SAS ~) that can be used to verify that two triangles are similar. Today you will continue to practice applying these conjectures and using flowcharts to organize your **reasoning**.

3-83.   Lynn wants to show that the triangles at right are similar.

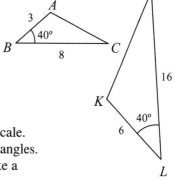

   a.   What similarity conjecture should Lynn use?

   b.   Make a flow chart showing that these triangles are similar.

3-84.   Below are six triangles, none of which is drawn to scale. Among the six triangles are three pairs of similar triangles. Identify the similar triangles, then for each pair make a flowchart **justifying** the similarity.

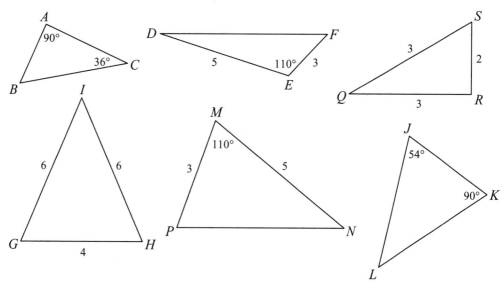

3-85.   Revisit the similar triangles from problem 3-84.

   a.   Which pair of triangles are congruent? How do you know?

   b.   Suppose that in problem 3-84, $AB = 3$ cm, $AC = 4$ cm, and $KJ = 12$ cm. Find all the other side lengths in $\triangle ABC$ and $\triangle JKL$.

3-86.   **Examine** the triangles at right.

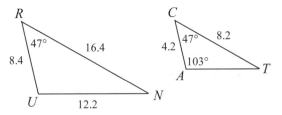

a.   Are these triangles similar? If so, make a flowchart **justifying** their similarity.

b.   Charles has $\triangle CAT \sim \triangle RUN$ as the conclusion of his flowchart. Leesa has $\triangle NRU \sim \triangle TCA$ as her conclusion. Who is correct? Why?

c.   Are $\triangle CAT$ and $\triangle RUN$ congruent? Explain how you know.

d.   Find all the missing side lengths and all the angle measures of $\triangle CAT$ and $\triangle RUN$.

3-87.   THE FAMILY FORTUNE, Part Two

In Lesson 1.1.4, you had to convince city officials that you were a relative of Molly "Ol' Granny" Marston, who had just passed away leaving a sizable inheritance. Below is the evidence you had available:

Family Portrait — a photo showing three young children. On the back you see the date 1968.

Newspaper Clipping — from 1972 titled "Triplets Make Music History." The first sentence catches your eye: "Jake, Judy, and Jeremiah Marston, all eight years old, were the first triplets ever to perform a six-handed piano piece at Carnegie Hall."

Jake Marston's Birth Certificate — showing that Jake was born in 1964, and identifying his parents as Phillip and Molly Marston.

Your Learner's Permit — signed by your father, Jeremiah Marston.

Wilbert Marston's Passport — issued when Wilbert was fifteen.

As their answer to this problem, one team wrote the following argument for the city officials:

> *The birth certificate shows that Jake Marston was Molly Marston's son. The newspaper clipping shows that Jeremiah Marston was Jake Marston's brother. Therefore, Jeremiah Marston was Molly Marston's son. The learner's permit shows that Jeremiah Marston is my father. Therefore, I am Molly Marston's grandchild.*

Your task: Make a flowchart showing the **reasoning** in this team's argument. This flowchart will have more levels than the ones you have made in the past, because certain conclusions will be used as facts to support other conclusions. So plan carefully before you start to draw your chart.

170   *Geometry Connections*

METHODS AND MEANINGS

Complete Conditions for Triangle Similarity

MATH NOTES

There are exactly three valid, non-redundant triangle similarity conjectures that use only sides and angles. (A conjecture is "redundant" if it includes more information than is necessary to establish triangle similarity.) They are abbreviated as: SSS ~, AA ~, and SAS ~. In the SAS ~ Conjecture, the "A" is placed between the two "S"s to indicate that the angle must be *between* the two sides used.

For example, $\triangle RUN$ and $\triangle CAT$ below are similar by SAS ~. $\frac{RU}{CA} = 2$ and $\frac{UN}{AT} = 2$, so two pairs of corresponding side lengths share a common ratio. The measure of the angle *between* $\overline{RU}$ and $\overline{UN}$, $\angle U$,

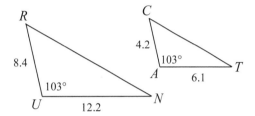

equals the measure of the angle between $\overline{CA}$ and $\overline{AT}$, $\angle A$, so the conditions for SAS ~ are met.

3-88.   Sketch each triangle described below, if possible. If not possible, explain why it is not possible.

   a.   Equilateral obtuse triangle

   b.   Right scalene triangle

   c.   Obtuse isosceles triangle

   d.   Acute right triangle

3-89.   Determine which similarity conjectures (AA ~, SSS ~, or SAS ~) could be used to
          establish that the following pairs of triangles are similar. List as many as you can.

a.

b.

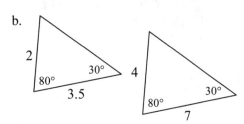

c.

3-90.   Solve each equation to find the value of x. Leave your answers in decimal form
          accurate to the thousandths place.

a. $\frac{3.2}{x} = \frac{7.5}{x^2}$ b. $4(x-2) + 3(-x+4) = -2(x-3)$

c. $2x^2 + 7x - 15 = 0$ d. $3x^2 - 2x = -1$

3-91.   On graph paper, sketch a rectangle with side lengths of 15 units and 9 units. Shrink
          the rectangle by a zoom factor of $\frac{1}{3}$. Make a table showing the area and perimeter of
          both rectangles.

3-92.   Susan lives 20 miles northeast of Matt. Simone lives 15 miles
          dues south of Susan. If Matt lives due west of Simone,
          approximately how many miles does he live from Simone?
          Draw a diagram and show all work.

3.2.6 What can I do with similar triangles?

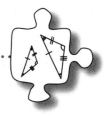

....

Applying Similarity

In previous lessons, you have learned methods for finding similar triangles. Once you find triangles are similar, how can that help you? Today you will apply similar triangles to analyze situations and solve new applications. As you work on today's problems, ask the following questions in your team:

What is the relationship?

Are any triangles similar? What similarity conjecture can we use?

3-93.   YOU ARE GETTING SLEEPY…

Legend has it that if you stare into a person's eyes in a special way, you can hypnotize them into squawking like a chicken. Here's how it works.

Place a mirror on the floor. Your victim has to stand exactly 200 cm away from the mirror and stare into it. The only tricky part is that you need to figure out where you have to stand so that when you stare into the mirror, you are also staring into your victim's eyes.

If your calculations are correct and you stand at the *exact* distance, your victim will squawk like a chicken!

a.   Choose a member of your team to hypnotize. Before heading to the mirror, first analyze this situation. Draw a diagram showing you and your victim standing on opposite sides of a mirror. Measure the heights of both yourself and your victim (heights to the eyes, of course), and label all the lengths you can on the diagram. (Remember: your victim will need to stand 200 cm from the mirror.)

b.   Are there similar triangles in your diagram? **Justify** your conclusion. (Hint: Remember what you know about how light reflects off mirrors.) Then calculate how far you will need to stand from the mirror to hypnotize your victim.

c.   Now for the moment of truth! Have your teammate stand 200 cm away from the mirror, while you stand at your calculated distance from the mirror. Do you make eye contact? If not, check your measurements and calculations and try again.

3-94.   **LESSONS FROM ABROAD**

Latoya was trying to take a picture of her family in front of the
Big Ben clock tower in London. However, after she
snapped the photo, she realized that the top of
her father's head exactly blocked the top
of the clock tower!

While disappointed with the picture,
Latoya thought she might be able to
estimate the height of the tower
using her math knowledge. Since
Latoya took the picture while
kneeling, the camera was 2 feet
above the ground. The camera was
also 12 feet from her 6-foot tall father, and he was standing about 930 feet from the
base of the tower.

a.   Sketch the diagram above on your paper and locate as many triangles as you
can. Can you find any triangles that must be similar? If so, explain how you
know they are similar.

b.   Use the similar triangles to determine the height of the Big Ben clock tower.

3-95.   **TRIANGLE CHALLENGE**

Use what you know about triangles and angle relationships to answer these questions
about the diagram below. As you work, make a careful record of your **reasoning**
(including a flowchart for any similarity arguments) and be ready to share it with the
class.

a.   First challenge: Find y.

b.   Second challenge: Is $\overline{HR}$ parallel to
$\overline{AK}$? How do you know?

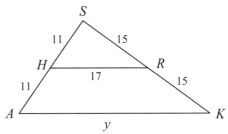

3-96.   Use the relationships in the diagram at right to
solve for x and y. **Justify** your solutions.

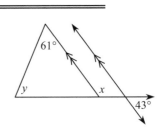

3-97. The area of the triangle at right is 25 square units. Find the value of h. Then find the perimeter of the entire triangle. Show all work.

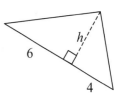

3-98. **Examine** the graph of the line at right.

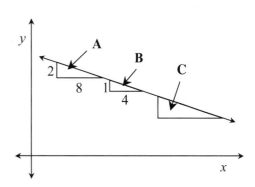

    a. Find the slope of this line using slope triangle A.

    b. Find the slope using slope triangle B.

    c. Without calculating, what does the slope ratio for slope triangle C have to be?

3-99. If $ABCD \sim WXYZ$, find x, y, and z.

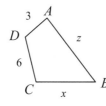
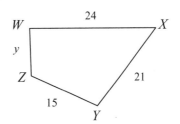

3-100. The area of the rectangle shown at right is 40 square units. Write and solve an equation to find x. Then find the dimensions of the rectangle.

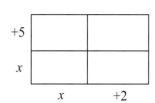

Chapter 3 Closure   What have I learned?

Reflection and Synthesis

The activities below offer you a chance to reflect on what you have learned in this chapter. As you work, look for concepts that you feel very comfortable with, ideas that you would like to learn more about, and topics you need more help with. Look for connections between ideas as well as connections with material you learned previously.

① TEAM BRAINSTORM

With your team, brainstorm a list for each of the following two subjects. Be as detailed as you can. How long can you make your list? Challenge yourselves. Be prepared to share your team's ideas with the class.

Topics: What have you studied in this chapter? What ideas and words were important in what you learned? Remember to be as detailed as you can.

Connections: How are the topics, ideas, and words that you learned in previous courses connected to the new ideas in this chapter? Again, make your list as long as you can.

② MAKING CONNECTIONS

The following is a list of the vocabulary used in this chapter. The words that appear in bold are new to this chapter. Make sure that you are familiar with all of these words and know what they mean. Refer to the glossary or index for any words that you do not yet understand.

AA ~	angle	conditional
congruent	conjecture	**corresponding sides**
dilation	**enlarge**	**flowchart**
hypotenuse	**logical argument**	original
perimeter	**proportional equation**	ratio
relationship	**SAS ~**	**sides**
similarity	**SSS ~**	**statement**
translate	vertex	**zoom factor**

Problem continues on next page →

Geometry Connections

② *Problem continued from previous page.*

Make a concept map showing all of the connections you can find among the key words and ideas listed above. To show a connection between two words, draw a line between them and explain the connection, as shown in the example below. A word can be connected to any other word as long as there is a justified connection. For each key word or idea, provide a sketch of an example.

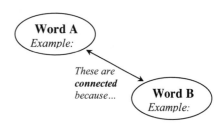

Your teacher may provide you with vocabulary cards to help you get started. If you use the cards to plan your concept map, be sure either to re-draw your concept map on your paper or to glue the vocabulary cards to a poster with all of the connections explained for others to see and understand.

While you are making your map, your team may think of related words or ideas that are not listed above. Be sure to include these ideas on your concept map.

③ SUMMARIZING MY UNDERSTANDING

This section gives you an opportunity to show what you know about certain math topics or ideas. Your teacher will direct you how to do this. Your teacher may give you a "GO" page to work on. "GO" stands for "Graphic Organizer," a tool you can use to organize your thoughts and communicate your ideas clearly.

④ WHAT HAVE I LEARNED?

This section will help you evaluate which types of problems you feel comfortable with and which types you need more help with. This section will appear at the end of every chapter to help you check your understanding. Even if your teacher does not assign this section, it is a good idea to try these problems and find out for yourself what you know and what you need to work on.

Solve each problem as completely as you can. The table at the end of this closure section has answers to these problems. It also tells you where you can find additional help and practice on problems like these.

CL 3-101. **Examine** the shape at right.

a. On graph paper, enlarge this shape by a factor of 3.

b. Now redraw the enlarged shape from part (a) using a zoom factor of $\frac{1}{2}$.

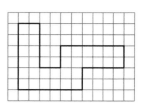

CL 3-102. Jermaine has a triangle with sides 8, 14, and 20. Sadie and Aisha both think that they have triangles that are similar to Jermaine's triangle. The sides of Sadie's triangle are 2, 3.5, and 5. The sides of Aisha's triangle are 4, 10, and 16. Decide who, if anyone, has a triangle similar to Jermaine's triangle. Be sure to explain how you know.

CL 3-103. Each pair of figures below is similar. Find the lengths of all unknown sides.

a.

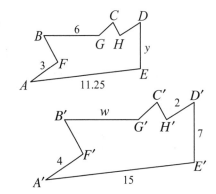

b.

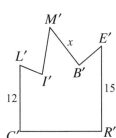

CL 3-104. Draw an example for each geometric term below. Diagrams should be clearly marked with all necessary information. You should also include a brief description of the qualities of each term.

a. supplementary angles

b. alternate interior angles

c. parallel lines

d. complementary angles

CL 3-105. Find the perimeter and area of the shape at right.

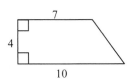

CL 3-106. Create a flowchart that represents the story:

Marcelle and Harpo live at Apt. I, 8 Logic St. Marcelle took his guitar to band practice across town and isn't back yet. Harpo hears guitar music in the hallway. He decides that someone else in the building also plays guitar.

CL 3-107. For each pair of triangles below, determine whether or not the triangles are similar. If they are similar, show your **reasoning** in a flowchart. If they are not similar, explain how you know.

a.   b.   c.

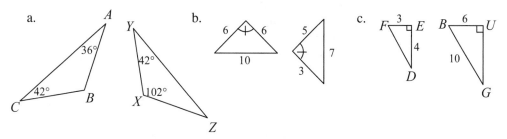

CL 3-108. Use the relationships in the diagrams below to find the values of the variables, if possible. The diagrams are not drawn to scale.

a.   b.   c.

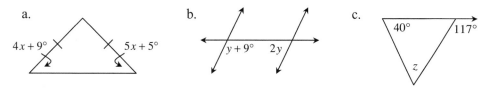

CL 3-109. Among the triangles below are pairs of similar triangles. Find the pairs of similar triangles and state the triangle similarity conjecture that you used to determine that the triangles are similar.

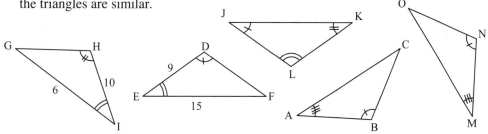

CL 3-110. Check your answers using the table at the end of this section. Which problems do you feel confident about? Which problems were hard? Have you worked on problems like these in math classes you have taken before? Use the table to make a list of topics you need help on and a list of topics you need to practice more.

This course focuses on five different **Ways of Thinking**: investigating, examining, reasoning and justifying, visualizing, and choosing a strategy/tool. These are some of the ways in which you think while trying to make sense of a concept or to solve a problem (even outside of math class). During this chapter, you have probably used each Way of Thinking multiple times without even realizing it!

This closure activity will focus on one of these Ways of Thinking: **reasoning and justifying**. Read the description of this Way of Thinking at right.

Think about the problems you have worked on in this chapter. When did you need to make sense out of multiple pieces of information? What helps you to determine what makes sense? Were there times when you made assumptions instead of relying on facts? How have you used **reasoning** in a previous math class? You may want to flip through the chapter to refresh your memory about the problems that you have worked on. Discuss any of the methods you have developed to help you **reason**.

Once your discussion is complete, think about the way you think as you answer the questions below.

Reasoning and Justifying
To use logical **reasoning** means to organize facts with the purpose of making and communicating a convincing argument. As you develop this way of thinking, you will learn what pieces of information are facts and how you can combine facts to make convincing (uncontestable) arguments. You think this way when you answer questions like *"Is that always true?"* When you catch yourself defending a statement or idea, you are using logical reasoning.

a.   At right is a crossword puzzle and a list of words that fit within the puzzle. Do you know where all of the words MUST go or is there more than one possible solution? Write an argument that will convince your teacher of your answer.

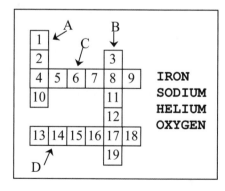

Problem continues on next page →

⑤   *Problem continued from previous page.*

b.   **Examine** the Venn diagram at right. What shape(s) can go in the intersection? **Justify** your statements so that they are convincing.

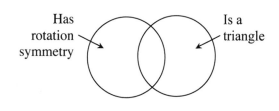

c.   Being able to **justify** your thinking is especially important when you are working in a study team. While you may have a correct idea, if you cannot convince your team that your ideas are valid, your teammates may not agree.

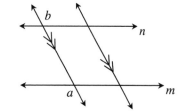

Consider the diagram above. Assume that your teammates think angles *a* and *b* must be congruent. Do you agree? How can you explain your **reasoning** so that your team is convinced? Use a diagram to support your argument.

d.   What if someone else is trying to convince <u>you</u> of something? How do you think as you follow someone else's argument? Consider this as you read Lila's **reasoning** below. Then decide if you agree with her statement or not. If you agree, what helped convince you?

I know that the sum of the angles of a triangle is 180°, but I don't think that is true for a quadrilateral. If I draw a diagonal, I split my quadrilateral into two triangles. I know that the angles of each of these triangles add up to 180°. Therefore, I think the angles of the quadrilateral must add up to 360°.

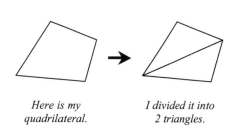

Here is my quadrilateral.   *I divided it into 2 triangles.*

Answers and Support for
Closure Activity #4
What Have I Learned?

Problem	Solutions	Need Help?	More Practice
CL 3-101.	a. b.	Lessons 3.1.1 and 3.1.2 Math Notes boxes, problem 3-11	Problems 3-12, 3-24, 3-26, 3-27, 3-28, 3-91
CL 3-102.	Sadie's triangle is similar to Jermaine by SSS ~ ; The ratio of the corresponding sides is $\frac{1}{4}$.	Lessons 3.2.1 and 3.2.5 Math Notes boxes	Problems 3-46, 3-24, 3-54, 3-55, 3-57, 3-66, 3-80, 3-83, 3-84, 3-86, 3-89
CL 3-103.	a. $w = 8$, $y = \frac{21}{4} = 5.25$ b. $x = 10$, $z = \frac{36}{5} = 7.2$	Lessons 3.1.2 and 3.1.3 Math Notes boxes	Problems 3-23, 3-35, 3-36, 3-38, 3-59, 3-65, 3-71, 3-99
CL 3-104.	a. b. c. d.	Lessons 2.1.1 and 2.1.4 Math Notes boxes, Problem 2-3	Problems 2-13, 2-16, 2-17, 2-18, 2-23, 2-24, 2-25, 2-62, 2-72, 3-20, 3-31, 3-60

Problem	Solutions	Need Help?	More Practice
CL 3-105.	Perimeter = 26 un., Area = 34 un.^{2}	Lessons 1.1.3, 2.2.4, and 2.3.3 Math Notes boxes	Problems 2-66, 2-75, 2-79, 2-90, 2-120, 3-5, 3-42, 3-51, 3-81
CL 3-106.	Harpo hears guitar music in the hallway. / Marcelle took his guitar to band practice across town and isn't back yet. → Harpo decides that someone else in the building also plays guitar.	Lesson 3.2.4 Math Notes box	Problems 3-62, 3-56, 3-67, 3-87
CL 3-107.	a. $\angle BAC = \angle XZY$ / $\angle ACB = \angle ZYX$ → $\triangle ABC \sim \triangle ZXY$ AA ~ b. Not similar because corresponding sides do not have the same ratio. c. $\frac{FD}{BG} = \frac{5}{10} = \frac{1}{2}$ / $\frac{DE}{GU} = \frac{4}{8} = \frac{1}{2}$ / $\frac{FE}{BU} = \frac{3}{6} = \frac{1}{2}$ → $\triangle FED \sim \triangle BUG$ SSS ~	Lessons 3.2.1 and 3.2.5 Math Notes Boxes	Problems 3-46, 3-49, 3-53, 3-54, 3-55, 3-57, 3-64, 3-66, 3-80, 3-83, 3-84, 3-89
CL 3-108.	a. $x = 4°$ b. Cannot determine because the lines are not marked parallel. c. $z = 77°$	Lessons 2.1.1 and 2.1.4 Math Notes boxes, Problem 2-3	Problems 2-13, 2-16, 2-17, 2-18, 2-23, 2-24, 2-25, 2-31, 2-32, 2-38, 2-49, 2-51, 2-62, 2-72, 2-111, 3-20, 3-31, 3-60, 3-70, 3-96
CL 3-109.	$\triangle ABC \sim \triangle MNO$, by AA ~ $\triangle EDF \sim \triangle IGH$, by SAS ~ $\triangle EDF \sim \triangle LJK$, by AA ~ $\triangle IGH \sim \triangle LJK$, by AA ~	Lessons 3.2.1 and 3.2.4 Math Notes boxes	Problems 3-46, 3-49, 3-53, 3-54, 3-55, 3-57, 3-64, 3-66, 3-80, 3-83, 3-84, 3-89

CHAPTER 4 Trigonometry and Probability

In Chapter 3, you **investigated** similarity and discovered that similar triangles have special relationships. In this chapter, you will discover that the side ratios in a right triangle can serve as a powerful mathematical tool that allows you to find missing side lengths and missing angle measures for any right triangle. You will also learn how these ratios (called trigonometric ratios) can be used in solving problems.

You will also develop additional skills in prediction as you extend your understanding of probability. You will **examine** different models to represent possibilities and to assist you in calculating probabilities.

Guiding Questions

Think about these questions throughout this chapter:

Are they similar?

What is the ratio?

What is the probability?

How can I represent it?

In this chapter, you will learn:

> how the tangent ratio is connected to the slope of a line.

> the trigonometric ratio of tangent.

> how to apply trigonometric ratios to find missing measurements in right triangles.

> how to model real world situations with right triangles and use trigonometric ratios to solve problems.

> several ways to model probability situations, such as tree diagrams and area models.

Chapter Outline

Section 4.1 Students will **investigate** the relationship between the slope of a line and the slope angle. The slope ratio will be used to find missing measurements of a right triangle and to solve real world problems.

Section 4.2 Students will continue with their study of probability by studying different models (systematic lists, tree diagrams and area models) used to represent situations involving chance.

4.1.1 What patterns can I use?

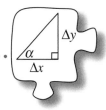

. . . .

Constant Ratios in Right Triangles

In Chapter 3, you looked for relationships and patterns among shapes such as triangles, parallelograms, and trapezoids. Now we are going to focus our attention on slope triangles, which were used in algebra to describe linear change. Are there geometric patterns within slope triangles themselves that we can use to answer other questions? In this lesson, you will look closely at slope triangles on different lines to explore their patterns.

4-1.   LEANING TOWER OF PISA

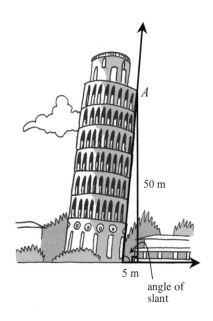

For centuries, people have marveled at the Leaning Tower of Pisa due to its slant and beauty. Ever since construction of the tower started in the 1100's, the tower has slowly tilted south and has increasingly been at risk of falling over. It is feared that if the angle of slant ever falls below 83°, the tower will collapse.

Engineers closely monitor the angle at which the tower leans. With careful measuring, they know that the point labeled *A* in the diagram at right is now 50 meters off the ground. Also, they determined that when a weight is dropped from point *A*, it lands five meters from the base of the tower, as shown in the diagram.

a.   With the measurements provided above, what can you determine?

b.   Can you determine the angle at which the tower leans? Why or why not?

c.   At the end of Section 4.1, you will know how to find the angle for this situation and many others. However, at this point, how else can you describe the "lean" of the leaning tower?

4-2.   PATTERNS IN SLOPE TRIANGLES

In order to find an angle (such as the angle at which the Leaning Tower of Pisa leans), you need to **investigate** the relationship between the angles and the sides of a right triangle. You will start by studying slope triangles. Obtain the Lesson 4.1.1 Resource Pages (two in all) from your teacher and find the graph shown below. Notice that one slope triangle has been drawn for you. **Note:** For the next several lessons angle measures will be rounded to the nearest degree.

a. Draw three new slope triangles on the line. Each should be a different size. Label each triangle with as much information as you can, such 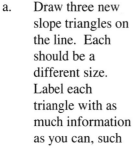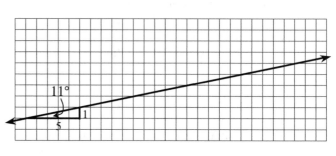 as its horizontal and vertical lengths and its angle measures.

b. What do these triangles have in common? How are these triangles related to each other?

c. Write the slope ratio for each triangle as a fraction, such as $\frac{\Delta y}{\Delta x}$. (Note: Δy represents the vertical change or "rise," while Δx represents the horizontal change or "run.") Then change the slope ratio into decimal form.

d. What do you notice about the slope ratios written in fraction form? What do you notice about the decimals?

4-3.   Tara thinks she sees a pattern in these slope triangles, so she decides to make some changes in order to **investigate** whether or not the patterns remain true.

a. She asks, "What if I draw a slope triangle on this line with $\Delta y = 6$? What would be the Δx of my triangle?" Answer her question and explain how you figured it out.

b. "What if Δx is 40?" she wonders. "Then what is Δy?" Find Δy, and explain your **reasoning**.

c. Tara wonders, "What if I draw a slope triangle on a different line? Can I still use the same ratio to find a missing Δx or Δy value?" Discuss this question with your team and explain to Tara what she could expect.

4-4.   CHANGING LINES

In part (c) of problem 4-3, Tara asked, "What if I draw my triangle on a different line?" With your team, **investigate** what happens to the slope ratio and slope angle when the line is different. Use the graph grids provided on your Lesson 4.1.1 Resource Pages to graph the lines described below. Use the graphs and your answers to the questions below to respond to Tara's question.

a.   On graph A, graph the line $y = \frac{2}{5}x$. What is the slope ratio for this line? What does the slope angle appear to be? Does the information about this line support or change your conclusion from part (c) of problem 4-3? Explain.

b.   On graph B, you are going to create $\angle QPR$ so that it measures 18°. First, place your protractor so that point P is the vertex. Then find 18° and mark and label a new point, R. Draw ray $\overrightarrow{PR}$ to form $\angle QPR$. Find an approximate slope ratio for this line.

c.   Graph the line $y = x + 4$ on graph C. Draw a slope triangle and label its horizontal and vertical lengths. What is $\frac{\Delta y}{\Delta x}$ (the slope ratio)? What is the slope angle?

4-5.   TESTING CONJECTURES

The students in Ms. Coyner's class are writing conjectures based on their work today. As a team, decide if you agree or disagree with each of the conjectures below. Explain your **reasoning**.

- All slope triangles have a ratio $\frac{1}{5}$.

- If the slope ratio is $\frac{1}{5}$, then the slope angle is approximately 11°.

- If the line has an 11° slope angle, then the slope ratio is approximately $\frac{1}{5}$.

- Different lines will have different slope angles and different slope ratios.

METHODS AND MEANINGS

Slope and Angle Notation

MATH NOTES

Slope is the ratio of the vertical distance to the horizontal distance in a slope triangle. The vertical part of the triangle is called Δy, (read "change in y"), while the horizontal part of the triangle is called Δx (read "change in x").

When we are missing a side length in a triangle, we often assign that length a variable from the English alphabet such as x, y, or z. However, sometimes we need to distinguish between an unknown side length and an unknown angle measure. With that in mind, mathematicians sometimes use Greek letters as variables for angle measurement. The most common variable for an angle is the Greek letter θ (*theta*), pronounced "THAY-tah." Two other Greek letters commonly used include α (*alpha*), and β (*beta*), pronounced "BAY-tah."

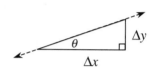

When a right triangle is oriented like a slope triangle, such as the one in the diagram above, the angle the line makes with the horizontal side of the triangle is called a **slope angle.**

4-6.   Use what you know about the angles of a triangle to find the value of x and the angles in each triangle below.

a.

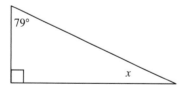

b.

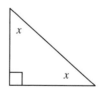

c.

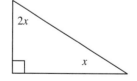

d.

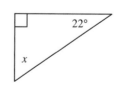

4-7.   Use the triangles at right to answer the following questions.

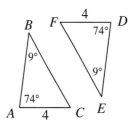

a. Are the triangles at right similar? How do you know? Show your **reasoning** in a flowchart.

b. **Examine** your work from part (a). Are the triangles also congruent? Explain why or why not.

4-8.   As Randi started to solve for x in the diagram at right, she wrote the equation $7^2 + x^2 = (x+1)^2$.

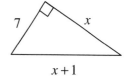

a.   Is Randi's equation valid? Explain your thinking.

b.   To solve her equation, first rewrite $(x+1)^2$ by multiplying $(x+1)(x+1)$. You may want to review the Math Notes box for Lesson 2.2.2.

c.   Now solve your equation for x.

d.   What is the perimeter of Randi's triangle?

4-9.   Assume that the shapes at right are similar. Find the values of x, y, and z.

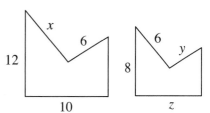

4-10.   Are the lines represented by the equations at right parallel? Support your **reasoning** with convincing evidence.

$$y = -\frac{3}{5}x + 2$$
$$y = -\frac{3}{5}x - 3$$

4.1.2 How important is the angle?

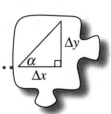

Connecting Slope Ratios to Specific Angles

In Lesson 4.1.1, you started **trigonometry**, the study of the measures of triangles. As you continue to **investigate** right triangles with your team today, use the following questions to guide your discussion:

What do we know about this triangle?

How does this triangle relate to other triangles?

Which part is Δx? Which part is Δy?

4-11. What do you know about this triangle? To what other triangles does it relate? Use any information you have to solve for *x*.

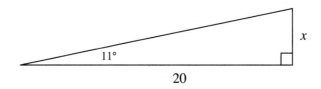

4-12. For each triangle below, find the missing angle or side length. Use your work from Lesson 4.1.1 to help you.

a.

b.
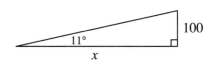

c.
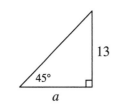

d.
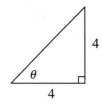

e.
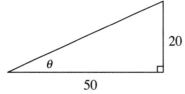

f.
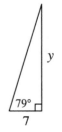

4-13. Sheila says the triangle in part (f) of problem 4-12 is the same as her drawing at right.

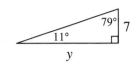

a. Do you agree? Use tracing paper to convince you of your conclusion.

b. Use what you know about the slope ratio of 11° to find the slope ratio for 79°.

c. What is the relationship of 11° and 79°? Of their slope ratios?

4-14. For what other angles can you find the slope ratios based on the work you did in Lesson 4.1.1?

a. For example, since you know the slope ratio for 22°, what other angle do you know the slope ratio for? Use tracing paper to find a slope ratio for the complement of each slope angle you know. Use tracing paper to help re-orient the triangle if necessary.

b. Use this information to find x in the diagram at right.

c. Write a conjecture about the relationship of the slope ratios for complementary angles. You may want to start with, "*If one angle of a right triangle has the slope ratio $\frac{a}{b}$, then ...*"

4-15. BUILDING A TRIGONOMETRY TABLE

So far you have looked at several similar slope triangles and their corresponding slope ratios. These relationships will be very useful for finding missing side lengths or angle measures of right triangles for the rest of this chapter.

Before you forget this valuable information, organize information about the triangles and ratios you have discovered so far in the table on the Lesson 4.1.2 ("Trig Table") Resource Page provided by your teacher or download from www.cpm.org. Keep it in a safe place for future reference. Include all of the angles you have studied up to this point. An example for 11° is filled in on the table to get you started.

MATH NOTES

In Lesson 4.1.1, you discovered that certain slope angles produce slope triangles with special ratios. Below are the triangles you have studied so far. Note that the angles below are rounded to the nearest degree.

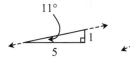 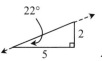

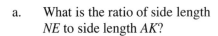

Review & Preview

4-16.   Use your Trig Table from problem 4-15 to help you find the value of each variable below.

a.

b.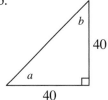

c.

4-17.   The triangles shown at right are similar.

a.   What is the ratio of side length *NE* to side length *AK*?

b.   Use a ratio to compare the perimeters of $\triangle ENC$ and $\triangle KAR$. How is the perimeter ratio related to the side length ratio?

c.   If you have not already done so, find the length of $\overline{EC}$.

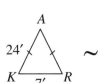 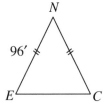

4-18.   **Examine** each pair of figures below. Are they similar? Explain how you know.

a.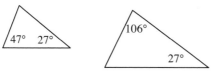

b.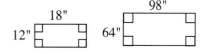

c.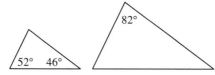

4-19.   Find the equation of the line with a slope of $\frac{1}{3}$ that goes through the point $(0, 9)$.

4-20.   **Examine** the figure at right, which is not drawn to scale. Which is longer, $\overline{AB}$ or $\overline{BC}$? Explain your answer.

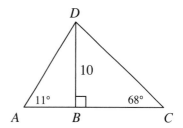

4.1.3   What if the angle changes?

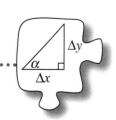

Expanding the Trig Table

In the last few lessons, you found the slope ratios for several angles. However, so far you are limited to using the slope angles that are currently in your Trig Table. How can we find the ratios for other angles? And how are the angles related to the ratio?

Today your goal is to determine ratios for more angles and to find patterns. As you work today, keep the following questions in mind:

What happens to the slope ratio when the angle increases? Decreases?

What happens to the slope ratio when the angle is 0°? 90°?

When is a slope ratio more than 1? When is it less than 1?

4-21. On your paper, draw a slope triangle with a slope angle of 45°.

a. Now **visualize** what would happen to the triangle if the slope angle increased to 55°. Which would be longer? Δy? Or Δx? Explain your **reasoning**.

b. Using your dynamic geometry tool, create a triangle with a slope angle measuring 55°. Then use the resulting slope ratio to help you solve for x in the triangle at right. (Note: the triangle at right is not draw to scale.)

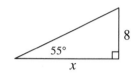

4-22. Copy each of the following triangles onto your paper. Decide whether or not the given measurements are possible. If the triangle is possible, find the value of x or θ. Use the dynamic geometry tool to find the appropriate slope angles or ratios. If technology is not available, your teacher will provide a Lesson 4.1.3 Resource Page with the needed ratios. Round angle measures to the nearest degree. If a triangle's indicated measurement is not possible, explain why.

a.

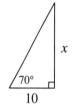

b.

c.

d.

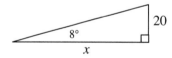

e.

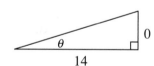

f.

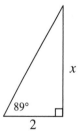

g.

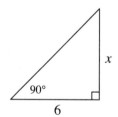

4-23. If you have not already, add these new slope ratios with their corresponding angles to your Trig Table. Be sure to draw and label the triangle for each new angle. Summarize your findings—which slope triangles did not work? Do you see any patterns?

4-24.   What statements can you make about the connections between slope
          angle and slope ratio? In your Learning Log, write down all of your
          observations from this lesson. Be sure to answer the questions given
          at the beginning of the lesson (reprinted below). Title this entry,
          "Slope Angles and Slope Ratios" and include today's date.

          What happens to the slope ratio when the angle increases? Decreases?

          What happens to the slope ratio when the angle is 0°? 90°?

          When is a slope ratio more than 1? When is it less than 1?

4-25.   Ben thinks that the slope ratio for this triangle is $\frac{7}{10}$.
          Carlissa thinks the ratio is $\frac{10}{7}$. Who is correct? Explain
          your thinking fully.

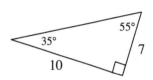

4-26.   Use your observations from problem 4-24 to answer the following
          questions:

          a.   Thalia did not have a tool to help her find the
               slope angle in the triangle at right. However,
               she claims that the slope angle has to be more
               than 45°. Do you agree with Thalia? Why?

          b.   Lyra was trying to find the slope ratio for the triangle at right,
               and she says the answer is $\frac{\Delta y}{\Delta x} = 2.675$. Isiah claims that
               cannot be correct. Who is right? How do you know?

          c.   Without finding the actual value, what information do you
               know about x in the diagram at right?

4-27.   An airplane takes off and climbs at an angle of 11°.
          If the plane must fly over a 120-foot tower with at
          least 50 feet of clearance, what is the minimum
          distance between the point where the plane leaves
          the ground and the base of the tower?

          a.   Draw and label a diagram for this situation.

          b.   What is the minimum distance between the point where the plane leaves the
               ground and the tower? Explain completely.

4-28.   Edwina has created her own Shape Bucket and has provided the clues below about
         her shapes. List one possible group of shapes that could be in her bucket.

$$P(equilateral) = 1$$

$$P(triangle) = \tfrac{1}{3}$$

4-29.   Use what you know about the sum of the
         angles of a triangle to find $m\angle ABC$ and
         $m\angle BAC$. Are these angles acute or obtuse?
         Find the sum of these two angles. How can we
         describe the relationship of these two angles?

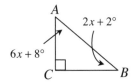

4.1.4 What about other right triangles?

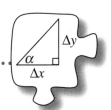

The Tangent Ratio

In Lesson 4.1.2 you started a Trig Table of angles and their related slope ratios. Unfortunately,
you only have the information for a few angles. How can you quickly find the ratios for other
angles when a computer is not available or when an angle is not on your Trig Table? Do you
have to draw each angle to get its slope ratio? Or is there another way?

4-30.   WILL IT TOPPLE?

         In problem 4-1, you learned that the Leaning
         Tower of Pisa is expected to collapse once its
         angle of slant is less than 83°. Currently, the
         top of the seventh story (point *A* in the
         diagram at right) is 50 meters above the
         ground. In addition, when a weight is dropped
         from point *A*, it lands 5 meters from the base
         of the tower, as shown in the diagram.

         a.   What is the slope ratio for the tower?

         b.   Use your Trig Table to determine the
               angle at which the Leaning Tower of
               Pisa slants. Is it in immediate danger of
               collapse?

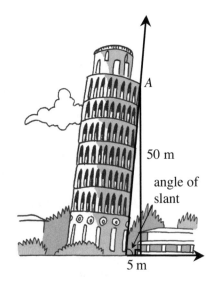

4-31.   Solve for the variables in the triangles below. It may be helpful to first orient the triangle (by rotating your paper or by using tracing paper) so that the triangle resembles a slope triangle. Use your Trig Table for reference.

a.

b.

c.
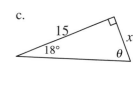

4-32.   MULTIPLE METHODS

Tiana, Mae Lin, Eddie, and Amy are looking at the triangle at right and trying to find the missing side length.

a.   Tiana declares, "Hey! We can rotate the triangle so that 18° looks like a slope angle, and then $\Delta y = 4$." Will her method work? If so, use her method to solve for a. If not, explain why not.

b.   Mae Lin says, "I see it differently. I can tell $\Delta y = 4$ without turning the triangle." How can she tell? Explain one way she could know.

c.   Eddie replies, "What if we use 72° as our slope angle? Then $\Delta x = 4$." What is he talking about? Discuss with your team and explain using pictures and words.

d.   Use Eddie's observation in part (c) to confirm your answer to part (a).

4-33.   USING A SCIENTIFIC CALCULATOR

Examine the triangle at right.

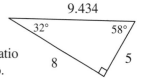

a.   According to the triangle at right, what is the slope ratio for 32°? Explain how you decided to set up the ratio. Write the ratio in both fraction and decimal form.

b.   What is the slope ratio for the 58° angle? How do you know?

c.   Scientific calculators have a button that will give the slope ratio when the slope angle is entered. In part (a), you calculated the slope ratio for 32° as 0.625. Use the "tan" button on your calculator to verify that you get ≈ 0.625 when you enter 32°. Does that button give you ≈ 1.600 when you enter 58°? Be ready to help your teammates find the button on their calculator.

d.   The ratio in a right triangle that you have been studying is referred to as the **tangent ratio**. When you want to find the slope ratio of an angle, such as 32°, it is written "tan 32°." So, an equation for this triangle can be written as $\tan 32° = \frac{5}{8}$. Read more about the tangent ratio in the Math Notes box for this lesson.

4-34.   For each triangle below, trace the triangle on tracing paper. Label its legs Δy and Δx based on the given slope angle. Then write an equation (such as $\tan 14° = \frac{x}{5}$), use your scientific calculator to find a slope ratio for the given angle, and solve for x.

a.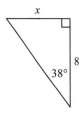

b.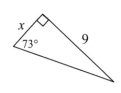

c.

4-35.   How do you set up a tangent ratio equation? How can you know which side is Δy? How can you use your scientific calculator to find a slope ratio? Write a Learning Log entry about what you learned today. Be sure to include examples or refer to your work from today. Title this entry "The Tangent Ratio" and include today's date.

⃝ℳETHODS AND MEANINGS

MATH NOTES

The Tangent Ratio

For any slope angle in a slope triangle, the ratio that compares the Δy to Δx is called the **tangent ratio**. The ratio for any angle is constant, regardless of the size of the triangle. It is written:

$$\tan\,(\text{slope angle}) = \frac{\Delta y}{\Delta x}$$

One way to identify which side is Δy and Δx is to first reorient the triangle so that it looks like a slope triangle, as shown at right.

For example, when the triangle at right is rotated, the resulting slope triangle helps to show that the tangent of θ is $\frac{p}{r}$, since θ is the slope angle, p is Δy and r is Δx. This is written:

$$\tan\,\theta = \frac{p}{r}$$

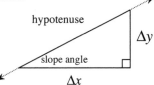

Whether the triangle is oriented as a slope triangle or not, you can identify Δy as the leg that is always opposite (across the triangle from) the angle, while Δx is the leg closest to the angle.

$$\tan\,\theta = \frac{opposite\ leg}{adjacent\ leg} = \frac{p}{r}$$

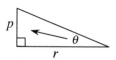

4-36.   Find the missing side length for each triangle. Use the tangent button on your calculator to help.

a.
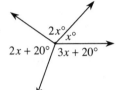

b.

c.
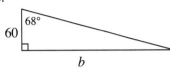

4-37.   Use the relationships in the diagrams below to write an equation and solve for *x*.

a.

b.
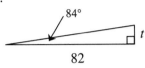

c.
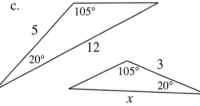

4-38.   What is the relationship of the triangles at right? **Justify** your conclusion.

4-39.   Mr. Singer made the flowchart at right about a student named Brian.

a.   What is wrong with Mr. Singer's flowchart?

b.   Rearrange the ovals so the flowchart makes more sense.

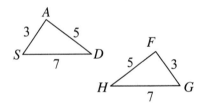

4-40.   When she was younger, Mary had to look up at a 68° angle to see into her father's eyes whenever she was standing 15 inches away. How high above the flat ground were her father's eyes if Mary's eyes were 32″ above the ground?

4.1.5 What if I can't measure it?

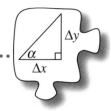

Applying the Tangent Ratio

In this section so far, you have learned how to find the legs of a right triangle using an angle. But how can you use this information? Today you and your team will use the tangent ratio to solve problems and answer questions.

4-41.   STATUE OF LIBERTY

Lindy gets nose-bleeds whenever she is 300 feet above ground. During a class fieldtrip, her teacher asked if she wanted to climb to the top of the Statue of Liberty. Since she does not want to get a nose-bleed, she decided to take some measurements to figure out how high the torch of the statue is. She found a spot directly under the torch and then measured 42 feet away and determined that the angle up to the torch was 82°. Her eyes are 5 feet above the ground.

Should she climb to the top or will she get a nose-bleed? Draw a diagram that fits this situation. **Justify** your conclusion.

4-42.   HOW TALL IS IT?

How tall is Mount Everest? How tall is the White House? Often we want to know a measurement of something we cannot easily measure with a ruler or tape measure. Today you will work with your team to measure the height of something inside your classroom or on your school's campus in order to apply your new tangent tool.

Your Task: Get a **clinometer** (a tool that measures a slope angle) and a meter stick (or tape measure) from your teacher. As a team, decide how you will use these tools to find the height of the object selected by your teacher. Be sure to record all measurements carefully on your Lesson 4.1.6A Resource Page and include a diagram of the situation.

Discussion Points

What should the diagram look like?

What measurements would be useful?

How can you use your tools effectively to get accurate measurements?

4-43. The trapezoids at right are similar.

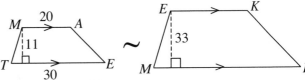

a. What is the ratio of the heights?

b. Compare the areas. What is the ratio of the areas?

4-44. For each diagram below, write an equation and solve for x, if possible.

a.

$3x + 3°$

$x + 7°$

b.

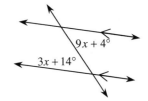

$9x + 4°$

$3x + 14°$

4-45. Joan and Jim are planning a dinner menu including a main dish and dessert. They have 4 main dish choices (steak, tuna casserole, turkey burgers, and lasagna) and 3 dessert choices (brownies, ice cream, and chocolate chip cookies.) How many different dinner menus do they have to choose from? List all of the possible menus.

4-46. Leon is standing 60 feet from a telephone pole. As he looks up, a red-tailed hawk lands on the top of the pole. Leon's angle of sight up to the bird is 22° and his eyes are 5.2 feet above the ground.

a. Draw a detailed picture of this situation. Label with all of the given information.

b. How tall is the pole? Show all of your work.

4-47. Find the value of x in the triangle at right. Refer to problem 4-8 for help. Show all work.

$x + 3$

26

10

4.2.1 How can I make a list?

Introduction to Probability Models

In Chapter 1 you began to study probability, which is a measure of the chance that a particular event will occur. In the next few lessons you will encounter a variety of situations that require probability calculations. In many cases you will have to develop new probability tools to help you analyze these situations. Today's lesson focuses on tools for listing *all* the possible outcomes of a probability situation.

4-48.   THE RAT RACE, Part One

Ryan has a pet rat Romeo that he boasts is the smartest rat in the county. Sammy overheard Ryan at the county fair claiming that Romeo could learn to run a particular maze and find the cheese at the end.

"I don't think Romeo's that smart!" Sammy declares, "I think the rat just chooses a random path through the maze."

Ryan has built a maze with the floor plan shown at right. Ryan places the cheese in an airtight container (so Romeo can't smell the cheese!) in room A.

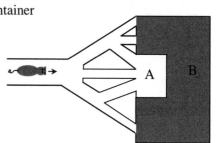

Ryan runs Romeo through the maze 100 times, and Romeo finds the cheese 66 times. "See how smart Romeo is?" Ryan asks, "He clearly learned something and got better at the maze as he went along." Sammy isn't so sure.

a.   Does the fact that Romeo ends up in Room A mean that he has learned the route and improved his ability to return to the same room over time? Or do you think he would achieve the same results by moving randomly throughout the maze? Discuss this question with your team.

b.   Predicting what will happen requires you to analyze the **probability** of possible outcomes. To accurately decide if Romeo is randomly guessing or if he is learning the route, you will need to learn more about probability. We will return to this problem at the end of Section 4.2. At this point, review the Math Notes box from Lesson 1.3.3 and then move on to problem 4-49.

4-49.   TOP OF THE CHARTS

Renae's MP3 player can be programmed to randomly play songs from her playlist without repeating a single song. Currently, Renae's MP3 player has 5 songs loaded on it, which are listed at right. As she walks between class, she only has time to listen to one song.

PLAYLIST

a. **I Love My Mama** (country)
   by the Strings of Heaven

b. **Don't Call Me Mama** (country)
   Duet by Sapphire and Hank Tumbleweed

c. **Carefree and Blue** (R & B)
   by Sapphire and Prism Escape

d. **Go Back To Mama** (Rock)
   Duet by Bjorn Free and Sapphire

e. **Smashing Lollipops** (Rock)
   by Sapphire

a.   What is the probability that her MP3 player will select a country song?

b.   What is the probability that Renae will listen to a song with "Mama" in the title?

c.   What is the probability she listens to a duet with Hank Tumbleweed?

d.   What is the probability she listens to a song that is not R & B?

4-50.   While waiting for a bus after school, Renae programmed her MP3 player to randomly play two songs. Assume that the MP3 player will not play the same song twice.

a.   List all the combinations of two songs that Renae could select. The order that she hears the songs does not matter for your list. How can you be sure that you listed all of the song combinations?

b.   Find the probability that Renae will listen to two songs with the name "Mama" in the title.

c.   What is the probability that at least one of the songs will have the name "Mama" in the title?

d.   Why does it make sense that the probability in part (d) is higher than the probability in part (c)?

4-51.   When a list is created by following a system (an orderly process), it is called a
          systematic list. Using a systematic list to answer questions involving probability
          can help you determine **all** of the possible outcomes. There are different **strategies**
          that may help you make a systematic list, but what is most important is that you
          methodically follow your system until it is complete. For the problem below, create
          a systematic list. Be prepared to share your **strategy**.

          To get home, Renae can take one of four
          busses: #41, #28, #55, or #81. Once she is on
          a bus, she will randomly select one of the
          following equally likely activities: listening to
          her MP3 player, writing a letter, or reading a
          book.

          a.   List all the possible ways Renae can get
                home. Use a systematic list to make
                sure you find all the combinations of a
                bus and an activity.

          b.   Use your list to find the following
                probabilities:

                (1)   P(Renae takes an odd-numbered bus)

                (2)   P(Renae does not write a letter)

                (3)   P(Renae catches the #28 bus and then reads a book)

          c.   Does her activity depend on which bus she takes? Explain why or why not.

4-52.   Sometimes writing an entire systematic list seems
          unnecessarily repetitive. However, creating a **tree
          diagram**, like the one started at right, is one way to
          avoid the repetition. This structure organizes the list
          by connecting each bus with each activity.

          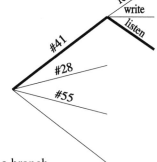

          In this tree, the first set of branches represents the bus
          options. At the end of each of these branches are
          branches representing the activities. This system
          reduces some of the writing repetition.

          Renae can then read the possible scenarios by following a branch
          across the diagram. For example, if you follow the bold
          branches, Renae will take the #41 bus and will listen to her MP3
          player.

          On your paper, complete this tree diagram to show all of the different travel options
          that Renae could take.

4-53.   For the evening, Renae has programmed her MP3 player to play all five songs in a random order.

PLAYLIST

a. **I Love My Mama** (country) by the Strings of Heaven

b. **Don't Call Me Mama** (country) Duet by Sapphire and Hank Tumbleweed

c. **Carefree and Blue** (R & B) by Sapphire and Prism Escape

d. **Go Back To Mama** (Rock) Duet by Bjorn Free and Sapphire

e. **Smashing Lollipops** (Rock) by Sapphire

a.   What is the probability that the first song is a country song?

b.   If the first song is a country song, does that affect the probability that the second song is a country song? Explain your thinking.

c.   As songs are playing, the number of songs left to play decreases. Therefore, the probability of playing each of the remaining songs is dependent on which songs that have played before it. This is an example of **dependent events.** If Renae has already listened to "Don't Call Me Mama," "Carefree and Blue," and "Smashing Lollipops," what is the probability that one of the singers of the fourth song will be Sapphire? Explain your **reasoning**.

d.   In problem 4-51, you considered a situation of **independent events**, when the bus that Renae took did not affect which activity she chose. For example, what is the probability that Renae writes a letter if she takes the #41 bus? What if she takes the #55 bus?

Review & Preview

4-54.   When Ms. Shreve randomly selects a student in her class, she has a $\frac{1}{3}$ probability of selecting a boy.

a.   If her class has 36 students, how many boys are in Ms. Shreve's class?

b.   If there are 11 boys in her class, how many girls are in her class?

c.   What is the probability that she will select a girl?

d.   Assume that Ms. Shreve's class has a total of 24 students. She selected one student (who was a boy) to attend a fieldtrip and then was told she needed to select one more student to attend. What is the probability that the second randomly selected student will also be a boy?

4-55.   What is wrong with the
         argument shown in the
         flow chart at right?
         What assumption does
         the argument make?

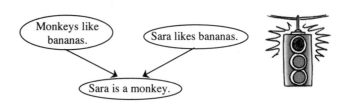

4-56.   For each diagram below, solve for x. Name the relationship(s) you used. Show all
         work.

         a.   b.

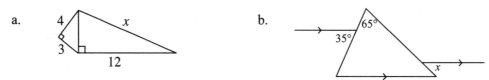

4-57.   Kamillah decided to find the height of the Empire State Building. She walked 1 mile
         away (5280 feet) from the tower and found that she had to look up 15.5° to see the
         top. Assuming Manhattan is flat, if Kamillah's eyes are 5 feet above the ground how
         tall is the Empire State Building?

4-58.   Renae wonders, "What if I program my MP3 player to randomly play 3 songs?"

         a.   Assuming that her MP3 player is still programmed as described in problem
              4-53, how many combinations of 3 songs could she listen to? Make a list of
              all the combinations. Remember that the order of the songs is not important.

         b.   Compare your answer to part (a) with the number of combinations that she
              could have with 2 songs in problem 4-50. Why does it make sense that these
              are the same?

4.2.2 How can I test my prediction?

Theoretical and Experimental Probability

In Lesson 4.2.1, you organized situations using systematic lists and tree diagrams in order to calculate the probability that an event would occur. How can you verify that your calculated probabilities are accurate? Today you will analyze a couple of games. For each game, you will calculate the chance of winning. Then you will play the game to decide if your probability analysis is accurate.

4-59.   Luis is going to spin the spinner at right.

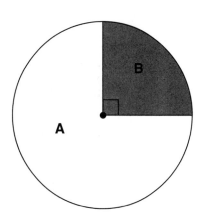

a.   What are the possible results of his spin?

b.   What do you predict Luis' result will be after one spin? **Justify** your prediction.

c.   If Luis spun this spinner four times, what do you predict his results would be?

d.   A paperclip with one end extended can be spun around the center of a spinner to collect data, as shown at right. With your team, spin a paperclip four times about the spinner above and record the results. Did your **experimental results** match your prediction in part (c)? Why or why not?

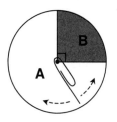

e.   What if the spinner is spun 40 times? 400 times? Would you predict the result to be closer to the expected results? Why or why not?

f.   What if Luis has spun the spinner 400 times and landed on region A every time? What would be the probability that the next spin would land on region B? Explain.

4-60.   Congratulations! You are going to be a
          contestant on a new game show with a chance to
          win some money. You will spin two spinners to
          see how much money you will win. The first
          spinner has an equal chance of coming out
          $100, $300, or $1500. The second spinner is
          equally divided between "Keep your winnings,"
          and "Double your winnings."

a.   Make a systematic list <u>and</u> tree
      diagram of all the possible outcomes.

b.   What is the probability that you will take
      home $200? What is the probability that
      you will take home more than $500?

c.   What is the probability that you will double
      your winnings? Does the probability that
      you will double your winnings depend on
      the result of the first spinner?

d.   Does your answer for part (c) mean that you are guaranteed to double your
      winnings half the time you play the game?

e.   What if the amounts on the first spinner were $100, $200, and $1500? What is
      the probability that you would take home $200? Justify your conclusion.

4-61.   ROLL AND WIN

          Now consider a game that is played with two regular dice, each numbered 1 through
          6. First, each player chooses a number. The two dice are rolled and the numbers
          that come up are **added** together. If the sum is the number you chose, you win a
          point. For example, if a 2 and a 5 are rolled, the sum is 7, so the person who chose
          the number 7 would get a point.

a.   First, use your intuition. Which number do you think will be the most common
      sum? In other words, which number would you pick to play?

b.   What are the possible results when two dice are rolled and their numbers
      added?

c.   Are the results from the separate dice dependent or
      independent? Does the result from one die affect the
      other? Explain.

d.   What about the separate rolls? When the two dice are
      rolled, does the sum depend on the previous rolls? Why
      or why not?

e.   Now play Roll and Win to see which sum occurs the most often. Have each
      member of your team choose a different number. Roll the dice 36 times and
      record the results. Which sum occurred most often?

Geometry Connections

4-62. One way to analyze this situation is to **make a table** like the one at right. Copy and complete this table of sums on your paper.

	Dice #1					
Dice #2	**1**	**2**	**3**	**4**	**5**	**6**
1			**5**			
2		**4**				
3						
4			**7**			
5						
6						

a. What is the probability the outcome will be odd? P(even)?

b. Which sum is the most likely result? What is the probability of rolling that sum?

c. What is the probability of rolling the sum of 2? 10? 15?

d. Compare your data from the table with that from problem 4-61. Did your experimental results match your theoretical probability? If not, explain what you could do to get results that are closer to the predicted results.

Review & Preview

4-63. What are the possible lengths for side $\overline{ML}$ in the triangle at right? Show how you know.

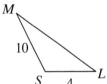

4-64. Alexis, Bart, Chuck, and Dariah all called in to a radio show to get free tickets to a concert. List all the possible orders in which their calls could have been received.

4-65. **Examine** the diagram at right. If $\overline{AC}$ passes through point *B*, then answer the questions below.

a. Are the triangles similar? If so, make a flowchart **justifying** your answer.

b. Are the triangles congruent? Explain how you know.

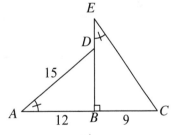

4-66. **Examine** pentagon *SMILE* at right. Do any of its sides have equal length? How do you know? Be sure to provide convincing evidence. You might want to copy the figure onto graph paper.

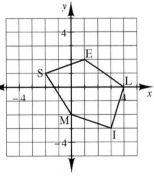

4-67. Find the area of the triangle at right. Show all work.

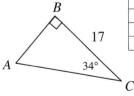

4.2.3 How can I represent it?

Using an Area Model

So far in this chapter, you have studied several different ways to represent situations involving chance. You have analyzed games using systematic lists, tree diagrams, and tables. In these games, each outcome you listed had an equal probability of occurring. But what if a game is biased so that some outcomes are more likely than others? How can you represent biased games? Today you will learn a new tool to analyze more complicated situations of chance, called an area model.

4-68.   IT'S IN THE GENES

Can you bend your thumb backwards at the middle joint to make an angle, like the example at right? Or does your thumb remain straight? The ability to bend your thumb back is thought to rely on a single gene.

What about your tongue? If you can roll your tongue into a "U" shape, you probably have a special gene that enables you to do this.

Example of a thumb that can bend backwards at the joint.

Assume that half of the U.S. population can bend their thumbs backwards and that half can roll their tongues. Also assume that these genes are independent (in other words, having one gene does not affect whether or not you have the other) and randomly distributed (spread out) throughout the population. Then the possible outcomes of these genetic traits can be organized in a table like the one below.

a.   According to this table, what is the probability that a random person from the U.S. has both special traits? That is, what is the chance that he or she can roll his or her tongue <u>and</u> bend his or her thumb back?

b.   According to this table, what is the probability that a random person has only one of these special traits? **Justify** your conclusion.

c.   This table is useful because every cell in the table is equally likely. Therefore, each possible outcome, such being able to bend your thumb but not roll your tongue, has a $\frac{1}{4}$ probability.

However, this table assumes that half the population can bend their thumbs backwards. In actuality, only a small percentage (about $\frac{1}{4}$) of the U.S. population has this special trait. It also turns out that a majority (about $\frac{7}{10}$) of the population can roll their tongues. How can this table be adjusted to represent these percentages? Discuss this with your team and be prepared to share your ideas with the class.

Geometry Connections

4-69.   USING AN AREA MODEL

One way to represent a probability situation that has outcomes with unequal
probabilities is by using an **area model**. An area model uses a square with an area
of 1 to represent all possible outcomes. The square is subdivided to represent the
different possible outcomes. The area of each outcome is the probability that the
outcome will occur.

For example, if $\frac{1}{4}$ of the U.S. population
can bend their thumbs back, then the
column representing this ability should
take only one-fourth of the square's
width, as shown at right.

BEND THUMB?

YES	NO
$\frac{1}{4}$	$\frac{3}{4}$

ROLL TONGUE?

a   How should the diagram be altered
     to that show that $\frac{7}{10}$ of the U.S. can
     roll their tongues? Copy this
     diagram on your paper and add two
     rows to represent this probability.

b.   The relative probabilities of the
     different outcomes are represented
     by the areas of the regions. For
     example, the portion of the area
     model representing people with both special traits is a rectangle with a width of
     $\frac{1}{4}$ and a height of $\frac{7}{10}$. What is the area of this rectangle? This area tells you
     the probability that a random person in the U.S. has both traits.

c.   What is the probability that a randomly selected person can roll his or her
     tongue but not bend his or her thumb back? Show how you got this
     probability.

4-70.   PROBABILITIES IN VEIN

You and your best friend may not only look
different, you may also have different types of
blood! For instance, members of the American
Navajo population can be classified into two
groups: 73% percent (73 out of 100) of the
Navajo population has type "O" blood, while
27% (27 out of 100) has type "A" blood.
(Blood types describe certain chemicals, called
"antigens," that are found in a person's blood.)

Navajo Person #1

Navajo Person #2	O $\frac{73}{100}$	A $\frac{27}{100}$
O $\frac{73}{100}$		
A $\frac{27}{100}$		

a.   Suppose you select two Navajo individuals at random. What is the probability
     that both individuals have type "A" blood? This time, drawing an area model
     that is exactly to scale would be challenging. Therefore, a **generic area model**
     (like the one above) is useful because it will still allow you to calculate the
     individual areas. Copy and complete this generic area model.

Problem continues on next page →

4-70.   *Problem continued from previous page.*

      b.   What is the probability that two Navajo individuals selected at random have the same blood type?

4-71.   SHIPWRECKED!

You and your best friend (both from the U.S.) are shipwrecked on a desert island! You have been injured and are losing blood rapidly, and your best friend is the only person around to give you a transfusion.

Unlike the Navajo you learned of in problem 4-70, most populations are classified into four blood types: O, A, B, and AB. For example, in the U.S., 45% of people have type O blood, 40% have type A, 11% have type B, and 4% have type AB (according to the American Red Cross, 2004). While there are other ways people's blood can differ, this problem will only take into account these four blood types.

      a.   Make a generic area model representing the blood types in this problem. List your friend's possible blood types along the top of the model and your possible blood types along the side.

      b.   What is the probability that you and your best friend have the same blood type?

      c.   Luckily, two people do not have to have the same blood type for the receiver of blood to survive a transfusion. Other combinations will also work, as shown in the diagram at right. Assuming that their blood is compatible in other ways, a donor with type O blood can donate to receivers with type O, A, B, or AB, while a donor with type A blood can donate to a receiver with A or AB. A donor with type B blood can donate to a receiver with B or AB, and a donor with type AB blood can donate only to AB receivers.

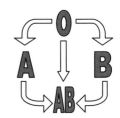

           Assuming that your best friend's blood is compatible with yours in other ways, determine the probability that he or she has a type of blood that can save your life!

4-72.   **Examine** the graph at right with slope triangles A, B, and C.

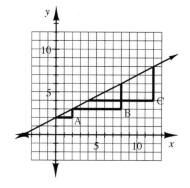

a.   Find the slope of the line using slope triangle A, slope triangle B, and then slope triangle C.

b.   Hernisha's slope triangle has a slope of $\frac{1}{2}$. What do you know about her line?

4-73.   Francis and John are racing. Francis is 2 meters in front of the starting line at time $t = 0$ and he runs at a constant rate of 1 meters per second. John is 5 meters in front of the starting line and he runs at a constant rate of 0.75 meters per second. After how long will Francis catch up to John?

4-74.   Can a triangle be made with sides of length 7, 10 and 20 units? **Justify** your answer.

4-75.   Solve each equation below for the given variable. Show all work and check your answer.

a. $\sqrt{x} - 5 = 2$ b. $-4(-2 - x) = 5x + 6$

c. $\frac{5}{x-2} = \frac{3}{2}$ d. $x^2 + 4x - 5 = 0$

4-76.   Find the perimeter of the shape at right. Clearly show all your steps.

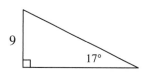

4.2.4 How can I represent it?

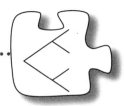

Choosing a Probability Model

In Lesson 4.2.3, you used area models to represent probability situations where some outcomes were more likely than others. Today you will consider how to represent these situations using tree diagrams. You will then return to a problem that was introduced in Lesson 4.2.1 to decide if a rat's behavior is random or not.

4-77. Your teacher challenges you to a spinner game. You spin the two spinners with the probabilities listed at right. The first letter should come from Spinner #1 and the second letter from Spinner #2. If the letters can form a two-letter English word, you win. Otherwise, your teacher wins.

$$P(I) = \tfrac{1}{2} \qquad P(T) = \tfrac{1}{4}$$
$$P(U) = \tfrac{1}{6} \qquad P(F) = \tfrac{3}{4}$$
$$P(A) = \tfrac{1}{3}$$

a. Make a generic area model and find the probability that you will win this game.

b. Is this game fair? If you played the game 100 times, who do you think would win more often, you or your teacher? Can you be sure this will happen?

4-78. Sinclair wonders how to model the spinner game in problem 4-77 using a tree diagram. He draws the tree diagram at right.

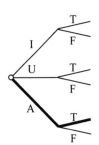

a. Sabrina says, "That can't be right. This diagram makes it look like all the words are equally likely." What is Sabrina talking about? Why is this tree diagram misleading?

b. To make the tree diagram reflect the true probabilities in this game, Sabrina writes numbers next to each letter showing the probability that the letter will occur. So she writes a "$\tfrac{1}{3}$" next to "A," a "$\tfrac{1}{4}$" next to each "T," etc. Following Sabrina's method, label the tree diagram with numbers next to each letter.

c. According to your area model from problem 4-77, what is the probability that you will spin the word "AT"? Now **examine** the bolded branch on the tree diagram shown above. How could the numbers you have written on the tree diagram be used to find the probability of spinning "AT"?

Problem continues on next page →

Geometry Connections

d.   Does this method work for the other combinations of letters? Similarly calculate the probabilities for each of the paths of the tree diagram. At the end of each branch, write its probability. (For example, write "$\frac{1}{12}$" at the end of the "AT" branch.) Do your answers match those from problem 4-77?

e.   Find all the branches with letter combinations that make words. Use the numbers written at the end of each branch to compute the total probability that you will spin a word. Does this probability match the probability you found with your area model?

4-79.   THE RAT RACE, Part Two

In problem 4-48, you read about Romeo, an amazing pet rat. As you previously learned, Sammy overheard Ryan at the county fair claiming that Romeo could learn to run a particular maze and find the cheese at the end.

"I don't think Romeo is that smart!" Sammy declares, "I think the rat just chooses a random path through the maze."

Ryan has built a maze with the floorplan shown at right. In addition, he has placed some cheese in an airtight container (so Romeo can't smell the cheese!) in room A.

a.   Suppose that every time Romeo reaches a split in the maze, he is equally likely to choose any of the paths in front of him. Choose a method and find the probability that Romeo will end up in each room. In a sentence or two, explain why you chose the method you did.

b.   If the rat moves through the maze randomly, how many out of 100 attempts would you expect Romeo to end up in room A? How many times would you expect him to end up in room B? Explain.

Problem continues on next page →

4-79.   *Problem continued from previous page.*

    c.   After 100 attempts, and Romeo finds the cheese 66 times. "See how smart Romeo is?" Ryan asks, "He clearly learned something and got better at the maze as he went along." Looking at your calculations, Sammy isn't so sure.

        Do you think Romeo learned and improved his ability to return to the same room over time? Or could he just have been moving randomly? Discuss this question with your team. Then, write an argument that would convince Ryan or Sammy.

4-80.   Always skeptical, Sammy says, "If Romeo really can learn, he ought to be able to figure out how to run this new maze I've designed." **Examine** Sammy's maze at right.

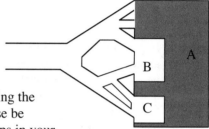

    a.   To give Romeo the best chance of finding the cheese, in which room should the cheese be placed? Choose a method, show all steps in your solution process, and **justify** your answer.

    b.

        If the cheese is in room C and Romeo finds the cheese 6 times out of every 10 tries, does he seem to be learning? Explain your conclusion.

4-81.   Make an entry in your Learning Log describing the various ways of representing probabilities you have learned in this chapter. Which seems easiest to use so far? Which is most versatile? Are there any conditions under which certain ones cannot be used? Label this entry "Probability Methods" and include today's date.

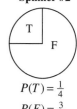ETHODS AND MEANINGS

Probability Models

When all the possible outcomes of a probabilistic event have the same probability, probabilities can be calculated by listing the possible outcomes in a **systematic list**. However, when some outcomes are more probable than others, a more sophisticated model is required to calculate probabilities.

Spinner #1	Spinner #2
$P(I) = \frac{1}{2}$	$P(T) = \frac{1}{4}$
$P(U) = \frac{1}{6}$	$P(F) = \frac{3}{4}$
$P(A) = \frac{1}{3}$	

One such model is an **area model**. In this type of model, the situation is represented by a square with area of 1 so that the areas of the parts are the probabilities of the different events that occur. For example, suppose you spin the two spinners shown above. The possible outcomes are represented in the area model at right. Notice that column "U" takes up $\frac{1}{6}$ of the width of the table since the "U" region is $\frac{1}{6}$ of Spinner #1. Similarly, the "T" row takes up $\frac{1}{4}$ of the height of the table, since the "T" region is $\frac{1}{4}$ of Spinner #2. Then the probability that the spinners come out "U" and "T" is equal to the area of the "UT" rectangle in the table: $\frac{1}{6} \cdot \frac{1}{4} = \frac{1}{24}$.

Spinner #1

	I $\left(\frac{1}{2}\right)$	**A** $\left(\frac{1}{3}\right)$	**U** $\left(\frac{1}{6}\right)$
T $\left(\frac{1}{4}\right)$	IT	AT	UT
F $\left(\frac{3}{4}\right)$	IF	AF	UF

Spinner #2

The situation can also be represented using a **tree diagram**. In this model, the branching points indicate probabilistic events, and the branches stemming from each event indicate the possible outcomes for the event. For example, in the tree diagram at right the first branching point represents spinning the first spinner. The first spinner can come out "I" "A" or "U", so each of those options has a branch. The numbers next to each letter represent the probability that that letter occurs.

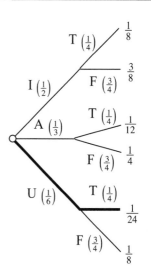

The numbers at the far right of the table represent the probabilities of various outcomes. For example, the probability of spinning "U" and "T" can be found at the end of the bold branch of the tree. This probability, $\frac{1}{24}$, can be found by multiplying the fractions that appear on the bold branches.

4-82.   Choose a method to represent the following problem and use it to answer the questions below.

Kiyomi has 4 pairs of pants (black, green, blue jeans, and plaid) and she has 5 shirts (white, red, teal, black, and brown).

a.   If any shirt can be worn with any pair of pants, how many outfits does she own? That is, how many different combinations of pants and shirts can she wear?

b.   What is the probability that she is wearing black?

4-83.   Are the triangles at right similar? If so, write a flowchart that **justifies** your conclusion. If not, explain how you know.

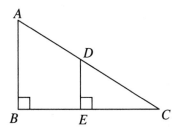

4-84.   You roll a die and it comes up a "6" three times in a row. What is the probability of rolling a "6" on the next toss?

4-85.   Find the values of θ and α in the diagram at right. State the relationships you used.

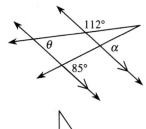

4-86.   Find x and y in the diagram at right. Show all of the steps leading to your answer.

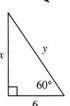

4.2.5 Which model should I use?

Optional: Applications of Probability Methods

Beginning with Lesson 4.2.1, you have been learning about various methods to help you determine probabilities of different outcomes. Today you will apply what you know to two probability problems. As you work in your teams today, keep the following questions in mind to guide your discussion:

> Which probability do we need to find?

> How can we find the probability?

> Which probability model should we use?

4-87. SHIFTY SHAUNA

Shauna has a bad relationship with the truth—she doesn't usually tell it! In fact, whenever Shauna is asked a question, she rolls a die. If it comes up 6, she tells the truth. Otherwise, she lies.

a. If Shauna flips a fair coin and you ask her how it came out, what is the probability that she says "heads" and is telling the truth? Choose a method to solve this problem and carefully record your work. Be ready to share your solution method with the class.

b. Suppose Shauna flips a fair coin and you ask her whether it came up heads or tails. What is the probability that she says "heads"? (Hint: The answer is not $\frac{1}{12}$!)

4-88. MIDNIGHT MYSTERY

Each year, the students at Haardvarks School randomly select a student to play a prank. Late last night, Groundskeeper Millie saw a student steal the school's National Curling Championship trophy from the trophy case. All Millie can tell the headmaster about the crime is that the student who stole the trophy looked like he or she had red hair.

Problem continues on next page →

4-88.   *Problem continued from previous page.*

Unfortunately, of the 100 students at Haardvarks, the only ones with red hair are your friend Don Measley and his siblings. Groundskeeper Millie insists that one of the 5 Measley children committed the crime and should be punished. Don is incensed: "We Measleys would never play such a stupid prank! Groundskeeper Millie claims to have seen someone with red hair, but it was so dark at the time and Millie's eyes are so bad, there is no way she could have identified the color of someone's hair!"

The headmaster isn't convinced, so he walks around with Millie at night and points to students one by one, asking Millie whether each one has red hair. Millie is right about the hair color 4 out of every 5 times.

This looks like bad news for Don, but Professor McMonacle agrees to take up his defense. "I still think," McMonacle says, "that the thief probably wasn't one of the Measleys."

Your task: Model this situation with a tree diagram or a generic area model. The two chance events in your model should be "The thief is/is not a redhead" and "Millie is/is not correct about the thief's hair color."

4-89.   Look back at your model from problem 4-88 and circle every outcome in which Millie would report that the thief had red hair. (There should be two!) Then use your model to answer the following questions:

a.   Suppose that to help make Don's case, you perform the following experiment repeatedly: you pick a Haardvarks student at random, show the student to Millie late at night, and see what Millie says about the student's hair color. If you performed this experiment with 100 students, how many times would you expect Millie to say the student had red hair?

b.   If you performed this experiment with 100 students, how many times would you expect Millie to say the student had red hair and be correct about it?

c.   If you performed this experiment with 100 students, what percentage of the students Millie *said* had red hair would *actually* have red hair?

d.   Can you use these calculations to defend the Measleys? Is it likely that a Measley was the thief?

4-90.   Have you *proven* that none of the Measleys stole the Curling Trophy?

4-91.   Based on the measurements provided for each triangle below, decide if the angle θ
         must be more than, less than, or equal to 45°. Assume the diagram is not drawn to
         scale. Show how you know.

a.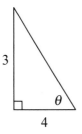

b.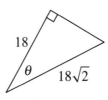

c.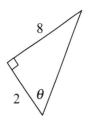

4-92.   Find the area and perimeter of the
         triangle at right.

4-93.   After doing well on a test, Althea's teacher
         placed a gold star on her paper. When Althea
         examined the star closely, she realized that it
         was really a regular pentagon surrounded by 5
         isosceles triangles, as shown in the diagram at
         right. If the star has the angle measurements
         shown in the diagram, find the sum of the
         angles inside the shaded pentagon. Show all
         work.

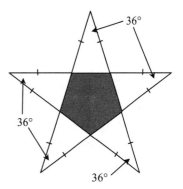

4-94.   Find the slope of the line through the points (–5, 86) and (95, 16). Then find at least
         one more point on the line.

4-95.   On graph paper, plot △ABC if A(–1, –1), B(3, –1), and C(–1, –2).

a.   Enlarge (dilate) △ABC from the origin so that the ratio of the side lengths is
       3. Name this new triangle △A′B′C′. List the coordinates of △A′B′C′.

b.   Rotate △A′B′C′ 90° clockwise (↻) about the origin to find △A″B″C″. List
       the coordinates of △A″B″C″.

c.   If △ABC is translated so that the image of A is located at (5, 3), where would
       the image of B lie?

Chapter 4 Closure   What have I learned?

Reflection and Synthesis

The activities below offer you a chance to reflect on what you have learned in this chapter. As you work, look for concepts that you feel very comfortable with, ideas that you would like to learn more about, and topics you need more help with. Look for connections between ideas as well as connections with material you learned previously.

①   TEAM BRAINSTORM

With your team, brainstorm a list for each of the following two subjects. Be as detailed as you can. How long can you make your list? Challenge yourselves. Be prepared to share your team's ideas with the class.

Topics: What have you studied in this chapter? What ideas and words were important in what you learned? Remember to be as detailed as you can.

Connections: How are the topics, ideas, and words that you learned in previous courses connected to the new ideas in this chapter? Again, make your list as long as you can.

The following is a list of the vocabulary used in this chapter. The words that appear in bold are new to this chapter. Make sure that you are familiar with all of these words and know what they mean. Refer to the glossary or index for any words that you do not yet understand.

α **(alpha)**	angle	**area model**
clinometer	conjecture	**dependent events**
hypotenuse	**independent events**	leg
orientation	probability	random
ratio	**slope angle**	slope ratio
slope triangle	**systematic list**	**tangent**
θ **(theta)**	**tree diagram**	**trigonometry**
Δx	Δy	

Make a concept map showing all of the connections you can find among the key words and ideas listed above. To show a connection between two words, draw a line between them and explain the connection, as shown in the example below. A word can be connected to any other word as long as there is a justified connection. For each key word or idea, provide a sketch of an example.

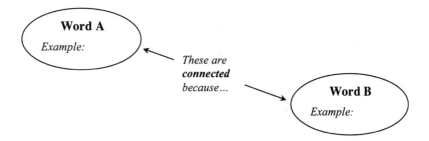

Your teacher may provide you with vocabulary cards to help you get started. If you use the cards to plan your concept map, be sure either to re-draw your concept map on your paper or to glue the vocabulary cards to a poster with all of the connections explained for others to see and understand.

While you are making your map, your team may think of related words or ideas that are not listed above. Be sure to include these ideas on your concept map.

SUMMARIZING MY UNDERSTANDING

This section gives you an opportunity to show what you know about certain math topics or ideas. Your teacher will direct you how to do this. Your teacher may give you a "GO" page to work on. "GO" stands for "Graphic Organizer," a tool you can use to organize your thoughts and communicate your ideas clearly.

④   WHAT HAVE I LEARNED?

This section will help you evaluate which types of problems you feel comfortable with and which types you need more help with. This section appears at the end of every chapter to help you check your understanding. Even if your teacher does not assign this section, it is a good idea to try these problems and find out for yourself what you know and what you need to work on.

Solve each problem as completely as you can. The table at the end of this closure section has answers to these problems. It also tells you where you can find additional help and practice on problems like these.

CL 4-96.   Solve for the missing side length or angle below.

a.
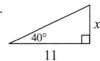

b.

c.

CL 4-97.   Use a flowchart to show how you know the triangles are similar. Then find the value of each variable.

a.
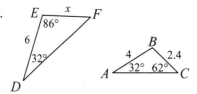

b.
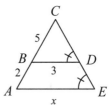

CL 4-98.   Salvador has a hot dog stand 58 meters from the base of the Space Needle in Seattle. He prefers to work in the shade and knows that he can calculate when his hotdog stand will be in the shade if he knows the height of the Space Needle. To measure its height, Salvador stands at the hotdog stand, gets out his clinometer, and measures the angle of sight to be 80°. Salvador's eyes are 1.5 meters above the ground. Assuming that the ground is level between the hotdog stand and the Space Needle, how tall is the Space Needle?

CL 4-99. Use the diagram at right to answer the questions below.

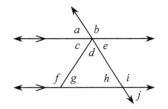

a. State the name of the geometric relationship between the angles below. Also describe the relationship between the angle measures, if one exists.

   i.   ∡a and ∡h

   ii. ∡b and ∡e

   iii. ∡c and ∡g

   iv. ∡g, ∡d, and ∡h

b. For each problem below, **justify** your answer. If $m\angle c = 32°$ and $m\angle e = 55°$, then find the measure of each angle below.

   i. $m\angle j$

   ii. $m\angle d$

   iii. $m\angle a$

   iv. $m\angle g$

CL 4-100. Draw a pair of axes in the center of a half sheet of graph paper. Then draw the figure to the right and perform the indicated transformations. For each transformation, label the resulting image $A'B'C'D'E'$.

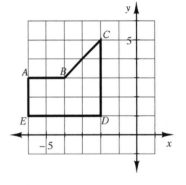

a. Rotate $ABCDE$ 180° ↻ around the origin.

b. Rotate $ABCDE$ 90° ↺ around the origin.

c. Reflect $ABCDE$ across the y-axis.

d. Translate $ABCDE$ up 5, left 7.

CL 4-101. In a certain town, 45% of the population has dimples and 70% has a widow's peak (a condition where the hairline above the forehead makes a "V" shape). Assuming that these physical traits are independently distributed, what is the probability that a randomly selected person has both dimples and a widow's peak? What is the probability that he or she will have neither? Use a generic area model or a tree diagram to represent this situation.

CL 4-102. Trace each figure onto your paper and label the sides with the given measurements.

  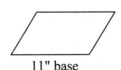 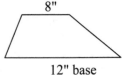

3" base   10" base   11" base   12" base

a. On your paper, draw a height that corresponds to the labeled base for each figure.

b. Assume that the height for each figure above is 7 inches. Add this information to your diagrams and find the area of each figure.

CL 4-103. Find the perimeter of each shape below. Assume the diagram in part (b) is a parallelogram.

a.

b.

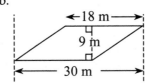

c.

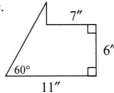

CL 4-104. For each equation below, solve for x:

a. $\frac{x}{23} = \frac{15}{7}$

b. $(x+2)(x-5) = 6x + x^2 - 5$

c. $x^2 + 2x - 15 = 0$

d. $2x^2 - 11x = -3$

CL 4-105. Check your answers using the table at the end of this section. Which problems do you feel confident about? Which problems were hard? Have you worked on problems like these in math classes you have taken before? Use the table to make a list of topics you need help on and a list of topics you need to practice more.

         Geometry Connections

This course focuses on five different **Ways of Thinking**: investigating, examining, reasoning and justifying, visualizing, and choosing a strategy/tool. These are some of the ways in which you think while trying to make sense of a concept or attempting to solve a problem (even outside of math class). During this chapter, you have probably used each Way of Thinking multiple times without even realizing it!

This closure activity will focus on one of these Ways of Thinking: **visualizing**. Read the description of this Way of Thinking at right.

Think about the problems you have worked on in this chapter. When did you need to think about what something looked like to make sense of it? What helps you to create mental pictures? Were there times when a problem didn't seem to make sense without seeing it? How have you used visualizing in a previous math class? You may want to flip through the chapter to refresh your memory about the problems that you have worked on. Discuss any of the methods you have developed to help you **visualize**.

Once your discussion is complete, think about the way you think as you answer the questions below.

Visualizing

To visualize means to make a picture in your mind that represents a situation or description. As you develop this Way of Thinking, you will learn how to turn a variety of situations into mental pictures. You think this way when you ask or answer questions like, "What does it look like when . . . ?" or "How can I draw . . . ?" When you catch yourself wondering what something might look like, you are visualizing.

How does it look when...?

a. **Visualization** is required when you imagine a situation and want to draw a diagram to represent it. Read the descriptions below and **visualize** what each situation looks like. Then draw a diagram for each. Label your diagrams appropriately with any given measurements.

   (1) *Karen is flying a kite on a windy day. Her kite is 80 feet above ground and her string is 100 feet long. Karen is holding the kite 3 feet above ground.*

   (2) *The bow of a rowboat (which is 1 foot above water level) is tied to a point on a dock that is 6 feet above the water level. The length of the rope between the dock and the boat is 8 feet.*

Problem continues on next page →

⑤ *Problem continued from previous page.*

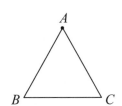

b. Sometimes, **visualization** requires you to think about how an object can move in relation to others. For example, consider equilateral △*ABC* at right.

(1) **Visualize** changing △*ABC* by stretching vertex *A* to point *D*, which is right above *A*. What does the new triangle look like? Do you have a name for it?

(2) What happened to $m\angle A$ as you stretched the triangle in part (1)? What happened to $m\angle B$ and $m\angle C$?

(3) Now **visualize** the result after vertex *A* stretched to point *E*. What type of triangle is the result? What happens to $m\angle B$ as the triangle is stretched? What happens to $m\angle C$?

c. An important use of **visualization** is to re-orient a right triangle to help you identify which leg is Δx and which leg is Δy. For each triangle below, **visualize** the triangle by rotating and/or reflecting it so that it is a slope triangle. Draw the result and label the appropriate legs Δx or Δy.

(1)

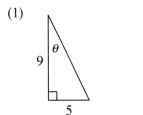

(2)

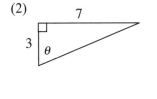

(3)

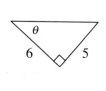

d. Finally, **visualization** can help you view an object from different perspectives. For example, consider the square-based pyramid at right. **Visualize** what you would see if you looked down at the pyramid from a point directly above the top vertex. Draw this view on your paper.

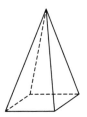

Answers and Support for
Closure Activity #4
What Have I Learned?

Problem	Solution	Need Help?	More Practice
CL 4-96.	a. $x \approx 9.23$ b. $x \approx 5.47$ c. $\theta = 45°$	Lessons 4.1.2 and 4.1.4 Math Notes boxes	Problems 4-11, 4-12, 4-16, 4-22, 4-25, 4-31, 4-34, 4-36, 4-46
CL 4-97.	a. $x \approx 3.6$ $\angle EDF = \angle BAC$ $\angle DEF = \angle ABC$ $\triangle EDF \sim \triangle BAC$ $AA \sim$ $\angle EDF = \angle BAC$ $\angle BCD = \angle ACE$ b. $x = 4.2$ $\triangle BCD \sim \triangle ACE$ $AA \sim$	Lessons 3.1.2, 3.1.3, 3.2.1, and 3.2.5 Math Notes boxes	Problems 3-23, 3-35, 3-36, 3-38, 3-46, 3-49, 3-53, 3-54, 3-55, 3-57, 3-64, 3-66, 3-80, 3-83, 3-84, 3-89, 4-7, 4-17, 4-18, 4-38, 4-39, 4-83
CL 4-98.	Total height ≈ 330.4 m 80° 58 m 1.5 m not to scale	Lesson 4.1.4 Math Notes box, Lesson 4.1.5	Problems 4-27, 4-30, 4-40, 4-41, 4-46, 4-57
CL 4-99.	a. *i.* corresponding angles, congruent *ii.* straight angle pair, supplementary *iii.* alternate interior angles, congruent *iv.* sum of the angles in a triangle, add up to 180° b. *i.* 55°: corresponding to *e* *ii.* 93°: straight angle with *e* and *c* *iii.* 55°: vertical to *e* *iv.* 32°: alternate interior to *c*	Lessons 2.1.1, 2.1.4, and 2.2.1 Math Notes boxes, Problem 2-3	Problems 2-13, 2-16, 2-17, 2-18, 2-23, 2-24, 2-25, 2-31, 2-32, 2-38, 2-49, 2-51, 2-62, 2-72, 2-111, 3-20, 3-31, 3-60, 3-70, 3-96, 4-6, 4-29, 4-44, 4-85, 4-93

Problem	Solution	Need Help?	More Practice
CL 4-100.	a. $A'(6, -3)$, $B'(4, -3)$, $C'(2, -5)$, $D'(2, -1)$, $E'(6, -1)$ b. $A'(-3, -6)$, $B'(-3, -4)$, $C'(-5, -2)$, $D'(-1, -2)$, $E'(-1, -6)$ c. $A'(6, 3)$, $B'(4, 3)$, $C'(2, 5)$, $D'(2, 1)$, $E'(6, 1)$ d. $A'(-13, 8)$, $B'(-11, 8)$, $C'(-9, 10)$, $D'(-9, 6)$, $E'(-13, 6)$	Lessons 1.2.2 and 1.2.3 Math Notes boxes	Problems 1-50, 1-51, 1-59, 1-60, 1-61, 1-64, 1-69, 1-73, 1-85, 1-96, 2-11, 2-19, 2-20, 2-33, 2-64, 2-113, 3-96
CL 4-101.	P(both) = 31.5% P(neither) = 16.5% Dimples? Widow's Peak? Yes: $\frac{45}{100}$ No: $\frac{55}{100}$ Yes $\frac{70}{100}$: 31.5% 38.5% No $\frac{30}{100}$: 13.5% 16.5%	Lesson 4.2.4 Math Notes box	Problems 4-52, 4-60, 4-62, 4-68, 4-69, 4-70, 4-71, 4-77, 4-78, 4-87
CL 4-102.	a. All heights should be 7". b. Areas in square inches: 10.5, 70, 77, 70	Lessons 1.1.3 and 2.2.4 Math Notes boxes	Problems 2-66, 2-68, 2-71, 2-75, 2-76, 2-78, 2-79, 2-82, 2-83, 2-90, 2-120, 3-51, 3-81, 3-97, 4-43, 4-67, 4-92
CL 4-103.	a. ≈ 23.899 mm b. $= 66$ m c. $\approx 32.93"$	Lessons 1.1.3, 4.1.2, and 4.1.4 Math Notes boxes	Problems 4-16, 4-20, 4-22, 4-36, 4-76, 4-92
CL 4-104.	a. $x = 49.286$ b. $x = \frac{-5}{9}$ c. $x = 3$ or $x = -5$ d. $x = \frac{11 \pm \sqrt{97}}{4} \approx 5.21$ or 0.29	Lessons 1.1.4 and 3.2.3 Math Notes box	Problems 1-17, 1-25, 1-32, 1-34, 1-45, 1-57, 1-74, 1-100, 1-124, 3-68, 3-90, 3-100, 4-8, 4-47, 4-75

CHAPTER 5 Completing the Triangle Toolkit

In Chapter 4, you **investigated** the powerful similarity and side ratio relationships in right triangles. In this chapter, you will learn about other side ratio relationships using the hypotenuse that will allow you to find missing side lengths and missing angle measures for any right triangles.

In addition, you will develop tools to complete your triangle toolkit so that you can find the missing angle measures and side lengths for any triangle, provided that enough information is given. You will then explore ways to **choose an appropriate tool** to solve new problems in unfamiliar contexts.

In this chapter, you will learn:

Guiding Questions

Think about these questions throughout this chapter:

Is there a shortcut?

How are the shapes related?

What information do I know?

Which tool(s) can I use?

➢ how to recognize and use special right triangles.

➢ the trigonometric ratios of sine and cosine as well as the inverses of these functions.

➢ how to apply trigonometric ratios to find missing measurements in right triangles.

➢ new triangle tools called the Law of Sines and the Law of Cosines.

➢ how to recognize when the information provided is not enough to determine a unique triangle.

Chapter Outline

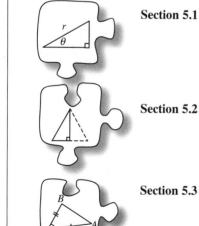

Section 5.1 Students will extend their understanding of trigonometric ratios to include sine, cosine, and inverse trigonometric functions and will use these tools to find missing measurements in right triangles.

Section 5.2 Students will apply the Pythagorean Theorem and similar triangles to find patterns in special right triangles, such as 30°- 60°- 90° and 45°- 45°- 90° triangles and those with side lengths that are Pythagorean Triples.

Section 5.3 Once students **investigate** all of the types of information that can be given about a triangle (Lesson 5.3.1), they will focus on developing tools to find missing side lengths and angle measures in non-right triangles.

234   *Geometry Connections*

5.1.1 What if I know the hypotenuse?

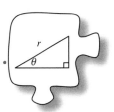

Sine and Cosine Ratios

In the previous chapter, you used the idea of similarity in right triangles to find a relationship between the acute angles and the lengths of the legs of a right triangle. However, we do not always work just with the legs of a right triangle—sometimes we only know the length of the hypotenuse. By the end of today's lesson, you will be able to use two new trigonometric ratios that involve the hypotenuse of right triangles.

5-1.   THE STREETS OF SAN FRANCISCO

While traveling around the beautiful city of San Francisco, Juanisha climbed several steep streets. One of the steepest, Filbert Street, has a slope angle of 31.5° according to her guidebook.

Once Juanisha finished walking 100 feet up the hill, she decided to figure out how high she had climbed. Juanisha drew the diagram below to represent this situation.

Can a tangent ratio be used to find Δy? Why or why not? Be prepared to share your thinking with the rest of the class.

Juanisha's Drawing

5-2.   In order to find out how high Juanisha climbed in problem 5-1, you need to know more about the relationship between the ratios of the sides of a right triangle and the slope angle.

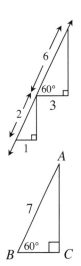

a.   Use **two different strategies** to find Δy for the slope triangles shown in the diagram at right.

b.   Find the ratio $\frac{\Delta x}{\text{hypotenuse}}$ for each triangle. Why must these ratios be equal?

c.   Find *BC* and *AC* in the triangle at right. Show all work.

NEW TRIG RATIOS

In problem 5-2, you used a ratio that included the
hypotenuse of $\triangle ABC$. There are several ratios that
you might have used. One of those ratios is known
as the **sine ratio** (pronounced "sign"). This is the
ratio of the length of the side **opposite** the acute
angle to the length of the **hypotenuse**.

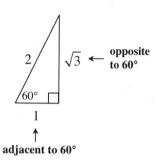

For the triangle shown at right, the sine of 60°
is $\frac{\sqrt{3}}{2} \approx 0.866$. This is written:

$$\sin 60° = \frac{\sqrt{3}}{2}$$

Another ratio comparing the length of the side **adjacent** to (which means "next to")
the angle to the length of the **hypotenuse**, is called the **cosine ratio** (pronounced
"co-sign"). For the triangle above, the cosine of 60° is $\frac{1}{2} = 0.5$. This is written:

$$\cos 60° = \frac{1}{2}$$

a.   Like the tangent, your calculator can give you both the sine and cosine ratios
for any angle. Locate the "sin" and "cos" buttons on your calculator and use
them to find the sine and cosine of 60°. Does your calculator give you the
correct ratios?

b.   Use a trig ratio to write an equation and solve for a
in the diagram at right. Does this require the sine
ratio or the cosine ratio?

c.   Likewise, write an equation and solve for b for the
triangle at right.

5-4.   Return to the diagram from Juanisha's
climb in problem 5-1. Juanisha still
wants to know how many feet she
climbed vertically when she walked
up Filbert Street. Use one of your
new trig ratios to find how high she
climbed.

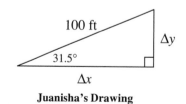

Juanisha's Drawing

Geometry Connections

5-5.   For each triangle below, decide which side is
        opposite and which is adjacent to the given
        acute angle. Then determine which of the three
        trig ratios will help you find x. Write and solve
        an equation.

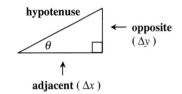

a.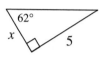

b.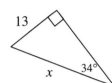

c.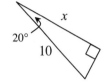

d.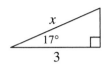

e.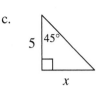

f.

5-6.   TRIANGLE TOOLKIT

        Obtain a Lesson 5.1.1 Resource Page ("Triangle Toolkit")
        from your teacher. This will be a continuation of the
        Geometry Toolkit you started in Chapter 1. Think about
        the tools you have developed so far to solve for the
        measure of sides and angles of a triangle. Then, in the
        space provided, add a diagram and a description of each
        tool you know. In later lessons, you will continue to add
        new triangle tools to this toolkit, so be sure to keep this
        resource page in a safe place. At this point, your toolkit
        should include:

        •   Pythagorean Theorem   •   Tangent

        •   Sine   •   Cosine

5-7.   You now have multiple trig tools to find missing side
        lengths of triangles. For the triangle at right, find the values
        of x and y. Your Triangle Toolkit might help. Which tools
        did you use?

5-8.   Lori has written the conjectures below. For each one, decide if it is true or not. If you believe it is not true, find a **counterexample** (an example that proves that the statement is false).

   a.   If a shape as four equal sides, it cannot be a parallelogram.

   b.   If $\tan \theta$ is more than 1, then θ must be more than 45°.

   c.   If two angles formed when two lines are cut by a transversal are corresponding, then the angles are congruent.

5-9.   Earl hates to take out the garbage and to wash the dishes, so he decided to make a deal with his parents: He will flip a coin once for each chore and will perform the chore if the coin lands on heads. What he doesn't know is that his parents are going to use a weighted coin that lands on heads 80% of the time!

   a.   What is the probability that Earl will have to do both chores?

   b.   What is the probability that Earl will have to take out the garbage, but will not need to wash the dishes?

5-10.   Copy the trapezoid at right on your paper. Then find its area and perimeter. Keep your work organized so that you can later explain how you solved it. (Note: The diagram is not drawn to scale.)

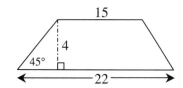

5-11.   Solve each of the equations below for the given variable. Be sure to check your answers.

   a. $4(2x+5)-11 = 4x - 3$ b. $\frac{2m-1}{19} = \frac{m}{10}$

   c. $3p^2 + 10p - 8 = 0$ d. $\sqrt{x+2} = 5$

5.1.2 Which tool should I use?

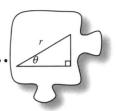

Selecting a Trig Tool

You now have several tools that will help you find the length of a side of a right triangle when given any acute angle and any side length. But how do you know which tool to use? And how can you identify the relationships between the sides and the given angle?

Today you will work with your team to develop **strategies** that will help you identify if cosine, sine, or tangent can be used to solve for a side of a right triangle. As you work, be sure to share any shortcuts you find that can help others identify which tool to use. As you work, keep the focus questions below in mind.

> Is this triangle familiar? Is there something special about this triangle?
>
> Which side is opposite the given angle? Which is adjacent?
>
> Which tool should I use?

5-12. Obtain the Lesson 5.1.2 Resource Page from your teacher. On it, find the triangles shown below. Note: the diagrams are not drawn to scale.

 $\cos \theta = \frac{adj}{hyp}$

 $\sin \theta = \frac{opp}{hyp}$

 $\tan \theta = \frac{opp}{adj}$

With your study team:

- Look through all the triangles first and see if any look familiar or are ones that you know how to answer right away without using a trig tool.

- Then, for all the other triangles, identify which tool you should use based on where the reference angle (the given acute angle) is located and which side lengths are involved.

- Write and solve an equation to find the missing side length.

a.

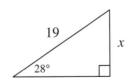

b.

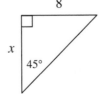

c.

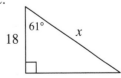

Problem continues on next page →

5-12.   *Problem continued from previous page.*

d.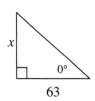

e.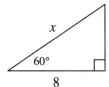

f.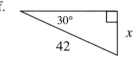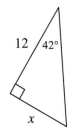

g.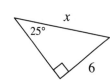

h.

i.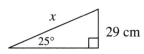

5-13.   Marta arrived for her geometry test only to find that she forgot her calculator. She decided to complete as much of each problem as possible.

a.   In the first problem on the test, Marta was asked to find the length x in the triangle shown at right. Using her algebra skills, she wrote and solved an equation. Her work is shown below. Explain what she did in each step.

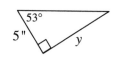

$$\sin 25° = \frac{29}{x}$$
$$x(\sin 25°) = 29$$
$$x = \frac{29}{\sin 25°}$$

b.   Marta's answer in part (a) is called an **exact answer**. Now use your calculator to help Marta find the **approximate** length of x.

c.   In the next problem, Marta was asked to find y in the triangle shown at right. Find an exact answer for y without using a calculator. Then use a calculator to find an approximate value for y.

5-14.   In problem 5-12, you used trig tools to find a side length. But do you have a way to find an angle? **Examine** the triangles below. Do any of them look familiar? How can you use information about the side lengths to help you figure out the reference angle (θ)? Your Trig Table from Chapter 4 may be useful.

a.

b.

c.

5-15.   Write a Learning Log entry explaining how you know which trig tool to use. Be sure to include examples with diagrams and anything else that would be useful to refer to later. Title this entry, "Choosing a Trig Tool" and include today's date.

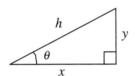
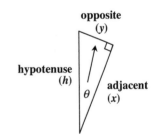

5-16.   For each triangle below, write an equation relating the **reference angle** (the given acute angle) with the two side lengths of the right triangle. Then solve your equation for x.

a.

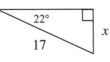

b.

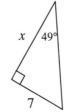

c.

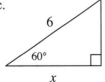

5-17.   While shopping at his local home improvement store, Chen noticed that the directions for an extension ladder state, "*This ladder is most stable when used at a 75° angle with the ground.*" He wants to buy a ladder to paint a two-story house that is 26 feet high. How long does his ladder need to be? Draw a diagram and set up an equation for this situation. Show all work.

5-18.   Kendra has programmed her cell phone to randomly show one of six photos when she turns it on. Two of the photos are of her parents, one is of her niece, and three are of her boyfriend, Bruce. Today, she will need to turn her phone on twice: once before school and again after school.

    a.   **Choose a model** (such as a tree diagram or generic area model) to represent this situation.

    b.   What is the probability that both photos will be of her boyfriend?

    c.   What is the probability that neither photo will be of her niece?

5-19.   Lori has written the conjectures below. For each one, decide if it is true or not. If you believe it is not true, find a **counterexample** (an example that proves that the statement is false).

    a.   If a triangle has a 60° angle, it must be an equilateral triangle.

    b.   To find the area of a shape, you always multiply the length of the base by the height.

    c.   All shapes have 360° rotation symmetry.

5-20.   **Examine** the triangles at right. Are the triangles similar? If so, show how you know with a flowchart. If not, explain how you know they cannot be similar.

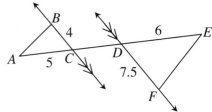

5.1.3 How can I find the angle?

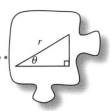

Inverse Trigonometry

You now know how to find the missing side lengths in a right triangle given an acute angle and the length of any side. But what if you want to find the measure of an angle? If you are given the lengths of two sides of a right triangle, can you work backwards to find the measurements of the unknown angles? Today you will work on "undoing" the different trigonometric ratios to find the angles that correspond to those ratios.

5-21.   Mr. Gow needs to build a wheelchair access ramp for the school's auditorium. The ramp must rise a total of 3 feet to get from the ground to the entrance of the building. In order to follow the state building code, the angle formed by the ramp and the ground cannot exceed 4.76°.

Mr. Gow has plans from the planning department that call for the ramp to start 25 feet away from the building. Will this ramp meet the state building code?

a.   Draw an appropriate diagram. Add all the measurements you can. What does Mr. Gow need to find?

b.   To find an angle from a trig ratio you need to "undo" the trig ratio, just like you can undo addition with subtraction, multiplication with division, or squaring by finding the square root. These examples are all pairs of **inverse** operations.

When you use a calculator to do this, you use inverse trig functions which look like "**sin⁻¹**", "**cos⁻¹**", and "**tan⁻¹**". These are pronounced, "inverse sine," "inverse cosine," and "inverse tangent." On many calculators, you must press the "inv" or "2nd" key first, then the "sin", "cos", or "tan" key.

Verify that your calculator can find an inverse trig value using the triangle at right from Lesson 5.1.2. When you find $\cos^{-1} \frac{8}{16}$, do you get 60°?

c.   Return to your diagram from part (a). According to the plan, what angle will the ramp make with the ground? Will the ramp be to code?

d.   At least how far from the building must the ramp start in order to meet the building code? If Mr. Gow builds the ramp exactly to code, how long will the ramp be? Show all work.

5-22. For the triangle at right, find the measures of $\angle A$ and $\angle B$. Once you have found the measure of the first acute angle (either $\angle A$ or $\angle B$), what knowledge about the angles in triangles could help you find the second acute angle?

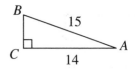

5-23. **Examine** the triangles below. Note: the diagrams are not drawn to scale.

$$\cos \theta = \frac{adj}{hyp} \qquad \sin \theta = \frac{opp}{hyp} \qquad \tan \theta = \frac{opp}{adj}$$

With your study team:

- Look through all the triangles first and see if any look familiar or are ones that you know how to answer right away without using a trig tool.

- Then, for all the other triangles, identify which tool to use based on where the reference angle (θ) is located and which side lengths are involved.

- Write and solve an equation to find the missing side length or angle.

a.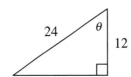

b.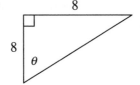

c.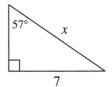

5-24. Peter cannot figure out what he did wrong. He wrote the equation below to find the missing angle of a triangle. However, his calculator gives him an error message each time he tries to calculate the angle.

Peter's work: $\cos \theta = \frac{8}{2.736}$

Jeri, his teammate, looked at his work and exclaimed, "Of course your calculator gives you an error! Your cosine ratio is impossible!" What is Jeri talking about? And how could she tell without seeing the triangle or checking on her calculator?

5-25. Write a Learning Log entry describing what you know about inverse trig functions. Be sure to include an example and a description of how to solve it. Title this entry, "Inverse Trig Functions" and include today's date.

Geometry Connections

5-26.   Solve the following equations for the given variable, if possible. Remember to check
         your answers.

   a. $6x^2 = 150$ b. $4m + 3 - m = 3(m + 1)$

   c. $\sqrt{5x - 1} = 3$ d. $(k - 4)^2 = -3$

5-27.   Mervin and Leela are in bumper cars. They are at opposite ends of a 100 meter track
         heading toward each other. If Mervin moves at a rate of 5.5 meters per second and
         Leela moves at a rate of 3.2 meters per second, how long does it take for them to
         collide?

5-28.   Donnell has a bar graph which shows the
         probability of a colored section coming up on
         a spinner, but part of the graph has been
         ripped off.

   a.   What is the probability of spinning red?

   b.   What is the probability of spinning yellow?

   c.   What is the probability of spinning blue?

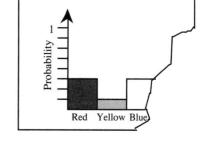

   d.   If there is only one color missing from the graph, namely green, what is the
         probability of spinning green? Why?

5-29.   Find the area and the perimeter of the figure at right.
         Be sure to organize your work so you can explain
         your method later.

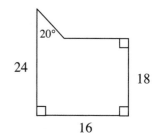

5-30.   While playing a board game, Mimi noticed that
         she could roll the dice 8 times in 30 seconds.
         How many minutes should it take her to roll the
         dice 150 times?

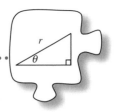

5.1.4 How can I use trig ratios?

Trigonometric Applications

Throughout this chapter, you have developed new tools to help you determine the length of any side or the measurement of any angle of a right triangle. Trig ratios, coupled with the Pythagorean Theorem, give you the powerful ability to solve problems involving right triangles. Today you will apply this knowledge to solve some real world problems.

As you are working with your team on the problems below, be sure to draw and label a diagram and determine which trig ratio to use before you start solving.

5-31.   CLIMBING IN YOSEMITE

David and Emily are climbing El Capitan, a big wall cliff in Yosemite National Park. David is on the ground holding the rope attached to Emily as she climbs. When Emily stops to rest, David wonders how high she has climbed. The rope is attached to his waist, about 3 feet off of the ground, and he has let out 40 feet of rope. This rope makes a 35° angle with the cliff wall.

a.   Assuming that the rope is taut (i.e., pulled tight), approximately how high up the wall has Emily climbed?

b.   How far away from the wall is David standing? Describe your method.

5-32.   The Bungling Brothers Circus is in town and you are part of the crew that is setting up its enormous tent. The center pole that holds up the tent is 70 feet tall. To keep it upright, a support cable needs to be attached to the top of the pole so that the cable forms a 60° angle with the ground.

a.   How long is the cable?

b.   How far from the pole should the cable be attached to the ground?

5-33.   Nathan is standing in a meadow, exactly 185 feet from the base of El Capitan. At 11:00 a.m., he observes Emily climbing up the wall, and determines that his angle of sight up to Emily is about 10°.

a.   If Nathan's eyes are about 6 feet above the ground, about how high is Emily at 11:00 a.m.?

b.   At 11:30 a.m., Emily has climbed some more, and Nathan's angle of sight to her is now 25°. How far has Emily climbed in the past 30 minutes?

c.   If Emily climbs 32 feet higher in ten more minutes, at what angle will Nathan have to look in order to see Emily?

5-34.   Forest needs to repaint the right side of his house because sunlight and rain have caused the paint to peel. Each can of paint states that it will cover 150 sq. feet. Help Forest decide how many cans of paint he should buy.

a.   Copy the shaded diagram below onto your paper. Work with your team to find the area that will be painted green.

b.   Assuming that Forest can only buy whole cans of paint, how many cans of paint should he buy? (Note: 1 square meter ≈ 10.764 square feet)

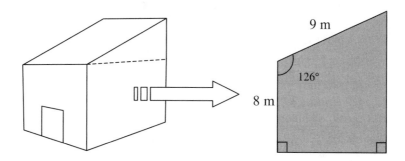

5-35.   TEAM CHALLENGE

It is 11:55 a.m., and Emily has climbed even higher. The rope now makes an angle of 6° with the cliff wall. If David is 18 feet away from the base of El Capitan, at what angle should Nathan (who is 185 feet from the base) look up to see Emily?

METHODS AND MEANINGS

Inverse Trigonometry

Just as subtraction "undoes" addition and multiplication "undoes" division, the inverse trigonometric functions "undo" the trig functions tangent, sine, and cosine. Specifically, **inverse trigonometric functions** are used to find the measure of an acute angle in a right triangle when a ratio of two sides is known. This is the **inverse**, or opposite, of finding the trig ratio from a known angle.

The inverse trigonometric functions that will be used in this course are $\sin^{-1}$, $\cos^{-1}$, and $\tan^{-1}$ (pronounced "inverse sine," "inverse cosine," and "inverse tangent"). Below is an example that shows how $\cos^{-1}$ may be used to find a missing angle, θ.

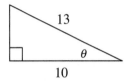

$$\cos \theta = \tfrac{10}{13}$$
$$\theta = \cos^{-1}\left(\tfrac{10}{13}\right)$$
$$\theta \approx 39.7°$$

To evaluate $\cos^{-1}\left(\tfrac{10}{13}\right)$ on a scientific calculator, most calculators require the "2nd" or "INV" button to be pressed before the "cos" button.

Review & Preview

5-36.   Which of the triangles below are similar to $\triangle LMN$ at right? How do you know? Explain.

a.

b.

c.
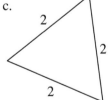

d.

5-37.   Find the equation of the line that has a 33.7° slope angle and a y-intercept at $(0, 7)$. Assume the line has a positive slope.

5-38.　For each triangle below, write a trigonometric equation relating a, b, and θ.

a.

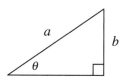

b.

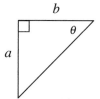

c.

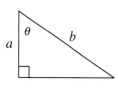

5-39.　Kendrick is frantic. He
remembers that several years ago
he buried his Amazing Electron
Ring in his little sister's sandbox,
but he cannot remember where.
A few minutes ago he heard that
someone is willing to pay $1000
for it. He has his shovel and is
ready to dig.

a.　The sandbox is rectangular, measuring 4 feet by 5
feet, as shown at right. If Kendrick only has time
to search in the 2 foot-square shaded region, what
is the probability that he will find the ring?

b.　What is the probability that he will not find the
ring? Explain how you found your answer.

c.　Kendrick decides instead to dig in the square
region shaded at right. Does this improve his
chances for finding the ring? Why or why not?

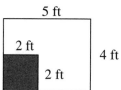

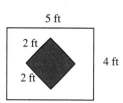

5-40.　Estelle is trying to find x in the triangle at right.
She lost her scientific calculator, but luckily her
teacher told her that $\sin 23° \approx 0.391$,
$\cos 23° \approx 0.921$, and $\tan 23° \approx 0.424$.

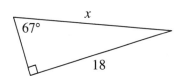

a.　Write an equation that Estelle could use to
solve for x.

b.　Without a calculator, how could Estelle find $\sin 67°$? Explain.

5.2.1 Is there a shortcut?

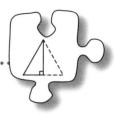

Special Right Triangles

You now know when triangles are similar and how to find missing side lengths in similar triangles. Today you will be using both of those ideas to **investigate** patterns within two types of special right triangles. These patterns will allow you to use a shortcut whenever you are finding side lengths in these particular types of right triangles.

5-41.   Darren wants to find the side lengths of the triangle at right. The only problem is that he left his calculator at home and he does not remember the value of cos 60°.

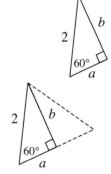

   a.   "That's okay," says his teammate, Jan. "I think I see a short-cut." Using tracing paper, she created the diagram at right by reflecting the triangle across the side with length b. What is the resulting shape? How do you know?

   b.   What if Jan had reflected across a different side? Would the result still be an equilateral triangle? Why or why not?

   c.   Use Jan's diagram to find the value of a without using a trig ratio.

   d.   Now find the length of b without a calculator. Leave your answer in **exact form**. In other words, do not approximate the height with a decimal.

5-42.   Darren's triangle is an example of a half-equilateral triangle, also known as a **30°- 60°- 90° triangle** because of its nice angle measures. Darren is starting to understand Jan's short-cut, but he still has some questions. Help Darren by answering his questions below.

   a.   "Will this approach work on all triangles? In other words, can you always form an equilateral triangle by reflecting a right triangle?" Explain your **reasoning**.

   b.   "What if the triangle looks different?" Use tracing paper to show how Darren can reflect the triangle at right to form an equilateral triangle. Then find the lengths of x and y without a calculator.

   c.   "Is the longer leg of a 30°- 60°- 90° triangle always going to be the length of the shorter leg multiplied by $\sqrt{3}$? Why or why not?"

Problem continues on next page →

5-42.   *Problem continued from previous page.*

d.   "What if I only know the length of the shorter leg?" Consider the triangle at right. **Visualize** the equilateral triangle. Then find the values of *n* and *m* without a calculator.

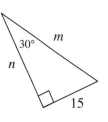

e.   Darren drew the triangle at right and is wondering if it also is a 30°- 60°- 90° triangle. What do you think? How do you know?

5-43.   Darren wonders if he can find a similar pattern in another special triangle he knows, shown at right.

a.   Use what you know about this triangle to help Darren find the lengths of *a* and *b* without a trig tool or a calculator. Leave your answer in exact form.

b.   What should Darren name this triangle?

c.   Use the fact that all 45°- 45°- 90° triangles are similar to find the missing side lengths in the right triangles below. Leave your answers in exact, radical form.

(1) 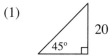   (2)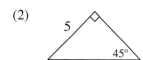

5-44.   Use your new 30°- 60°- 90° and 45°- 45°- 90° triangle patterns to quickly find the lengths of the missing sides in each of the triangles below. Do not use a calculator. Leave answers in exact form. Note: The triangles are not necessarily drawn to scale.

a. 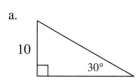   b.   c.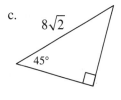

d. 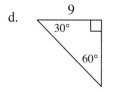   e. 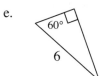   f.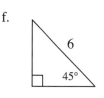

5-45.   In your Learning Log, explain what you know about 30°- 60°- 90° and 45°- 45°- 90° triangles. Include diagrams of each. Label this entry "Special Right Triangles" and include today's date.

LOOKING DEEPER

Rationalizing a Denominator

In Lesson 5.2.1, you developed some shortcuts to help find the lengths of the sides of a 30°- 60°- 90° and 45°- 45°- 90° triangle. This will enable you to solve similar problems in the future without a calculator or a Trig Table.

However, sometimes using the shortcuts leads to some strange looking answers. For example, when finding the length of a in the triangle at right, you will get the expression $\frac{6}{\sqrt{2}}$.

A number with a radical in the denominator is difficult to estimate. Therefore, it is sometimes beneficial to **rationalize the denominator** so that no radical remains in the denominator. Study the example below.

Example: Simplify $\frac{6}{\sqrt{2}}$.

First, multiply the numerator and denominator by the radical in the denominator. Since $\frac{\sqrt{2}}{\sqrt{2}} = 1$, this does not change the value of the expression.

Example

$$\frac{6}{\sqrt{2}} \cdot \frac{\sqrt{2}}{\sqrt{2}} = \frac{6\sqrt{2}}{2}$$
$$= 3\sqrt{2}$$

After multiplying, notice that the denominator no long has a radical, since $\sqrt{2} \cdot \sqrt{2} = 2$.

Often, the product can be further simplified. Since 2 divides evenly into 6, the expression $\frac{6\sqrt{2}}{2}$ can be rewritten as $3\sqrt{2}$.

5-46.   For each triangle below, use your triangle shortcuts from this lesson to find the missing side lengths. Then find the area and perimeter of the triangle.

a.
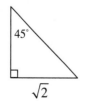

b.
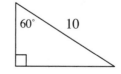

5-47.   Use the relationships found in each of the diagrams below to solve for *x* and *y*.
Assume the diagrams are not drawn to scale. State which geometric relationships
you used.

a.   b.

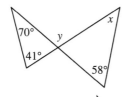   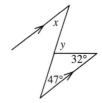

c.   d.

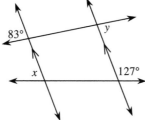   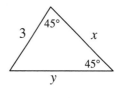

5-48.   On graph paper, graph $\overline{AB}$ if A(1, 6) and B(5, 2).

a.   Find *AB* (the length of $\overline{AB}$). Leave your answer in **exact form**. That is, do
not approximate with a decimal. Explain your method.

b.   Reflect $\overline{AB}$ across the *y*-axis to create $\overline{A'B'}$. What type of shape is *ABB'A'* if
the points are connected in order? Then find the area of *ABB'A'*.

5-49.   Fill in the blank ovals below so that each flowchart is correct.

a.   b.

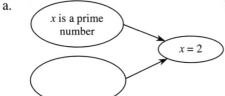   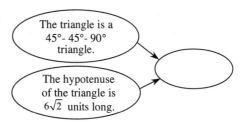

5-50.   Decide if each pair of triangles below are similar. If they are similar, show a flowchart that organizes your **reasoning**. If they are not similar, explain how you know.

a.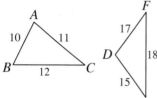

b.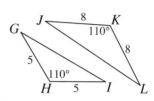

c.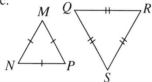

d.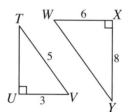

5.2.2   How can I use similar triangles?

Pythagorean Triples

In Lesson 5.2.1, you developed shortcuts that will help you quickly find the lengths of the sides of certain right triangles. What other shortcuts can be helpful? As you work today with your study team, look for patterns and connections between triangles.

5-51.   Use the tools you have developed to find the lengths of the missing sides of the triangles below. If you know of a shortcut, share it with your team. Be ready to share your strategies with the class.

a.

b.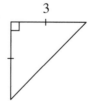

c.

d.

e.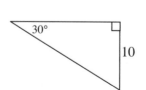

f.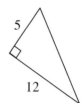

Problem continues on next page →

5-51.   *Problem continued from previous page.*

g.   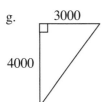   h.   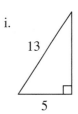   i.

5-52.   Karl noticed some patterns as he was finding the sides of the triangles in problem 5-46. He recognized that the triangles in parts (c), (d), (f), (g), (h), and (i) are integers. He also noticed that knowing the triangle in part (c) can help find the hypotenuse in parts (d) and (g).

a.   Groups of numbers like 3, 4, 5 and 5, 12, 13 are called **Pythagorean Triples**. Why do you think they are called Pythagorean Triples?

b.   What other sets of numbers are also Pythagorean Triples? How many different sets can you find?

5-53.   With your team, find the missing side lengths for each triangle below. Try to use your new shortcuts when possible.

a.   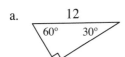   b.   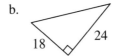   c.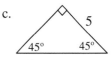

d.   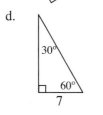   e.   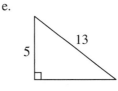   f.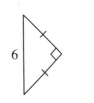

5-54.   Diana looked at the next problems and thought she could not do them. Erik pointed out that they are the same as the previous triangles, but instead of numbers, each side length is given in terms of a variable. Use Erik's idea to help you find the missing side lengths of the triangles below:

a.   b.   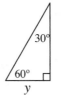   c.

d.   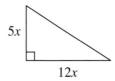   e.

5-55.   Use your shortcuts to find the area and perimeter of each shape below:

a.

b.

c.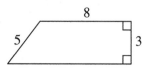

Review & Preview

5-56.   The sides of each of the triangles below can be found using one of the shortcuts from Section 5.2. Try to find the missing lengths using your patterns. Do not use a calculator.

a.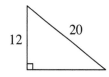

b.

c.

d.

5-57.   Copy the diagram at right onto your paper.

a.   Find the measures of all the angles in the diagram.

b.   Make a flowchart showing that the triangles are similar.

c.   Cheri and Roberta noticed their similarity statements for part (b) were not the same. Cheri had stated $\triangle ABC \sim \triangle DEC$, while Roberta maintained that $\triangle ABC \sim \triangle EDC$. Who is correct? Or are they both correct? Explain your **reasoning**.

5-58.   Using the triangle at right, write an expression representing $\cos 52°$. Then write an expression for $\tan 52°$ and $\cos 38°$.

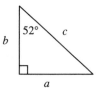

5-59.   Hadrosaurs, a family of duck-billed, plant-eating dinosaurs, were large creatures with thick, strong tails. It has recently been determined that hadrosaurs probably originated in North America.

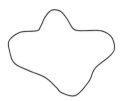

Example of a hadrosaur footprint.

Scientists in Alaska recently found a hadrosaur footprint like the one at right that measured 14 inches across. It is believed that the footprint was created by a young dinosaur that was approximately 27 feet long. Adult hadrosaurs have been known to be 40 feet long. How wide would you expect a footprint of an adult hadrosaur to be? Show your **reasoning**.

5-60.   Jeynysha has a Shape Bucket with a trapezoid, right triangle, scalene triangle, parallelogram, square, rhombus, pentagon, and kite. If she reaches in the bucket and randomly selects a shape, find:

   a.   P(at least one pair of parallel sides)   b.   P(hexagon)

   c.   P(not a triangle)   d.   P(has at least 3 sides)

5.3.1 What triangle tools do I still need?

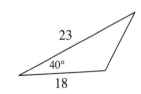

Finding Missing Parts of Triangles

When do you have enough information to find all of the angle measures and side lengths of a triangle? For example, can you find all of the side lengths if you are only informed about the three angles? Does it matter if the triangle has a right angle or not? Today you will organize your triangle knowledge so that you know what tools you have and for which triangles you can accurately find all of the angle measures and side lengths.

5-61.   How many ways can three pieces of information about a triangle be given? For example, the three given measurements could be one angle and two side lengths, as shown in the triangle at right. List as many other combinations of three pieces of information about a triangle as you can.

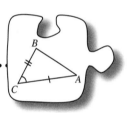

5-62.   HOW MUCH INFORMATION DO YOU NEED?

So far in this course you have developed several tools to find missing parts of triangles. But how complete is your Triangle Toolkit? Are there more tools that you need to develop?

Your Task: With your team, find the missing angles and sides of the triangles below. (Also printed on the Lesson 5.3.1 Resource Page). Notice that each triangle has three given pieces of information about its angles and sides. If there is not enough information or if you do not yet have the tools to find the missing information, explain why.

Discussion Points

- Are there any triangles that look familiar or that you already have a **strategy** for?

- What tools do you have to solve for parts of triangles? For what types of triangles do these tools work?

- Would it be helpful to subdivide any of the triangles into right triangles?

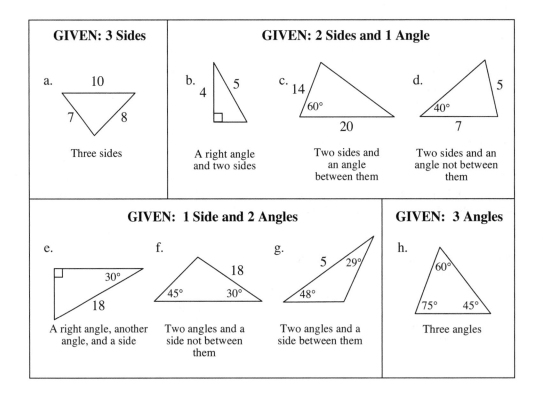

Geometry Connections

5-63.   While looking for different strategies to use, advice from a teammate can often help.

a.   Angelo thinks that the Pythagorean Theorem is a useful tool. For which types of triangles is the Pythagorean Theorem useful? Look for these types of triangles in problem 5-62 and use the theorem to solve for any missing sides.

b.   Tomas remembers using trigonometric ratios to find the missing sides and angles of a triangle. Which triangles from problem 5-62 can be analyzed using this **strategy**?

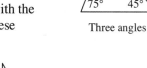

Three angles

c.   Ngoc thinks that more than one triangle exists with the angles at right. Is she correct? If so, how are these triangles related?

d.   Does it matter if the triangle is a right triangle? For example, both of these triangles (from parts (b) and (d)) give an angle and two sides. Can you use the same tool for both? Why or why not?

A right angle and two sides

Two sides and an angle not between them

_____   *Further Guidance*   _____
section ends here.

5-64.   WHAT IF IT DOES NOT HAVE A RIGHT ANGLE?

If your team needs help on parts (c) and (f) of problem 5-62, Leila has an idea. She knows that she has some tools to use with right triangles but noticed that some of the triangles in problem 5-62 are <u>not</u> right triangles. Therefore, she thinks it is a good idea to split a triangle into two right triangles.

a.   Discuss with your team how to change the diagram at right so that the triangle is divided into two right triangles. Then use your right triangle tools to solve for the missing sides and angles.

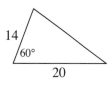

b.   Leila wonders if her method would work for other triangles too. Test her method on the triangle from part (f) of problem 5-62 (also shown at right). Does her method work?

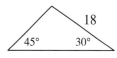

5-65.   Ryan liked Leila's idea so much, he looked for a way to create a right triangle in the triangle from part (g) of problem 5-62. He decided to draw a height <u>outside</u> the triangle, forming a large right triangle. Use the right triangle to help you find the missing side lengths of the original triangle.

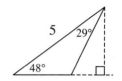

5-66.   LEARNING LOG

Return to problem 5-62 and **examine** all of the ways three pieces of information can be given about a triangle. For which triangle(s) were you able to find missing side lengths and angles? For which triangle(s) do you not have enough given information? For which triangle(s) do you need a new **strategy**? Reflect on the strategies you have developed so far. Title this entry "Strategies to find Sides and Angles of a Triangle" and include today's date.

$\bigcirc\!\!\!\!\mathbf{M}$ ETHODS AND $\mathbf{M}$ EANINGS

MATH NOTES

Special Right Triangles

So far in this chapter, you have learned about several special right triangles that will reappear throughout the rest of this course. Being able to recognize these triangles will enable you to quickly find the lengths of the sides and will save you time and effort in the future.

The half-equilateral triangle is also known as the **30°- 60°- 90° triangle**. The sides of this triangle are always in the ratio $1:\sqrt{3}:2$, as shown at right.

Another special triangle is the **45°- 45°- 90° triangle.** This triangle is also commonly known as an isosceles right triangle. The ratio of the sides of this triangle is always $1:1:\sqrt{2}$.

You also discovered several **Pythagorean triples**. A Pythagorean triple is any set of 3 positive integers a, b, and c for which $a^2 + b^2 = c^2$. Two of the common Pythagorean triples that you will see throughout this course are shown at right.

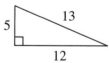

5-67.   To paint a house, Travis leans a ladder against the wall. If the ladder is 16 feet long and it makes contact with the house 14 feet above ground, what angle does the ladder make with the ground? Draw a diagram of this situation and show all work.

5-68.   WACKY DIAGRAMS

After drawing some diagrams on his paper, Jason thinks there is something wrong. **Examine** each diagram below and decide whether or not the triangle could exist. If it cannot exist, explain why not.

a.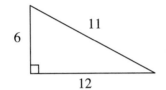

b.

5-69.   William thinks that the hypotenuse must be the longest side of a right triangle, but Chad does not agree. Who is correct? Support your answer with an explanation and a counterexample, if possible.

5-70.   Plot $\triangle ABC$ on graph paper with points $A(3, 3)$, $B(1, 1)$, and $C(6, 1)$.

a.   Reflect $\triangle ABC$ across the x-axis. Then translate the result to the left 6 and down 3. Name the coordinates of the result.

b.   Rotate $\triangle ABC$ 90° counterclockwise (↺) about the origin. Then reflect the result across the y-axis. Name the coordinates of the result.

5-71.   Solve the equations below, if possible. If there is no solution, explain why.

a. $\frac{8-x}{x} = \frac{3}{2}$

b. $-2(5x-1) - 3 = -10x$

c. $x^2 + 8x - 33 = 0$

d. $\frac{2}{3}x - 12 = 180$

5-72.   **Multiple Choice:** Based on the relationships provided in the diagram, which of the equations below is correct? **Justify** your solution.

a. $4x + 2° + 9x - 3° = 90°$

b. $4x + 2° = 9x - 3°$

c. $4x + 2° + 9x - 3° = 180°$

d. $(4x + 2°)(9x - 3°) = 90°$

5.3.2   Is there a faster way?

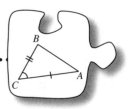

Law of Sines

In problem 6-64, you used a complicated **strategy** to find the lengths of sides and measures of angles for a non-right triangle. Is there a tool you can use to find angles and side lengths of non-right triangles directly, using fewer steps? Today you will explore the relationships that exist among the sides and angles of triangles and will develop a new tool called the Law of Sines.

5-73. Is there a relationship between a triangle's side and the angle opposite to it? For example, assume that the diagram for $\triangle ABC$, shown at right, is not drawn to scale. Based on the angle measures provided in the diagram, which side must be longest? Which side must be shortest? How do you know?

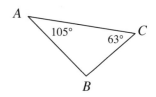

5-74. When Madelyn **examined** the triangle at right, she said, "I don't think this diagram is drawn to scale because I think the side labeled x has to be longer than 4."

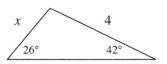

a. Do you agree with Madelyn? Why or why not?

b. Leila thinks that x can be found by using right triangles. Review what you learned in Lesson 5.3.1 by finding the value of x.

c. Find the area of the triangle.

5-75. Thui and Ivan came up with two different ways to find the height of the triangle at right.

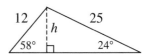

- Using the right triangle on the left, Thui wrote: $\sin 58° = \frac{h}{12}$.

- Ivan also used the sine function, but his equation looked like this: $\sin 24° = \frac{h}{25}$.

a. Which triangle did Ivan use?

b. Calculate h using Thui's equation and again using Ivan's equation. How do their answers compare?

Geometry Connections

5-76.   LAW OF SINES

Edwin wonders if Thui's and Ivan's methods can
help find a way to relate the sides and angles of a
non-right triangle. To find the height, Ivan and
Thui each used the sine ratio with an acute angle
and the hypotenuse of a right triangle.

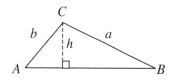

a.   Use the triangle above to find **two expressions** for h using the individual right
triangles like you did in problem 5-75.

b.   Use your expressions from part (a) to show that $\frac{\sin(m\angle B)}{b} = \frac{\sin(m\angle A)}{a}$.

c.   Describe where $\angle B$ is located in relation to the side labeled b. How is $\angle A$
related to the side labeled a?

d.   The relationship $\frac{\sin(m\angle B)}{b} = \frac{\sin(m\angle A)}{a}$ is called
the **Law of Sines**. Read the Math Notes box
for this lesson to learn more about this
relationship. Then use this relationship to solve
for x in the triangle at right.

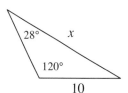

5-77.   EXTENSION

Does the Law of Sines work for a right triangle as well?
Test this idea by solving for x in the triangle at right <u>twice</u>:
once using the Law of Sines and again using right-triangle
trigonometry (such as sine, cosine, or tangent). What
happened? If it worked, do you think it will work for all
right triangles?

5-78.   LEARNING LOG

Reflect on what you have learned during this lesson about the sides
and angles of a triangle. What is the Law of Sines and when can it be
used? Include an example. Title this entry "Law of Sines" and
include today's date.

METHODS AND MEANINGS

Law of Sines

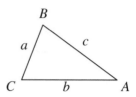

For any $\triangle ABC$, the ratio of the sine of an angle to the length of the side opposite the angle is constant. This means that:

$$\frac{\sin(m\angle A)}{a} = \frac{\sin(m\angle B)}{b},$$

$$\frac{\sin(m\angle B)}{b} = \frac{\sin(m\angle C)}{c}, \text{ and}$$

$$\frac{\sin(m\angle A)}{a} = \frac{\sin(m\angle C)}{c}.$$

This property is called the **Law of Sines**. This is a powerful tool because it allows you to use the sine ratio to solve for measures of angles and lengths of sides of *any* triangle, not just right triangles. The law works for angle measures between 0° and 180°.

Review & Preview

5-79.   Lizzie noticed that two angles in $\triangle DEF$, shown at right, have the same measure. Based on this information, what statement can you make about the relationship between $\overline{ED}$ and $\overline{EF}$?

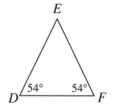

5-80.   Find the length of $\overline{DF}$ in the diagram from problem 5-79 if $DE = 9$ units.

5-81.   Find the area of the triangle at right. Show all work.

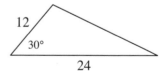

5-82.   In the diagram at right, $\triangle ABC$ and $\triangle ADE$ are similar. If $AB = 5$, $BD = 4$, and $BC = 7$, then what is DE?

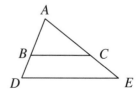

5-83. A particular spinner only has two regions: green and purple. If the spinner is randomly spun twice, the probability of it landing on green twice is 16%. What is the probability of the spinner landing on purple twice?

5-84. Solve the system of equations below. Write your solution as a point in (x, y) form. Check your solution.

$$y = -3x - 2$$
$$2x + 5y = 16$$

5·3·3 How can I complete my triangle toolkit?

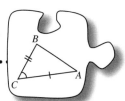

Law of Cosines

So far, you have three tools that will help you solve for missing sides and lengths of a triangle. In fact, one of those tools, the Law of Sines, even helps when the triangle is not a right triangle. There are still two triangles from our exploration in Lesson 5.3.1 that you cannot solve directly with any of your existing tools. Today you will develop a tool to help find missing side lengths and angle measures for triangles such as those shown at right.

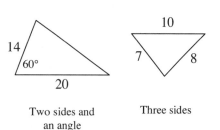

Two sides and an angle between them

Three sides

By the end of this lesson, you will have a complete set of tools to help solve for the side lengths and angle measures of *any* triangle, as long as enough information is given.

5-85. LAW OF COSINES

Leila remembers that in problem 5-64, she solved for the side lengths and missing angles of the triangle at right by dividing the triangle into two right triangles. She thinks that using two right triangles may help find a tool that works for any triangle with two given sides and a given angle between them. Help Leila generalize this process by answering the questions below.

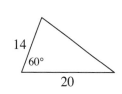

    a.   **Examine** $\triangle ABC$ at right. Assume that you know the lengths of sides a and b and the measure of $\angle C$. Notice how the side opposite $\angle A$ is labeled a and the side opposite $\angle B$ is labeled b, and so on. Line segment $\overline{BD}$ is drawn so that $\triangle ABC$ is divided into two right triangles. If $CD = x$, then what is DA?

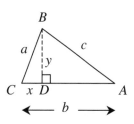

Problem continues on next page →

5-85.   *Problem continued from previous page.*

   b.   Write an equation relating *a*, *x*, and *y*. Likewise, write an equation relating the side lengths of △*BDA* .

   c.   Leila noticed that both expressions from part (b) have a y^2-term. "Can we combine these equations so that we have one equation that links sides *a*, *b*, and *c*?" she asked. With your team, use algebra to combine these two equations so that y^2 is eliminated. Then simplify the resulting equation as much as possible.

   d.   The equation from part (c) still has an *x*-term. Using only *a* and $m\angle C$, find an expression for *x* using the left-hand triangle. Solve your equation for *x*, then substitute this expression into your equation from part (c) for *x*.

   e.   Solve your equation from part (d) for c^2 . You have now found an equation that links the lengths of two sides and the measure of the angle between them to find the length of the side opposite the angle. This relationship is called the **Law of Cosines**. Read the Math Notes box for this lesson to learn more about the Law of Cosines.

   f.   Use the Law of Cosines to solve for *x* in the triangle at right.

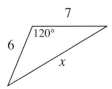

5-86.   You now have many tools to solve for missing side lengths and angle measures. Decide which tool to use for each of the triangles below and solve for *x*. Decide if your answer is reasonable based on the diagram.

   a.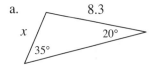

   b.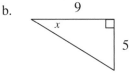

   c.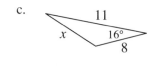

5-87.   **EXTENSION**

Not only can the Law of Cosines be used to solve for side lengths, but it can also be used to solve for angles. Consider the triangle from Lesson 5.3.1, shown at right.

a.   Write an equation that relates the three sides and the angle α. Then solve the equation for α.

b.   Now solve for the other two angles using any method. Be sure to name which tool(s) you used!

5-88.   You have now completed your Triangle Toolkit and can find the missing side lengths and angle measures for *any* triangle, provided that enough information is given. Add the Law of Sines and Law of Cosines to your Triangle Toolkit for reference later in this course.

MATH NOTES

METHODS AND MEANINGS

Law of Cosines

Just like the Law of Sines, the **Law of Cosines** represents a relationship between the sides and angles of a triangle.

Specifically, when given the lengths of any two sides, such as a and b, and the angle between them, $\angle C$, the length of the third side, in this case c, can be found using this relationship:

$$c^2 = a^2 + b^2 - 2ab\cos C$$

Similar equations can be used to solve for a and b.

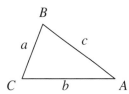

5-89.   Eugene wants to use the cosine ratio to find y on this triangle.

a.   Which angle should he use for his slope angle? Why?

b.   Set up an equation, and solve for y using cosine.

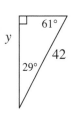

5-90.   Copy the graph at right onto graph paper.

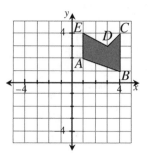

   a.   If the shape *ABCDE* were rotated about the
         origin 180°, where would point *A′* be?

   b.   If the shape *ABCDE* were reflected across the
         x-axis, where would point *C′* be?

   c.   If the shape *ABCDE* were translated so that
         each point (x, y) corresponds to $(x - 1, y + 3)$,
         where would point *B′* be?

5-91.   Jerry was trying to use a flowchart to describe how his friend Marcy
         feels about Whizzbangs candy. **Examine** his flowchart below.

   a.   How do you know Jerry's flowchart is incorrect?

   b.   Make a flowchart on your paper
         with the same three ovals, but with
         arrows drawn in so the flowchart
         makes sense. Explain why your
         flowchart makes more sense than
         the one at right.

5-92.   On graph paper, graph the line $y = \frac{3}{4}x + 6$. Then find the slope angle (the acute
         angle the line makes with the *x*-axis).

5-93.   The spinners at right are spun and the results are
         added.

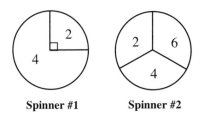

   a.   Find P(sum is 4).

   b.   Find P(sum is 8).

   Spinner #1   Spinner #2

5-94.   **Examine** the diagram at right. Which angle below is
         another name for $\measuredangle ABC$? Note: More than one
         solution is possible.

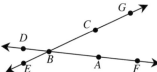

   a. $\measuredangle ABE$ b. $\measuredangle GBD$

   c. $\measuredangle FBG$ d. $\measuredangle EBC$

   e.   None of these

5.3.4 Is there more than one possible triangle?

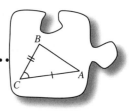

Optional: Ambiguous Triangles

Now that you have completed your Triangle Toolkit, you can solve for the missing angles or sides of any triangle, provided that enough information is given. But how do you know if you have enough information? What if there is more than one possible triangle? Today you will explore situations where your tools may not be adequate to solve for the missing side lengths and angle measures of a triangle.

5-95. **Examine** the triangle at right from problem 5-62. Notice that two side lengths and an angle measure not between the labeled sides are given. (This situation is sometimes referred to as **SSA**.)

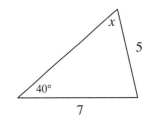

a. Assuming that the triangle is not drawn to scale, what do you know about x? Is it more than 40° or less than 40°? **Justify** your conclusion.

b. Solve for x. Was your conclusion from part (a) correct?

c. "Hold on!" proclaims your teammate, Missy. "That's not what I got. I found out that $x \approx 115.9°$." She then drew the triangle at right. Do you agree with Missy? Use the Law of Sines to test her answer. That is, find out if the ratios of $\frac{\sin(angle)}{\text{opposite side}}$ are equal for each angle and its opposite side.

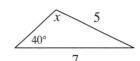

d. What happened? How can there be two possible answers for x? **Examine** the diagram at right for clues.

e. What is the relationship between the two solution angles? Do you think this relationship always exists? **Examine** the diagram above, which shows the two angle solutions, and use it to explain how the solutions are related.

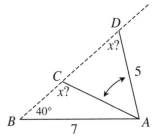

5-96.   In problem 5-95, you determined that it was possible to create two triangles because you were given only two side lengths and an angle not between them. When this happens, we call this **triangle ambiguity** since we cannot tell which triangle was the one we were supposed to find. Will there always be two possible triangles? Can there ever be more than two possible triangles? Think about this as you answer the questions below.

   a.   Obtain the Lesson 5.3.4 Resource Page and some linguini (or other flat manipulative) from your teacher. Prepare pieces of linguini that are 1 inch, 1.5 inches, 2 inches, 2.5 inches, and 3 inches long.

   b.   For each length of linguini, place one end at point A in the diagram on the resource page. Determine if you can form a triangle by connecting the linguini with the dashed side to close the triangle. If you can make a triangle, label the third vertex C and label $\overline{AC}$ with its length. Can you form more than one triangle with the same side length? Is a triangle always possible? Record any conjectures you make.

   c.   If the technology is available, test your conjectures from part (b) with a dynamic geometry tool. Try to learn everything you can about SSA triangles. Use the questions below to guide your **investigation**.

   •   Can you find a way to create three possible triangles with one set of SSA information?

   •   Is it ever impossible to form a triangle?

   •   Is it possible to choose SSA information that will create only one triangle? How?

Geometry Connections

5-97.   Now that Alex knows that SSA (two sides
and an angle not between them) can result
in more than one possible triangle, he
wants to know if other types of given
information can also create ambiguous
results. For example, when given three
side lengths, is more than one triangle
possible?

Examine each of the diagrams below
(from problem 5-62) and determine if any
other types of triangles are also
ambiguous. You may want to imagine
building the shapes with linguini.
Remember that the given information
cannot change – thus, if a side length is
given, it cannot be lengthened or
shortened.

a.

Three sides (SSS)

b.
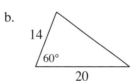
Two sides and an angle
between them (SAS)

c.

Two angles and a side not
between them (AAS)

d.

60°

75°   45°

Three angles (AAA)

5-98.   You now have many tools to use when solving for missing sides lengths and angle
measures. Decide which tool to use for each of the triangles below and solve for *x*.
If there is more than one solution, find both. Name the tool you use.

a.

b.
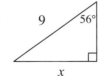

c.
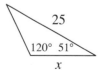

5-99. **EXTENSION**

While examining the triangle at right, Alex stated, "Well, I know that there can be at most one solution."

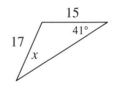

a. **Examine** the information given in the triangle. What do you know about x? Is it more than 41° or less? How can you tell?

b. Alex remembered that if there were two solutions, then they had to be supplementary. Explain why this triangle cannot have two different values for x.

5-100. Farmer Jill has a problem. She lives on a triangular plot of land that is surrounded on all three sides by a fence. Yesterday, one side of the fence was torn down in a storm. She wants to determine the length of the side that needs to be rebuilt so she can purchase enough lumber. Since the weather is still too poor for her to go outside and measure the distance, she decides to use the lengths of the two sides that are still standing (116 feet and 224 feet) and the angle between them (58°).

a. Draw a diagram of this situation. Label all of the sides and angles that Farmer Jill has measurements for.

b. Find the length of the fence that needs to be replaced. Show all work. Which tool did you use?

5-101. Mr. Miller has informed you that two shapes are similar.

a. What does this tell you about the angles in the shapes?

b. What does this tell you about the lengths of the sides of the shapes?

5-102. Find the equation of the line that has a slope angle of 25° and a y-intercept of $(0, 4)$. Sketch a graph of this line. Assume the slope of the line is positive.

5-103. Two sides of a triangle have lengths 9 and 14 units. Describe what you know about the length of the third side.

5-104. Tehachapi High School has 839 students and is increasing by 34 students per year. Meanwhile, Fresno High School has 1644 students and is decreasing by 81 students per year. In how many years will the two high schools have the same number of students? Be sure to show all work.

5-105. **Multiple Choice:** In the triangle at right, x must be:

a.   42°   b.   69°   c.   21°

d.   138°   e.   none of these

5·3·5   Which tool should I use?

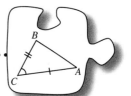

Choosing a Tool

During this section, you have developed new tools such as the Law of Sines and the Law of Cosines to find the lengths of sides and the measures of angles of a triangle. These strategies are very useful because they work with all triangles, not just right triangles. But when is each **strategy** the best one to use? Today you will focus on which **strategy** is most effective to use in different situations. You will also apply your **strategies** to triangles in different contexts.

As you work on these problems, keep in mind that good communication and a joint brainstorming of ideas will greatly enhance your team's ability to **choose a strategy** and to solve these problems.

5-106. LAKE TOFTEE, Part One

A bridge is being designed to connect two towns along the shores of Lake Toftee in Minnesota (one at point A and the other at point C). Lavanne has been given the responsibility of determining the length of the bridge.

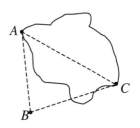

Since he could not accurately measure across the lake (AC), he measured the only distance he could by foot (AB). He drove a stake into the ground at point B and found that $AB = 684$ feet. He also used a protractor to determine that $m\angle B = 79°$ and $m\angle C = 53°$. How long will the bridge need to be?

5-107.   LAKE TOFTEE, Part Two

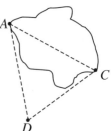

Lavanne was not convinced that his measurements from
problem 5-106 were correct. He decided to measure the
distance between towns A and C again using a different
method to verify his results.

This time, he decided to drive a stake in the ground at point
D, which is 800 feet from town A and 694 feet from town C.
He also determined that $m\angle D = 68°$. Using these
measurements, how wide is the lake between points A and C? Does this confirm the
results from problem 5-106?

5-108.   The lid of a grand piano is propped open by a
supporting arm, as shown in the diagram at
right. Carson knows that the supporting arm
is 2 feet long and makes a 60° angle with the
piano. He also knows that the piano lid
makes a 23° angle with the piano. How wide
is the piano?

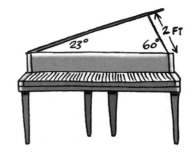

5-109.   PENNANT RACE

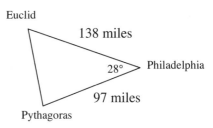

Your basketball team has made it to the semi-
finals and now needs to win only two more
games to go to the finals. Your plan is to
leave Philadelphia, travel 138 miles to the
town of Euclid, and then play the team there.
Then you will leave Euclid, travel to
Pythagoras, and play that team. Finally, you
will travel 97 miles to return home.

Your team bus can travel only 300 miles on one tank of gas. Assuming that all of
the roads connecting the three towns are straight and that the two roads that connect
in Philadelphia form a 28° angle, will one tank of gas be enough for the trip? **Justify**
your solution.

5-110.   Shonte is buying pipes to install sprinklers for her
front lawn. She needs to fit the pipes into the bed of
her pickup truck to get them home from the store.
She knows that the longest dimension in her truckbed
is from the top of one corner to the bottom of the bed
at the opposite corner. After measuring the truckbed,
she drew the diagram at right. What is the longest
pipe she can buy?

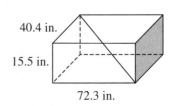

5-111.   In problem 5-105 from homework, you used your intuition to state that if a triangle has two angles that are equal, then the triangle has two sides that are the same length. Now use your triangle tools to show that your intuition was correct. For example, for $\triangle PQR$, show that if $m\angle Q = m\angle R$, then $PQ = PR$.

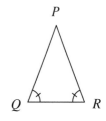

5-112.   As Freda gazes at the edge of the Grand Canyon, she decides to try to determine the height of the wall opposite her. Using her trusty clinometer, she determines that the top of the wall is at a 38° angle above her, while the bottom is at a 46° angle below her, as shown in the diagram at right. If the base of the wall is 253 feet from the point on the ground directly below Freda, determine the height of the wall opposite her.

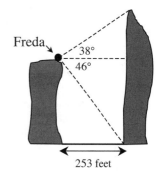

5-113.   While facing north, Lisa and Aaron decide to hike to their campsite. Lisa plans to hike 5 miles due north to Lake Toftee before she goes to the campsite. Aaron plans to turn 38° east and hike 7.4 miles directly to the campsite. How far will Lisa have to hike from the lake to meet Aaron at the campsite? Start by drawing a diagram of the situation, then calculate the distance.

5-114.   Solve for the missing side lengths and angles in the triangle at right.

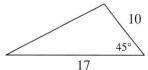

5-115.   **Examine** the triangle shown at right. Solve for x twice, using two different methods. Show your work for each method clearly.

5-116.   In Chapter 1 you learned that all rectangles are parallelograms because they all have two pairs of opposite parallel sides. Does that mean that all parallelograms are rectangles? Why or why not? Support your statements with **reasons**.

5-117.   Find the area and perimeter of each shape below. Show all work.

a.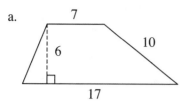

b.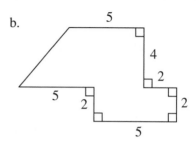

5-118.   △*ABC* was reflected across the *x*-axis, and then that result was rotated 90° clockwise about the origin to result in △*A′B′C′*, shown at right. Find the coordinates of points *A*, *B*, and *C* of the original triangle.

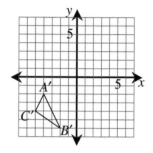

5-119.   Find the equation of a line parallel to the line $y = \frac{3}{4}x - 5$ that passes through the point (−4, 1). Show how you found your answer.

5-120.   **Examine** trapezoid *ABCD* at right.

a.   Find the measures of all the angles in the diagram.

b.   What is the sum of the angles that make up the trapezoid *ABCD*? That is, what is $m\angle A + m\angle ABC + m\angle BCD + m\angle D$?

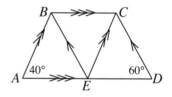

5-121.   Use your triangle tools to solve the problems below.

a.   Find *PR* in the diagram at right.

b.   Find the perimeter of quadrilateral *PQRS*.

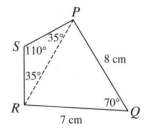

5-122. Find the area and perimeter of $\triangle ABC$ at right. Give approximate (decimal) answers, not exact answers.

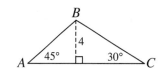

5-123. Earl still hates to wash the dishes and take out the garbage. (See problem 5-9.) He found his own weighted coin, one that would randomly land on heads 30% of the time. He will flip a coin once for each chore and will perform the chore if the coin lands on heads.

   a.   What is the probability that Earl will get out of doing both chores?

   b.   What is the probability that Earl will have to take out the garbage, but will not need to wash the dishes?

5-124. On graph paper, draw $\triangle ABC$ if $A(3, 2)$, $B(-1, 4)$, and $C(0, -2)$.

   a.   Find the perimeter of $\triangle ABC$.

   b.   Dilate $\triangle ABC$ from the origin by a factor of 2 to create $\triangle A'B'C'$. What is the perimeter of $\triangle A'B'C'$?

   c.   If $\triangle ABC$ is rotated 90° clockwise ($\circlearrowright$) about the origin to form $\triangle A''B''C''$, name the coordinates of C''.

5-125. Solve each equation below for x. Check your solution if possible.

   a. $\frac{4}{5}x - 2 = 7$ b. $3x^2 = 300$

   c. $\frac{4x-1}{2} = \frac{x+5}{3}$ d. $x^2 - 4x + 6 = 0$

Chapter 5 Closure   What have I learned?

Reflection and Synthesis

The activities below offer you a chance to reflect on what you have learned in this chapter. As you work, look for concepts that you feel very comfortable with, ideas that you would like to learn more about, and topics you need more help with. Look for connections between ideas as well as connections with material you learned previously.

① TEAM BRAINSTORM

With your team, brainstorm a list for each of the following two subjects. Be as detailed as you can. How long can you make your list? Challenge yourselves. Be prepared to share your team's ideas with the class.

Topics: What have you studied in this chapter? What ideas and words were important in what you learned? Remember to be as detailed as you can.

Connections: How are the topics, ideas, and words that you learned in previous courses connected to the new ideas in this chapter? Again, make your list as long as you can.

②   MAKING CONNECTIONS

The following is a list of the vocabulary used in this chapter. The words that appear in bold are new to this chapter. Make sure that you are familiar with all of these words and know what they mean. Refer to the glossary or index for any words that you do not yet understand.

adjacent	**ambiguous**	angle
approximate	**cosine**	equilateral
exact answer	hypotenuse	**inverse sin, cos, tan**
law of cosines	**law of sines**	leg
opposite	**Pythagorean triple**	ratio
right triangle	**sine**	slope
tangent	theta (θ)	

Make a concept map showing all of the connections you can find among the key words and ideas listed above. To show a connection between two words, draw a line between them and explain the connection, as shown in the example below. A word can be connected to any other word as long as there is a justified connection. For each key word or idea, provide a sketch of an example.

Your teacher may provide you with vocabulary cards to help you get started. If you use the cards to plan your concept map, be sure either to re-draw your concept map on your paper or to glue the vocabulary cards to a poster with all of the connections explained for others to see and understand.

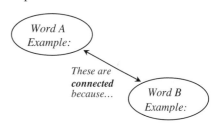

While you are making your map, your team may think of related words or ideas that are not listed above. Be sure to include these ideas on your concept map.

③   SUMMARIZING MY UNDERSTANDING

This section gives you an opportunity to show what you know about certain math topics or ideas. Your teacher will direct you how to do this. Your teacher may give you a "GO" page to work on. "GO" stands for "Graphic Organizer," a tool you can use to organize your thoughts and communicate your ideas clearly.

④ WHAT HAVE I LEARNED?

This section will help you recognize those types of problems you feel comfortable with and those you need more help with. This section will appear at the end of every chapter to help you check your understanding. Even if your teacher does not assign this section, it is a good idea to try these problems and find out for yourself what you know and what you need to work on.

Solve each problem as completely as you can. The table at the end of this closure section has answers to these problems. It also tells you where you can find additional help and practice on problems like these.

CL 5-126. For each diagram, write an equation and solve to find the value for each variable.

a.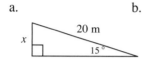
b.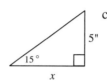
c.
d.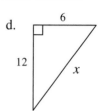

CL 5-127. Copy the diagram at right onto your paper.

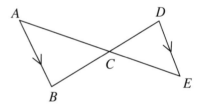

a. Are the triangles similar? If so, show your **reasoning** with a flowchart.

b. If $m\angle B = 80°$, $m\angle ACB = 29°$, $AB = 14$, and $DE = 12$, find CE.

CL 5-128. Cynthia is planning a party. For entertainment, she has designed a game that involves spinning two spinners. If the sum of the numbers on the spinners is 10 or greater, the guests can choose a prize from a basket of candy bars. If the sum is less than 10, then the guest will be thrown in the pool. She has two possible pairs of spinners, shown below. For each pair of spinners, determine the probability of getting tossed in the pool. Assume that Spinners B, C, and D are equally subdivided.

a.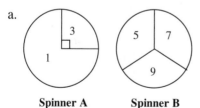
Spinner A   Spinner B

b.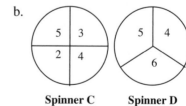
Spinner C   Spinner D

CL 5-129. While working on homework, Zachary was finding the value of each variable in the diagrams below. His first step for each problem is shown under the diagram. If his first step is correct, continue solving the problem to find the solution. If his first step is incorrect, explain his mistake and solve the problem correctly.

a.

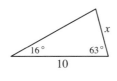

$$\sin 16° = \frac{x}{10}$$

b.

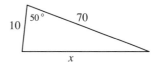

$$x^2 = 10^2 + 70^2 - 2(10)(70)\cos 50°$$

CL 5-130. **Examine** the diagram at right.

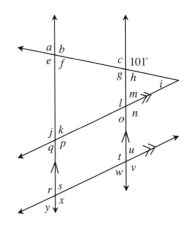

a. Find the measures of each of the angles below, if possible. If it is not possible, explain why it is not possible. If it is possible, state your **reasoning**.

(1) $m\angle b$ (2) $m\angle f$

(3) $m\angle m$ (4) $m\angle g$

(5) $m\angle h$ (6) $m\angle i$

b. If $m\angle p = 130°$, can you now find the measures of any of the angles from part (a) that you couldn't before? Find the measures for all that you can. Be sure to justify your **reasoning**.

CL 5-131. Trace each figure at right onto your paper. The side labeled $\overline{AB}$ is the base in each figure. Then:

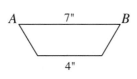

Trapezoid

Parallelogram

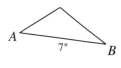

Triangle

a. Draw a height for each figure to side $\overline{AB}$.

b. Find the area of each figure assuming that the height of each shape is 4" long.

CL 5-132. Bob is hanging a swing from a pole high off the ground so that it can swing a total angle of 120°. Since there is a bush 5 feet in front of the swing and a shed 5 feet behind the swing, Bob wants to ensure that no one will get hurt when they are swinging. What is the maximum length of chain that Bob can use for the swing?

    a.   Draw a diagram of this situation.

    b.   What is the maximum length of chain that Bob can use? State what tools you used to solve this problem.

CL 5-133. **Examine** the triangle at right. Solve for x <u>twice</u> using two different methods.

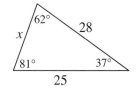

CL 5-134. Graph the points $(3, -4)$ and $(7, 2)$ on graph paper and draw the line segment and a slope triangle that connects the points. Find:

    a.   The length of the segment   b.   The slope of the line segment

    c.   The area of the slope triangle   d.   The measure of the slope angle

CL 5-135. Trace the figure at right onto your paper and then perform all of the transformations listed below on the same diagram. Then find the perimeter of the final shape.

    •   Reflect $\triangle ABC$ across $\overline{AB}$.

    •   Rotate $\triangle ABC$ 180° around the midpoint of $\overline{BC}$.

    •   Reflect $\triangle ABC$ across $\overline{AC}$.

CL 5-136. Check your answers using the table at the end of this section. Which problems do you feel confident about? Which problems were hard? Have you worked on problems like these in math classes you have taken before? Use the table to make a list of topics you need help on and a list of topics you need to practice more.

⑤   HOW AM I THINKING?

This course focuses on five different **Ways of Thinking**: investigating, examining, reasoning and justifying, visualizing, and choosing a strategy/tool. These are some of the ways in which you think while trying to make sense of a concept or to solve a problem (even outside of math class). During this chapter, you have probably used each Way of Thinking multiple times without even realizing it!

This closure activity will focus on one of these Ways of Thinking: **choosing a strategy/tool**. Read the description of this Way of Thinking at right.

Think about the problems you have worked on in this chapter. When did you need to think about what method you would use to solve a problem? What helped you decide how to approach a problem? Were there times when more than one **strategy** seemed most useful? You may want to flip through the chapter to refresh your memory about the problems that you have worked on. Discuss these ideas with the class.

Once your discussion is complete, think about the way you think as you answer the questions below.

a.   List all the triangle tools that you have learned so far in this course. For each tool, find or create a problem that can be solved with this **strategy**.

b.   Sometimes, the key to being able to **choose a strategy or tool** is to recognize that different tools can be used on the same problem, but that sometimes some tools are more efficient than others.

> ### Choosing a Strategy/Tool
>
> To choose a strategy means to think about what you know about a problem and match that information with methods and processes for solving problems. As you develop this way of thinking you will learn how to choose ways of solving problems based in given information. You think this way when you ask/answer questions like "What strategy might work for…?" or, "How can I use this information to answer…?" When you catch yourself looking for a method to answer a problem, you are choosing a strategy/tool.
>
>

Solve for x in each diagram below <u>twice</u>. Each time, use a different **strategy** or tool. Then decide which method was easiest for that problem or state that both methods were of equal value.

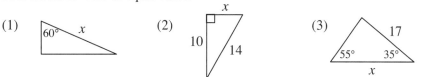

(1) $60°$ x

(2) x 10 14

(3) 17 $55°$ $35°$ x

Problem continues on next page →

⑤ *Problem continued from previous page.*

c. While you have focused most of this chapter on developing strategies for measuring triangles, you have developed strategies for several other topics so far in this course. Consider your **strategy** options as you answer the questions below.

(1) **Examine** the diagram at right. Assume you know the measure of $\angle b$. How could you find $m\angle h$? Describe two different **strategies**.

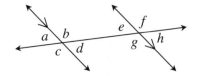

(2) Now consider the trapezoid below. Find the area of the shape twice, using two different **strategies**.

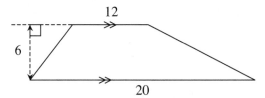

(3) While shopping at the Two Tired Bike Shop, Barry notices that he may choose from many bicycles. Of the bikes at the store, $\frac{1}{4}$ were mountain bikes, while the rest were racing bikes. Also, $\frac{1}{2}$ of the bikes were blue, $\frac{1}{3}$ of the bikes were red, and $\frac{1}{6}$ of the bikes were purple. Barry decides that he will randomly choose a bicycle from the store.

Choose a probability model to represent this situation. Then find the probability that he chooses a purple racing bike. [**Students can use a scaled area model, generic area model, or a tree diagram;** $\frac{1}{6} \cdot \frac{3}{4} = \frac{1}{8}$]

d. What about multiple strategies that you have learned in an earlier class? For example, you have multiple ways to approach a problem involving a system of equations. Consider the system below. Solve the system <u>twice</u>: once by graphing and again algebraically. Which **strategy** seemed most efficient? Why? [**(10, –2)**]

$$y = -x + 8$$
$$y = \tfrac{1}{2}x - 7$$

Answers and Support for
Closure Activity #4
What Have I Learned?

Problem	Solutions	Need Help?	More Practice
CL 5-126.	a. $\sin 15° = \frac{x}{20}$; $x \approx 5.176$ m b. $\tan 15° = \frac{5}{x}$; $x \approx 18.66$ in c. $\cos\theta = \frac{10}{11}$; $\theta \approx 24.62°$ d. $6^2 + 12^2 = x^2$; $x \approx 13.416$ un	Lessons 2.2.3, 4.1.4, 5.1.2, and 5.1.4 Math Notes boxes	Problems 4-11, 4-12, 4-16, 4-22, 4-25, 4-31, 4-34, 4-36, 4-46, 5-5, 5-7, 5-12, 5-16, 5-23, 5-89
CL 5-127.	a. $\measuredangle ACB \cong \measuredangle ECD$ $\measuredangle B \cong \measuredangle D$ $\triangle ACB \sim \triangle ECD$ AA $\sim$ b. $x \approx 24.38$	Lessons 3.1.2, 3.1.3, 3.2.1, 3.2.5, and 5.3.2 Math Notes boxes	Problems 3-64, 3-65, 3-83, 3-84, 3-89, 4-7, 4-17, 4-38, 4-39, 4-83, 5-20, 5-49, 5-50, 5-57, 5-76, 5-86, 5-91
CL 5-128.	a. $\frac{7}{12}$ b. $\frac{9}{12} = \frac{3}{4}$	Lesson 4.2.4 Math Notes box	Problems 4-52, 4-60, 4-62, 4-68, 4-69, 4-70, 4-71, 4-77, 4-78, 4-87, 5-9, 5-18, 5-28, 5-39, 5-83, 5-123
CL 5-129.	a. Cannot use sine in this manner in a non-right triangle. Use Law of Sines instead: $\frac{x}{\sin(16°)} = \frac{10}{\sin(101°)}$; $x \approx 2.807$ b. $x \approx 64.03$ un.	Lessons 5.3.2 and 5.3.3 Math Notes boxes	Problems 5-76, 5-77, 5-80, 5-81, 5-86, 5-98, 5-106, 5-107, 5-109, 5-114, 5-121
CL 5-130.	a. (1) $b = 101°$, corres. angles are equal (2) $f = 79°$, supplementary angles (3) not enough information (4) $g = 101°$, vertical angles are equal (5) $h = 79°$, supplementary angles (6) not enough information b. $m = 50°$; $v = 130°$; $i = 51°$	Lessons 2.1.1, 2.1.4, and 2.2.1 Math Notes boxes, Problem 2-3	Problems 2-13, 2-16, 2-17, 2-23, 2-24, 2-25, 2-31, 2-32, 2-38, 2-49, 2-51, 2-62, 2-72, 2-111, 3-20, 3-31, 3-60, 3-70, 3-96, 4-6, 4-29, 4-44, 4-85, 4-93, 5-47

Problem	Solutions	Need Help?	More Practice
CL 5-131.	Trapezoid Area: 22 square in. Parallelogram Area: 20 square in. Triangle Area: 14 square in.	Lessons 1.1.3 and 2.2.4 Math Notes boxes	Problems 2-66, 2-68, 2-71, 2-75, 2-76, 2-78, 2-79, 2-82, 2-83, 2-90, 2-120, 3-51, 3-81, 3-97, 4-43, 4-67, 4-92, 5-10, 5-29, 5-34, 5-55, 5-81, 5-117, 5-122
CL 5-132.	a. 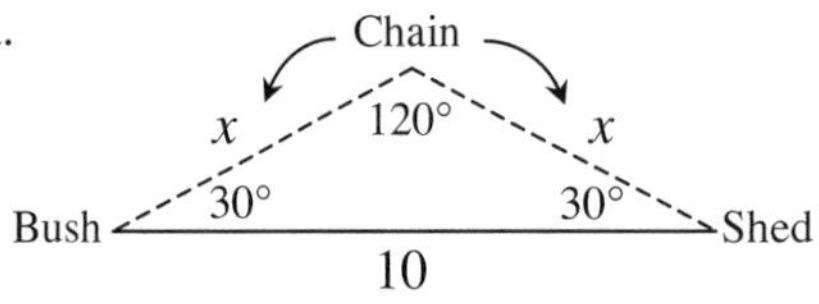 b. $x = \frac{10\sqrt{3}}{3} \approx 5.77$ ft.	Lessons 5.1.2, 5.3.2, and 5.3.3 Math Notes boxes	Problems 5-17, 5-76, 5-77, 5-80, 5-81, 5-86, 5-98, 5-106, 5-107, 5-109, 5-112, 5-114, 5-121
CL 5-133.	Both the Law of Sines and the Law of Cosines will work, as does dividing the triangle into two right triangles; $x \approx 17.06$ un.	Lessons 5.1.2, 5.3.2, and 5.3.3 Math Notes boxes	Problems 5-16, 5-17, 5-23, 5-76, 5-77, 5-80, 5-81, 5-86, 5-98, 5-106, 5-107, 5-109, 5-112, 5-114, 5-121
CL 5-134.	a. $\sqrt{52} \approx 7.21$ un. b. $\frac{3}{2}$ c. 12 sq. un. d. 56.31°	Lessons 1.2.5, 2.3.3, 4.1.1, 4.1.4, 5.1.4 Math Notes boxes	Problems 2-116, 5-2, 5-37, 5-92, 5-102
CL 5-135.	The final shape is shown below: Perimeter: 24 feet 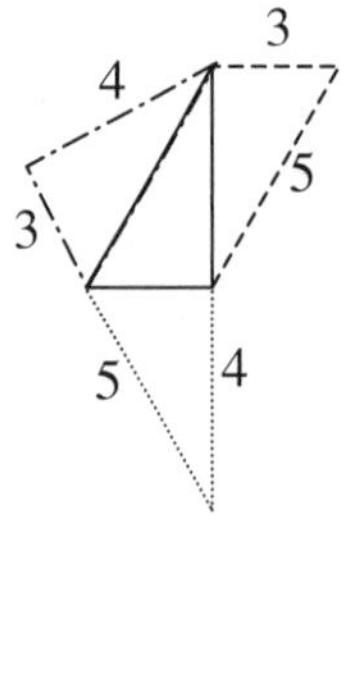 	Lessons 1.1.3, 1.2.2 and 1.2.3 Math Notes boxes	Problems 1-50, 1-51, 1-64, 1-69, 1-73, 1-85, 1-96, 2-11, 2-19, 2-20, 2-33, 2-64, 2-113, 3-96, 5-41, 5-70, 5-90, 5-118, 5-124

CHAPTER 6

Congruent Triangles

In Chapter 5, you completed your work with the measurement of triangles, so you can now find the missing side lengths and angles of a triangle when sufficient information is given. Earlier, you developed ways to determine if two triangles are similar, and can use the ratios of similarity to learn more about the sides and angles of similar figures. But what if two triangles are congruent? What information can congruent triangles provide? In this chapter, you will find ways to determine whether two triangles are congruent.

In addition, Section 6.2 offers several projects and activities that will help you synthesize your understanding and make connections between different concepts you have learned so far. You will consolidate what you know, apply it in new ways, and identify what you still need to learn.

Guiding Questions

Think about these questions throughout this chapter:

How are the triangles related?

What is the connection?

What information do I need?

What do I still need to learn?

In this chapter, you will learn:

> what information is needed in order to conclude that two triangles are congruent.

> what a converse of a conditional statement is and how to recognize whether or not the converse is true.

> how to organize a flowchart that helps conclude that two triangles are congruent.

Chapter Outline

Section 6.1 This section turns the focus to congruent triangles. You will develop strategies to directly conclude that two triangles are congruent without first concluding that they are similar.

Section 6.2 This section includes several big problems and activities to help you learn how different threads of geometry are connected and to help you assess what you know and what you still need to learn.

Geometry Connections

6.1.1 Are the triangles congruent?

Congruent Triangles

In Chapter 3, you learned how to identify similar triangles and used them to solve problems. But what can be learned when triangles are congruent? In today's lesson, you will practice identifying congruent triangles using what you know about similarity. As you search for congruent triangles in today's problems, focus on these questions:

What do I know about these triangles?

How can I show similarity?

What is the common ratio?

6-1.   **Examine** the triangles at right.

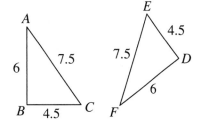

a.   Make a flowchart showing that these triangles are similar.

b.   Are these triangles also congruent? Explain how you know.

c.   While the symbol for similar figures is "~", the symbol for congruent figures is "≅". How is the congruence symbol related to the similarity symbol? Why do you think mathematicians chose this symbol for congruence?

d.   Luis wanted to write a statement to convey that these two triangles are congruent. He started with "$\triangle CAB$ ...", but then got stuck because he did not know the symbol for congruence. Now that you know the symbol for congruence, complete Luis's statement for him.

6-2.   The diagrams below are not drawn to scale. For each pair of triangles:

   - Determine if the two triangles are congruent.

   - If you find congruent triangles, write a congruence statement (such as $\triangle PQR \cong \triangle XYZ$).

   - If the triangles are not congruent or if there is not enough information to determine congruence, write "cannot be determined."

a. $\overline{AC}$ is a straight segment:
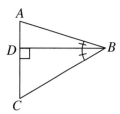

b.
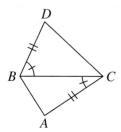

c.
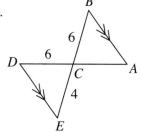

d.
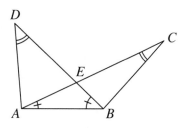

6-3.   Consider square *MNPQ* with diagonals intersecting at *R*, as shown at right.

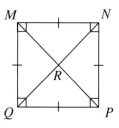

a.   How many triangles are there in this diagram? (Hint: There are more than 4!)

b.   How many lines of symmetry does *MNPQ* have? On your paper, trace *MNPQ* and indicate the location of each line of symmetry with a dashed line.

c.   Write as many triangle congruence statements as you can that involve triangles in this diagram. Be prepared to **justify** each congruence statement you write.

d.   Write a similarity statement for two triangles in the diagram that are not congruent. **Justify** your similarity statement with a flowchart.

METHODS AND **M**EANINGS

Congruent Shapes

MATH NOTES

The information below is from Chapter 3 and is reprinted here for your convenience.

If two figures have the same shape and are the same size, they are **congruent**. Since the figures must have the same shape, they must be similar.

Two figures are congruent if they meet both the following conditions:

- the two figures are similar, and
- their side lengths have a common ratio of 1.

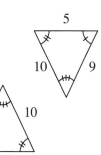

Review & Preview

6-4.   Use the diagram at right to answer the following questions.

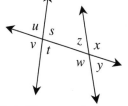

a.   What type of angle pair is ∡z and ∡t? That is, what is their geometric relationship?

b.   What type of angle pair do ∡s and ∡v form?

c.   Name all pairs of corresponding angles in the diagram. Hint: There are four different pairs.

6-5.   **Examine** the triangles at right.

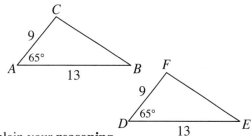

a.   Are these triangles similar? If so, use a flowchart to show how you know. If they are not similar, explain how you know.

b.   Are the triangles congruent? Explain your **reasoning**.

6-6.   Using the diagram at right, write an equation and find x. Check your answer.

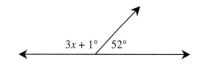

6-7. For each of the triangles below, find *x*.

a.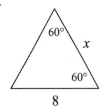

b.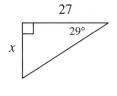

c.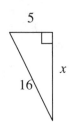

6-8. Joey is in charge of selling cupcakes at the basketball games. The game is today at 5:00 p.m. At 3:00 p.m., Joey put some cupcakes into the oven and started working on his homework. He fell asleep and did not wake up until 3:55 p.m. What do you think happened?

Joey knows that the following four statements are facts:

   i.   If the cupcakes are burned, then the fans that attend the varsity basketball games will not buy them.

   ii. If cupcakes are in the oven for more than 50 minutes, they will burn.

   iii. If the fans do not buy the cupcakes, then the team will not have enough money for new uniforms next year.

   iv. The cupcakes are in the oven from 3:00 p.m. to 3:55 p.m.

Copy the flowchart at right and decide how to organize the facts into the ovals. You will need to fill in one of the ovals with your own logical conclusion. Be sure that your ovals are in the correct order and that the arrows really show connections between the ovals. What conclusion should your flowchart make?

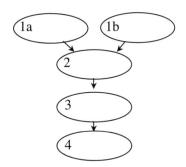

6-9. **Multiple Choice:** $\triangle ABC$ is defined by points A(3, 2), B(4, 9), and C(6, 7). Which triangle below is the image of $\triangle ABC$ when it is rotated 90° counter-clockwise (↺) about the origin?

   a. $A'(-2, 3)$, $B'(-9, 4)$, $C'(-7, 6)$ b. $A'(-3, 2)$, $B'(-4, 9)$, $C'(-6, 7)$

   c. $A'(-2, 3)$, $B'(-7, 6)$, $C'(-9, 4)$ d. $A'(2, -3)$, $B'(9, -4)$, $C'(7, -6)$

   e. None of these

6.1.2 What information do I need?

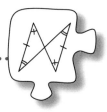

Conditions for Triangle Congruence

In Lesson 6.1.1, you identified congruent triangles by looking for similarity and a common side length ratio of 1. Must you go through this two-step process every time you want to argue that triangles are congruent? Are there shortcuts for establishing triangle congruence? Today you will develop Triangle Congruence Shortcuts in order to quickly determine if two triangles are congruent.

6-10. Review your work from problem 6-5, which required you to determine if two triangles are congruent. Then work with your team to answer the questions below.

   a. Derek wants to find general shortcuts that can help determine if triangles are congruent. To help, he draws the diagram at right to show the relationships between the triangles in problem 6-5.

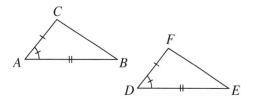

   If two triangles have the relationship shown in the diagram, do they have to be congruent? How do you know?

   b. Write a conjecture in the form of a conditional statement or arrow diagram based on this relationship. If you write a conditional statement, it should look like, "If two triangles …, then the triangles are congruent." What would be a good name (abbreviation) for this shortcut?

TRIANGLE CONGRUENCE SHORTCUTS

Derek wonders, "What other types of information can determine that two triangles are congruent?"

Your Task: Examine the pairs of triangles below to decide what other types of information force triangles to be congruent. Notice that since no measurements are given in the diagrams, you are considering the general case of each type of pairing. For each pair of triangles below that you can prove are congruent, enter the appropriate Triangle Congruence Conjecture on your Lesson 6.1.2 Resource Page with an explanation defending your decision. An entry for SAS ≅ (the conjecture you developed in problem 6-10) is already created on the resource page as an example. Be prepared to share your results with the class.

Discussion Points

- In what ways can you show that triangles are similar?

- What must be true in order for triangles to be congruent?

- Which of the conditions below DO NOT assure congruence?

a.
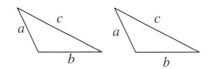

Side-Side-Side: SSS

b.

Hypotenuse-Leg: HL

c.
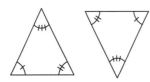

Angle-Angle-Angle: AA

d.
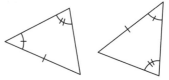

Angle-Angle-Side: AAS

e.
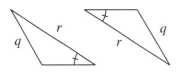

Side-Side-Angle: SSA

f.
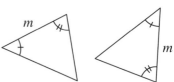

Angle-Side-Angle: ASA

Geometry Connections

6-12.   Use your triangle congruence conjectures to determine if the following pairs of triangles must be congruent. Note: the diagrams are not necessarily drawn to scale.

a.
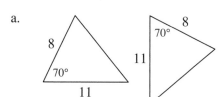

b.
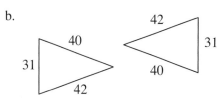

c.
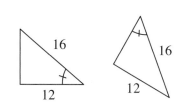

d.
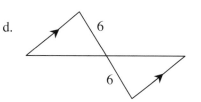

e.
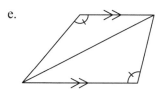

f.
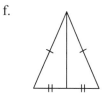

g.
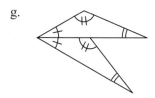

h.
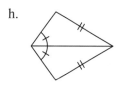

6-13.   **Examine** the relationships that exist in the diagram at right. Find the measures of angles *a*, *b*, *c*, and *d*. Remember that you can find the angles in any order, depending on the angle relationships you use.

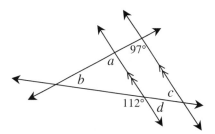

6-14.   **Examine** the triangles below. For each one, solve for *x* and name which **tool** you used. Show all work.

a.
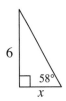

b.
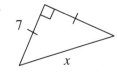

c.
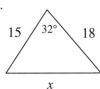

6-15.   The two shapes at right are similar.

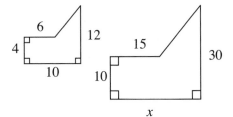

a.   Find the value of *x*. Show all work.

b.   Find the area of each shape.

6-16.   On graph paper, graph △*ABC* if A(–2, 7), B(–5, 8), and C(–3, 1).

a.   Reflect △*ABC* across the *x*-axis to form △*A′B′C′*. Name the coordinates of each new vertex.

b.   Now rotate △*A′B′C′* from part (a) 180° about the origin (0, 0) to form △*A″B″C″*. Name the coordinates of each new vertex.

c.   Describe a single transformation that would change △*ABC* to △*A″B″C″*.

6-17.   Solve the problem below using any method. Show all work.

Angle *A* of △*ABC* measures 5° more than 3 times the measure of angle *B*. Angle *C* measures 20° less than angle *B*. Find the measure of angles *A*, *B*, and *C*.

6-18.   **Multiple Choice:** Which list of side lengths below could form a triangle?

a.   2, 6, 7   b.   3, 8, 13   c.   9, 4, 2   d.   10, 20, 30

Geometry Connections

6.1.3 How can I organize my reasoning?

Flowcharts for Congruence

Now that you have shortcuts for establishing triangle congruence, how can you organize information in a flowchart to show that triangles are congruent? Consider this as you work today with your team.

6-19. In problem 6-1, you determined that the triangles at right are congruent.

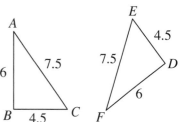

a. Which triangle congruence conjecture shows that these triangles are congruent?

b. Make a flowchart showing your argument that these triangles are congruent.

c. How is a flowchart showing congruence different from a flowchart showing similarity? List every difference you can find.

6-20. In Don's congruence flowchart for problem 6-19, one of the ovals said "$\frac{AB}{FD} = 1$". In Phil's flowchart, one of the ovals said, "$AB = FD$". Discuss with your team whether these ovals say the same thing. Can equality statements like Phil's always be used in congruence flowcharts?

6-21. Make a flowchart showing that the triangles below are congruent.

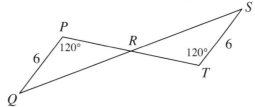

6-22.   In each diagram below, determine whether the triangles are congruent, similar but not congruent, or not similar. If you claim triangles are similar or congruent, make a flowchart **justifying** your answer.

a.
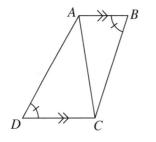

b.
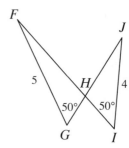

c.
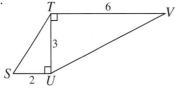

d.
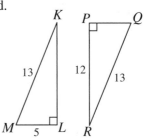

6-23.   Suppose you are working on a problem involving the two triangles $\triangle UVW$ and $\triangle XYZ$, and you know that $\triangle UVW \cong \triangle XYZ$. What can you conclude about the sides and angles of $\triangle UVW$ and $\triangle XYZ$? Write down every equation involving side lengths or angle measures that must be true.

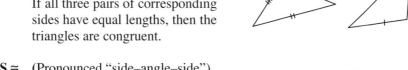

METHODS AND MEANINGS

Triangle Congruence Shortcuts

MATH NOTES

   To show that triangles are congruent, you can show that the triangles are similar and that the common ratio between side lengths is 1. However, you can also use certain combinations of congruent, corresponding parts as shortcuts to determine if triangles are congruent. These combinations, called **triangle congruence conjectures**, are:

SSS ≅ (Pronounced "side–side–side") If all three pairs of corresponding sides have equal lengths, then the triangles are congruent.

SAS ≅ (Pronounced "side–angle–side") If two pairs of corresponding sides have equal lengths *and* the angles between them (the included angle) are equal, then the triangles are congruent.

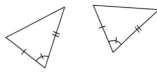

ASA ≅ (Pronounced "angle–side–angle") If two angles and the side between them in a triangle are congruent to the corresponding angles and side in another triangle, then the triangles are congruent.

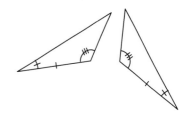

AAS ≅ (Pronounced "angle–angle–side") If two pairs of corresponding angles *and* a pair of corresponding sides that are not between them have equal measures, then the triangles are congruent.

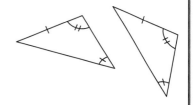

HL ≅ (Pronounced "hypotenuse–leg") If the hypotenuse and a leg of one right triangle have the same lengths as the hypotenuse and a leg of another right triangle, the triangles are congruent.

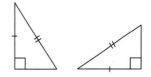

6-24. Use your conjectures about parallel lines and the angles formed with a third line to find the measures of the labeled angles below. Show each step you use and be sure to **justify** each with an angle conjecture from your Angle Relationships Toolkit. Be sure to write down your **reasoning** in the order that you find the angles.

a.

b.
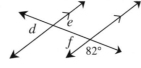

c.
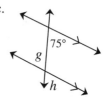

6-25. In a basketball game, a player who gets fouled is awarded a free-throw called a "one-and-one." When this happens, the player gets a "free shot" from the free-throw line. If the player makes the free-throw, then he or she gets one point and gets to attempt another one-point shot. If the player misses the first free-throw, then no additional attempt is provided.

a. If Vicki makes a free-throw 60% of the time, what is the probability that she will score no points? How did you get your answer?

b. Is there a greater probability that she will make only one free-throw or two? Create a diagram or chart to show your **reasoning**.

6-26. Solve the systems of equations below using any method, if possible. Show all work and check your solution. If there is no solution, explain why not.

a. $y = 2x + 8$
$3x + 2y = -12$

b. $2x - 5y = 4$
$5y - 2x = 10$

6-27. **Examine** the triangles at right. Solve for x.

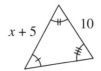

6-28.   **Examine** the triangle and line of reflection at right.

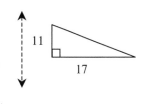

a.   On your paper, trace the triangle and the line of reflection. Then draw the image of the triangle after it is reflected across the line. Verify your reflection is correct by folding your paper along the line of reflection.

b.   Find the perimeter of the image. How does it compare with the perimeter of the original figure?

c.   Find the measure of both acute angles in the original triangle.

6-29.   **Multiple Choice:** Given the diagram at right, which of the statements below is not necessarily true?

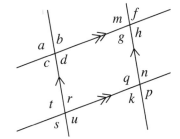

a. $a = d$ b. $d + r = 180°$

c. $u = n$ d. $t = m$

e.   These all must be true.

6.1.4 What's the relationship?

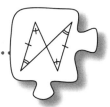

· ·

Converses

So far in this chapter, you have completed several problems in which you were given certain information and had to determine whether triangles were congruent. But what if you already know triangles are congruent? What information can you conclude then? Thinking this way requires you to take your Triangle Congruence Conjectures and reverse them. Today you will look more generally at what happens when you reverse a conjecture.

6-30.   Jorge is working with the diagram at right, and concludes that $\overleftrightarrow{AB} \parallel \overleftrightarrow{CD}$. He writes the following conditional statement to justify his reasoning:

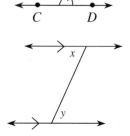

If alternate interior angles are equal, then lines are parallel.

a.   Margaret is working with a different diagram, shown at right. She concludes that $x = y$. Write a conditional statement or arrow diagram that justifies her reasoning.

b.   How are Jorge's and Margaret's statements related? How are they different?

c.   Conditional statements that have this relationship are called **converses**. Write the converse of the conditional statement below.

If lines are parallel, then corresponding angles are equal.

6-31.   In problem 6-30, you learned that each conditional statement has a converse. Are all converses true? Consider the arrow diagram of a familiar conjecture below.

Triangles congruent ➔ *corresponding sides are congruent.*

a.   Is this arrow diagram true?

b.   Write the converse of this arrow diagram as an arrow diagram or as a conditional statement. Is this converse true? **Justify** your answer.

c.   Now consider another true congruence conjecture below. Write its converse and decide if it is true.

Triangles congruent ➔ *corresponding angles are congruent.*

d.   Write the converse of the arrow diagram below. Is this converse true? **Justify** your answer.

A shape is a rectangle ➔ *the area of the shape is $b \cdot h$.*

Geometry Connections

6-32.   CRAZY CONVERSES

For each of these problems below, make up a conditional statement or arrow diagram that meets the stated conditions. You must use a different example each time, and none of your examples can be about math!

a.   A true statement whose converse is true.

b.   A true statement whose converse is false.

c.   A false statement whose converse is true.

d.   A false statement whose converse is false.

6-33.   INFORMATION OVERLOAD

Raj is solving a problem about three triangles. He is trying to find the measure of $\angle H$ and the length of $\overline{HI}$. Raj summarizes the relationships he has found so far in the diagrams below:

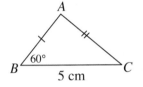 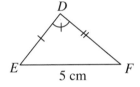 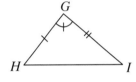

a.   Help Raj out! Assuming everything marked in the diagram is true, find $m\angle H$ and the length of $\overline{HI}$. Make sure to **justify** all your claims—don't make assumptions based on how the diagram looks

b.   Raj is still confused. Write a careful explanation of the **reasoning** you used to find the values in part (a). Whenever possible, use arrow diagrams or conditional statements in explaining your **reasoning**.

6-34.   Write an entry in your Learning Log about the converse relationship. Explain what a converse is, and give an example of a conditional statement and its converse. Also discuss the relationship between the truth of a statement and its converse. Title this entry "Converses" and include today's date.

Methods and Meanings

Converses

When conditional statements (also called "If ..., then ..." statements) are written backwards so that the condition (the "if" part) is switched with the conclusion (the "then" part), the new statement is called a **converse**. For example, examine the conjecture and its converse below:

Conjecture: If two parallel lines are cut by a transversal, then pairs of corresponding angles are equal.

Reversal: If two corresponding angles are equal, then the lines cut by the transversal are parallel.

The second statement (reversal) is called the **converse** of the first statement. Note that just because a conjecture is true does not mean that its converse must be true. For example, if the conditional statement, "If the dog has a meaty bone, then the dog is happy," is true, but its converse, "If the dog is happy, then the dog has a meaty bone," is not necessarily true. The dog could be happy for other reasons, such as going for a walk.

6-35. Copy the diagram at right onto your paper. If $a = 53°$ and $g = 125°$, find the measures of each labeled angle. Explain how you find each angle, citing definitions and conjectures from your toolkit that support your steps. Remember that you can find the angles in any order, depending on the angle relationships you use.

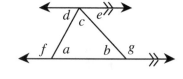

6-36. On graph paper, graph the line $y = \frac{-3}{2}x + 6$. Name the x- and y-intercepts.

304   *Geometry Connections*

6-37.   As Samone looked at the triangles at right, she said, "I think these triangles are congruent." Her teammate, Darla, said, "But they don't look the same. How can you tell?" Samone smiled and said, "Never trust the picture! Look at the angles and the sides. The measures are all the same."

a.   Solve for the missing side of each triangle. How do they compare?

b.   Are you convinced that Samone is correct? Explain.

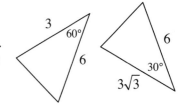

6-38.   Write a converse for each conditional statement below. Then, assuming the original statement is true, decide if the converse must be true or not.

a.   If it rains, then the ground is wet.

b.   If a polygon is a square, then it is a rectangle.

c.   If a polygon is a rectangle, then it has four 90° angles.

d.   If the shape has three angles, it is a triangle.

e.   If two lines intersect, then vertical angles are congruent.

6-39.   In the diagram at right, $\triangle ABC \sim \triangle ADE$.

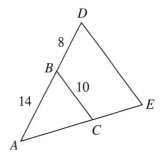

a.   Draw each triangle separately on your paper. Be sure to include all measurements in your diagrams.

b.   Find the length of $\overline{DE}$.

6-40.

One measurement used in judging kite-flying competitions is the size of the angle formed by the kite string and the ground. This angle can be used to find the height of the kite. Suppose the length of the string is 600 feet and the angle at which the kite is flying measures 40°. Calculate the height, h, of the kite.

6.2.1 How can I use it? What's the connection?

Angles on a Pool Table

The activities in this section review several big topics you have studied so far. Work with your team to decide which combination of **tools** you will need for each problem. As you work together, think about which skills and tools you are comfortable using and which ones you need more practice with.

As you work on this activity, keep in mind the following questions:

What mathematical concepts did you use to solve this problem?

What do you still want to know more about?

What connections did you find?

6-41. TAKE IT TO THE BANK

Ricky just watched his favorite pool player, "Montana Mike," make a double bank shot in a trick-shot competition. Montana bounced a ball off two rails (sides) of the table and sank it in the corner pocket. "That doesn't look too hard," Ricky says, "I just need to know where to put the ball and which direction to hit it in."

A diagram of Montana's shot is shown at right. The playing area of a tournament pool table is 50 inches by 100 inches. Along its rails, a pool table is marked with a diamond every 12.5 inches. Montana started the shot with the ball against the top rail and the ball hit the bottom rail three diamonds from the right rail.

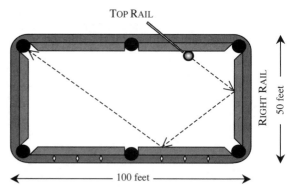

TOP RAIL

RIGHT RAIL

50 feet

100 feet

Your task: Figure out where on the top rail Ricky needs to place his ball and where he needs to aim to repeat Montana Mike's bank shot. Write instructions that tell Ricky how to use the diamonds on the table to place his ball correctly, and at what angle from the rail to hit the ball.

6-42. **EXTENSIONS**

The algebraic and geometric **tools** you have developed so far will enable you to answer many questions about the path of a ball on a pool table. Work with your team to analyze the situations below.

a. Ricky decided he wants to alter Montana's shot so that it hits the right rail exactly in at its midpoint. Where would Ricky need to place the ball along the top rail so that his shot reflects off the right rail, then the bottom rail, and enters the upper left pocket? At what angle with the top rail would he need to hit the ball?

b. During another shot, Ricky noticed that Montana hit the ball as shown in the diagram at right. He estimated that the ball traveled 18 inches before it entered the pocket. Before the shot, the announcers noted that the distance of the ball to the top rail was 2 inches more than the distance along the top rail, as shown in the diagram. Where was the ball located before Montana hit it?

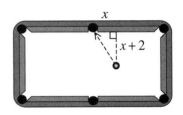

c. Ricky wants to predict how Montana's next shot will end. The ball is placed at the second diamond from the left along the bottom rail, as shown at right. Montana is aiming to hit the ball toward the second diamond from the right along the top rail. Assuming he hits the ball very hard so that the ball continues traveling indefinitely, will the ball ever reach a pocket? If so, show the path of the ball. If not, explain how you know.

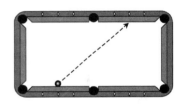

d. After analyzing the path in part (c), Montana decided to start his ball from the third diamond from the left along the bottom rail, as shown at right. He is planning to aim at the same diamond as he did in part (c). If he hits the ball sufficiently hard, will the ball eventually reach a pocket? If so, show the path of the ball. If not, explain how you know.

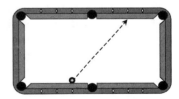

6-43.   Use your triangle **tools** to solve for x in the triangles below.

a.

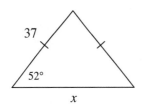

b.

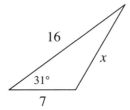

c.

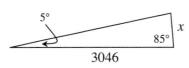

d.

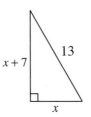

6-44.   Penelope measured several sides and heights of
$\triangle ABC$, as shown in the diagram at right. Find
the area of $\triangle ABC$ <u>twice</u>, using two different
methods.

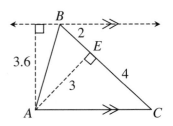

6-45.   The shaded figures at right are similar.

a.   Solve for m and n.

b.   Find the area and
perimeter of each
figure.

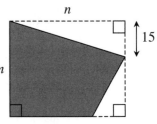

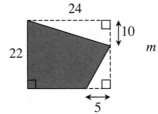

6-46.   Decide if each triangle below is congruent to $\triangle ABC$ at
right, similar but not congruent to $\triangle ABC$, or neither.
Justify each answer.

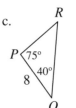

a.

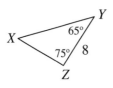

b.

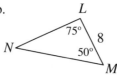

c.

6-47.   On graph paper, graph the line $y = 3x + 1$.

   a.   What is this line's slope angle? That is, what is the acute angle the line makes with the *x*-axis?

   b.   Find the equation of a new line that has a slope angle of 45° and passes through the point (0, 3). Assume that the slope is positive.

   c.   Find the intersection of these two lines using any method. Write your solution as a point in the form (*x*, *y*).

6-48.   **Multiple Choice:** Listed below are the measures of several different angles. Which angle is obtuse?

   a.   0°   b.   52°   c.   210°   d.   91°

6.2.2   How can I use it? What's the connection?

Investigating a Triangle

The activities in this section review several big topics you have studied so far. Work with your team to decide which combination of **tools** you will need for each problem. As you work together, think about which skills and tools you are comfortable using and which ones you need more practice with.

As you work on this activity, keep in mind the following questions:

What mathematical concepts did you use to solve this problem?

What do you still want to know more about?

What connections did you find?

6-49.   GETTING TO KNOW YOUR TRIANGLE

   a.   If you were asked to give every possible measurement of a triangle, what measurements could you include?

   b.   Consider a triangle on a coordinate grid with the following vertices:

   $A(2, 3)$, $B(32, 15)$, $C(12, 27)$

   Your Task: On graph paper, graph $\triangle ABC$ and find all of its measurements. Be sure to find every measurement you listed in part (a) of this problem and show all of your calculations. With your team, be prepared to present your method for finding the area.

6-50.   **Examine** the diagram at right. Based on the information
         below, what statement can you make about the relationships
         between the lines? Be sure to **justify** each conclusion.
         Remember: each part below is a separate problem.

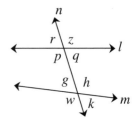

   a. $p = h$ b. $w = k$

   c. $r = q$ d. $z + k = 160°$

6-51.   Determine if each pair of triangles below are congruent, similar but not congruent, or
         if they are neither. If they are congruent, organize your **reasoning** into a flowchart.

   a.   b.   c.

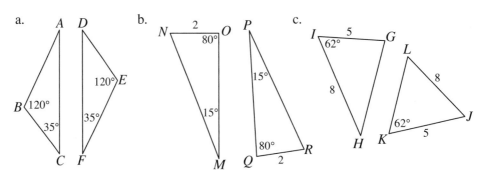

6-52.   On graph paper, plot and connect the points to form quadrilateral $WXYZ$ if $W(3, 7)$,
         $X(3, 4)$, $Y(9, 1)$, and $Z(5, 6)$.

   a.   What is the shape of quadrilateral $WXYZ$? **Justify** your conclusion.

   b.   Find the perimeter of quadrilateral $WXYZ$.

   c.   If quadrilateral $WXYZ$ is reflected across the y-axis to form quadrilateral
         $W'X'Y'Z'$, then where is Y' ?

   d.   Rotate quadrilateral $WXYZ$ about the origin 90° clockwise (↻) to form
         quadrilateral $W''X''Y''Z''$. What is the slope of $\overline{W''Z''}$?

6-53.   As Ms. Dorman looked from the window of her third-story classroom, she noticed
         Pam in the courtyard. Ms. Dorman's eyes were 52 feet above ground and Pam was
         38 feet from the building. Draw a diagram of this situation. What is the angle at
         which Ms. Dorman had to look down? That is, what is the angle of depression?
         (Assume that Ms. Dorman was looking at the spot on the ground below Pam.)

6-54.   **Examine** the diagram at right. Find two equivalent
         expressions that represent the area of the <u>inner</u> square.

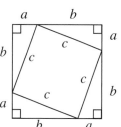

6-55.   **Multiple Choice:** To decide if his class would take a
         quiz today, he will flip a coin three times. If all three
         results are heads or all three results are tails, he will
         give the quiz. Otherwise, his students will not be
         tested. What is the probability that his class will take the quiz?

   a. $\frac{1}{8}$ b. $\frac{1}{4}$ c. $\frac{1}{2}$ d.   1

6.2.3 How can I use it? What's the connection?

Creating a Mathematical Model

The activities in this section review several big topics you have studied so far. Work with your
team to decide which combination of **tools** you will need for each problem. As you work
together, think about which skills and tools you are comfortable using and which ones you need
more practice with.

As you work on this activity, keep in mind the following questions:

What mathematical concepts did you use to solve this problem?

What do you still want to know more about?

What connections did you find?

6-56.   AT YOUR SERVICE

Carina, a tennis player, wants to make her serve a truly powerful part of her game. She wants to hit the ball so hard that it appears to travel in a straight path. Unfortunately, the ball always lands beyond the service box. After a few practice serves, she realizes that the height at which you hit the ball determines where the ball lands. Before she gets tired from serving, she sits down to figure out how high the ball must be when she hits it so that her serve is legal.

In the game of tennis, every point begins with one player serving the ball. For a serve to be legal, the player must stand outside the court and hit the ball so that it crosses over the net and lands within the service box (shown shaded below). It can be difficult to make the ball land in the service box because the ball is often hit too low and touches the net or is hit too high and lands beyond the service box.

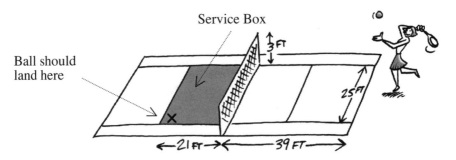

A tennis court is 78 feet long with the net located at the center. The distance from the net to the back of the service box is 21 feet, and the net is 3 feet tall.

Your task: Assuming Carina can hit the ball so hard that its path is linear, from what height must she hit the ball to have the serve just clear the net and land in the service box? Decide whether or not it is reasonable for Carina to reach this height if she is 5'7" tall. Also, at what angle does the ball hit the ground?

Your solution should include:

- a labeled diagram that shows a birds' eye view of the path of the ball.

- a labeled diagram that shows the side view of Carina, the ideal height of the tennis racket, the ideal path of the tennis ball, and the measurements that are needed from the birds' eye view diagram.

Discussion Points

What would you see if you were a bird looking down
on the court as Carina served?

Which distances do you know and which do you need to find?

Further Guidance

6-57.   To help solve this problem, copy the diagram at right, which shows the tennis court from above (called a "birds-eye" view). On your diagram, locate the position of Carina and the spot where the ball should land. Be sure to include the path of the ball.

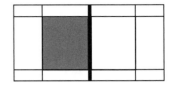

a.   Label all distances you already know. Do you see any triangles?

b.   What distance(s) can you find? What geometric **tool(s)** can you use?

6-58.   Next **visualize** the situation from the side and draw the path of the ball.

a.   As you draw this diagram, be sure to include Carina, the net, and the spot where the ball should land.

b.   Are there any similar triangles? Be careful to include any measurements that might help you determine the height of the serve.

c.   What triangle **tool** can you use to find the angle of depression of the path of the ball? Find the acute angle the path of the ball makes with the ground.

6-59.   How high must the ball be hit so that it just clears the net and lands in the service box? If Carina is 5'7" tall, is it possible for her to accomplish this serve? Assume that a tennis racquet is 2 feet long.

Further Guidance
section ends here.

6-60.   You've been hired as a consultant for the National Tennis Association. They are considering raising the net to make the serve even more challenging. They want players to have to jump to make a successful serve (assuming that the powerful serves will be hit so hard that the ball will travel in a straight line). Determine how high the net should be so that the players must strike the ball from at least 10 feet.

6-61.   **Examine** the triangles in the diagram at right.

a.   Are the triangles similar? **Justify** your conclusion.

b.   Solve for x. Show all work.

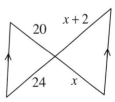

6-62.   For each diagram below, use geometric relationships
         to solve for the given variable(s). Check your answer.

   a.   b.
                            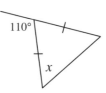

   c.   d.
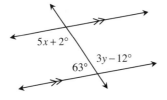   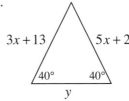

6-63.   Find the area and perimeter of the trapezoid at right.

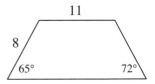

6-64.   A map of an island is shown at right. Each unit of length
         on the grid represents 32 feet.

   a.   Find the actual dimensions of the island (the overall
         width and length).

   b.   Find the area of the shape at right and the actual area
         of the island.

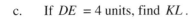

6-65.   **Examine** the two triangles at right.

   a.   Are the triangles congruent? **Justify** your
         conclusion. If they are congruent, complete
         the congruence statement $\triangle DEF \cong$ _____.

   b.   What transformation(s) are needed to change
         $\triangle DEF$ to $\triangle JKL$?

   c.   If $DE = 4$ units, find KL.

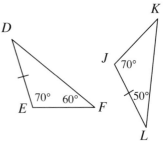

6-66.   **Multiple Choice:** The length of a rectangle is three units shorter than twice its
         width. Which expression below could represent the area of the rectangle?

   a. $2x^2 - 3$ b. $2x^2 - 6x$ c. $2x^2 - 3x$ d. $(2x - 3)^2$

6.2.4 How can I use it? What's the connection?

Analyzing a Game

The activities in this section review several big topics you have studied so far. Work with your team to decide which combination of **tools** you will need for each problem. As you work together, think about which skills and tools you are comfortable using and which ones you need more practice with.

As you work on this activity, keep in mind the following questions:

> What mathematical concepts did you use to solve this problem?
>
> What do you still want to know more about?
>
> What connections did you find?

6-67. THE MONTY HALL PROBLEM

Wow! Your best friend, Lee, has been selected as a contestant in the popular "Pick-A-Door" game show. The game show host, Monty, has shown Lee three doors and, because he knows what is behind each door, has assured her that behind one of the doors lies a new car! However, behind each of the other two doors is a goat.

"Which door do you pick?" Monty asks.

Lee's original choice

"I pick Door #1," Lee replies confidently.

"Okay. Now, before I show you what is behind Door #1, let me show you what is behind Door #3. It is a goat! Now, would you like to change your mind and choose Door #2 instead?" Monty asks.

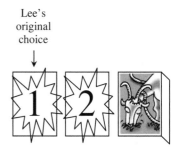

What should Lee do? Should she stay with Door #1 or should she switch to Door #2? Does she have a better chance of winning if she switches, or does it not matter? Discuss this situation with the class and make sure you provide **reasons** for your statements.

6-68.   Now test your prediction from problem 6-67 by simulating this game
         with a partner using either a computer or a programmable calculator.
         If no technology is available, collect data by playing the game with a
         partner as described below.

         Choose one person to be the contestant and one person to be the game show host.
         As you play, carefully record information about whether the contestant switches
         doors and whether the contestant wins. Play as many times as you can in the time
         allotted, but be sure to record at least 10 results from switching and 10 results from
         not switching. Be ready to report your findings with the class.

         If playing this game without technology, the host should:

         •   Secretly choose the winning door. Make sure that the contestant has no way
             of knowing which door has been selected.

         •   Ask the contestant to choose a door.

         •   "Open" one of the remaining two doors that does not have the winning prize.

         •   Ask the contestant if he or she wants to change his or her door.

         •   Show if the contestant has won the car and record the results.

6-69.   **Examine** the data the class collected in problem
         6-68.

         a.   What does this data tell you? What should
              Lee do in problem 6-67 to maximize her
              chance of winning?

         b.   Your teammate, Kaye, is confused. "Why
              does it matter? At the end, there are only
              two doors left. Isn't there a 50-50 chance
              that I will select the winning door?" Explain
              to Kaye why switching is better.

         c.   Gerald asks, "What if there are 4 doors? If Monty now reveals two doors with
              a goat, is it still better to switch?" What do you think? Analyze this problem
              and answer Gerald's question.

6-70.   One of the topics you studied during Chapters 1 through 6 was
         probability. You investigated what made a game fair and how to
         predict if you would win or lose. Reflect on today's activity and
         write a Learning Log entry about the mathematics you used today
         to analyze the "Monty Hall" game. Title this entry "Game
         Analysis" and include today's date.

6-71.   Determine whether or not the triangles in each pair below are congruent. **Justify** your conclusion.

a.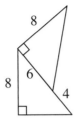

b.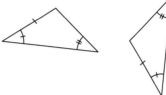

c.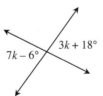

d.

6-72.   **Examine** the geometric relationships in each of the diagrams below. For each one, write and solve an equation to find the value of the variable. Name all geometric relationships or conjectures that you use.

a.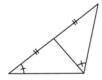

b.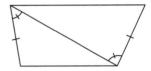

c.

d.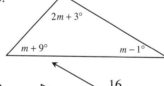
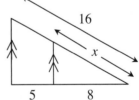

6-73.   In the diagram at right, $\overrightarrow{BD}$ **bisects** $\angle ABC$. This means that $\overrightarrow{BD}$ divides the angle into two equal parts. If $m\angle ABD = 3x + 24°$ and if $m\angle CBD = 5x + 2°$, solve for x. Then find $m\angle ABC$.

Chapter 6: Congruent Triangles

6-74.   Assume that 25% of the student body at your school is male and that 40% of the students walk to school. If a student from this school is selected at random, find the following probabilities.

   a.   P(student is female)

   b.   P(student is male and does not walk to school)

   c.   P(student walks to school or does not walk to school)

6-75.   If $b + 2a = c$ and $2a + b = 10$, then what is c? Explain how you know.

6-76.   **Multiple Choice:** The measure of $\angle ABD$ at right is:

   a.   27°   b.   161°

   c.   118°   d.   153°

   e.   None of these

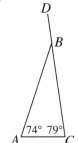

$6.2.5$ How can I use it? What's the connection?

Using Transformations and Symmetry to Create a Snowflake

The activities in this section review several big topics you have studied so far. Work with your team to decide which combination of **tools** you will need for each problem. As you work together, think about which skills and tools you are comfortable using and which ones you need more practice with.

As you work on this activity, keep in mind the following questions:

What mathematical concepts did you use to solve this problem?

What do you still want to know more about?

What connections did you find?

6-77.   THE PAPER SNOWFLAKE

You have volunteered to help the decorating committee
make paper snowflakes for the upcoming winter school
dance. A paper snowflake is made by folding and
cutting a square of paper in such a way that when the
paper is unfolded, the result is a beautiful design with
symmetric patterns similar to those of a real snowflake.

Looking through your drawer of craft projects, you find the directions for how to
fold the paper snowflake (see below). However you cannot find any directions for
how to cut the folded paper to make specific designs in the final snowflake.

DESIGNING A PAPER SNOWFLAKE: Cut out a 20 unit by 20 unit square
from a piece of graph paper. Making sure that the gridlines are visible on the
outside of the shape, fold the square three times as shown in the diagram
below. Your final shape should be a 45°- 45°- 90° triangle that has folds
along the hypotenuse.

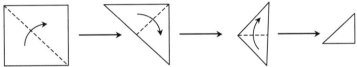

Once you have folded your paper, correctly label the
sides of your folded triangle "hypotenuse," "folded
leg," and "open leg" (the leg comprised of your
original square's edges) after orienting your triangle as
shown in the diagram at right.

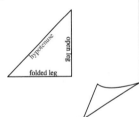

Your Task: You want to be fully prepared to help the decorating committee for the
school dance. Explore and be ready to explain the relationships between the shapes
that are cut out and the design that appears after unfolding the paper. For each
possible location of a cutout, use what you know about symmetry and
transformations to describe the shapes that result when you unfold the paper.

Discussion Points

What are your goals for this task?

What **tools** would be useful to complete this task?

Visualize the result when a shape is cut along the hypotenuse.
What qualities will the result have?

6-78.   Since there are so many cuts that are possible, it is helpful to begin by considering a simple cutout along one side of the triangle.

    a.   Sketch triangles onto your folded paper according to these directions, making sure that none of the shapes share a side or overlap:

- A 45°-45°-90° triangle with its hypotenuse along the paper triangle's hypotenuse. Each leg should be 3 units long.

- A 45°-45°-90° triangle with a 3-unit leg along the paper triangle's folded leg.

- A 45°-45°-90° triangle with a 3-unit leg along the paper triangle's open leg.

    b.   Cut out the shapes you sketched from part (a). Then unfold the paper to view your snowflake. Identify the three different kinds of shapes that resulted from the triangles you cut. Describe them with as much specific vocabulary as you can.

    c.   Find each pair of shapes shown below on your own snowflake. For each pair, describe <u>two different ways</u> to transform one shape into the other.

i.

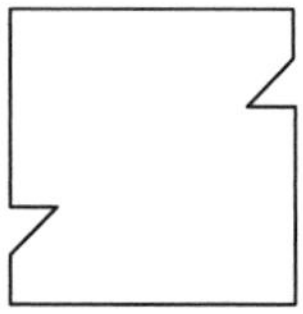

ii.

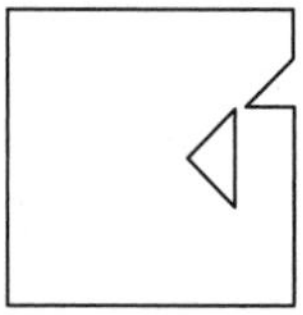

iii.

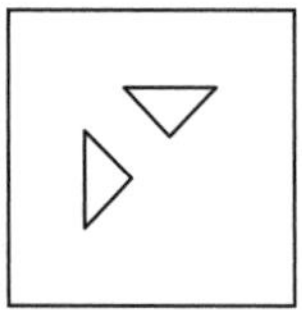

iv.

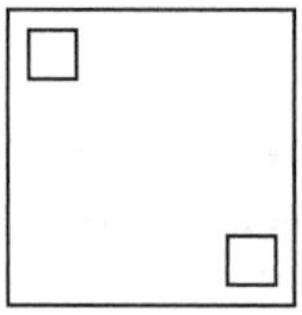

*Further Guidance
section ends here.*

Geometry Connections

6-79.   Some possible folded triangles with cutouts are shown below. What would each of these snowflakes look like when unfolded? Draw the resulting designs on the squares provided on the Activity 6.2.5 Resource Page (or draw the result on your paper.)

a.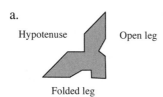

Hypotenuse   Open leg

Folded leg

b.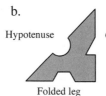

Hypotenuse   Open leg

Folded leg

6-80.   What shape must you cut out along your folded triangle hypotenuse to get the following shapes on your snowflake when your paper is unfolded? Sketch and label diagrams to show that you can accomplish each result.

a.   Rectangle with a length that is twice its width

b.   Kite

c.   Rhombus

d.   Pentagon

6-81.   EXTENSION

Get a final piece of grid paper from your teacher. Make one more snowflake that includes at least four of the following shapes. Make sure you sketch all of your planned cuts before cutting the paper with scissors. After you cut out your shapes and unfold your snowflake, answer questions (a) through (d) below.

- Draw a shape at the vertex where the hypotenuse and folded leg intersect that results in a shape with 8 lines of symmetry. Note that you will actually have to cut the vertex off to do this.

- Many letters, such as E, H, and I have reflection symmetry. Pick a letter (maybe one of your initials) and draw a shape along the folded triangle's hypotenuse or folded leg that results in a box letter when cut out and the snowflake is unfolded.

- Draw a shape along the folded triangle's hypotenuse or folded leg that results in a regular hexagon when cut out and the snowflake is unfolded.

- Draw a shape along the folded triangle's hypotenuse or folded leg that results in a star when cut out and the snowflake is unfolded.

- Draw curved shapes along the open leg so that there is at least one line of symmetry that is perpendicular to the open leg.

Problem continues on next page →

6-81.   *Problem continued from previous page.*

    a.   What has to be true about the shape you cut out in order for your unfolded shape to have 8 lines of symmetry?

    b.   There are two different shapes you could have drawn that would result in a regular hexagon. Sketch both shapes. Why do both shapes work?

    c.   Some shapes are impossible to create by cutting along the hypotenuse of the folded paper triangle. Sketch several different examples of these shapes. What is the common characteristic of these impossible shapes?

    d.   How is cutting along the open leg different from cutting along the folded leg or hypotenuse? Try to describe the difference in terms of symmetry. Use a diagram to help make your description clear.

6-82.   **Examine** the angles formed by parallel lines at right.

    a.   If $r = 5x + 3°$ and $k = 4x + 9°$, solve for x. **Justify** your answer.

    b.   If $c = 114°$, what is q? **Justify** your answer.

    c.   If $g = 88°$, then what is q? **Justify** your answer.

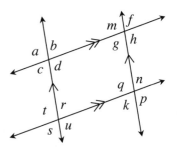

6-83.   **Examine** the triangles below. For each, solve for x and name which **tool** you use. Show all work.

    a.

    b.

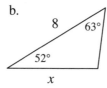

    c.

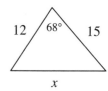

6-84.　For each part below, decide if the triangles are similar. If they are similar, use their similarity to solve for *x*. If they are not similar, explain why not.

a.

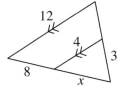

b.

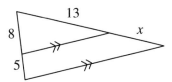

c.

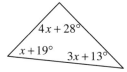

d.

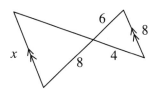

6-85.　For each shape below, use the geometric relationships to solve for the given variable(s). Show all work. Name the geometric relationships you used.

a.

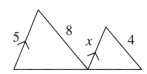

b.

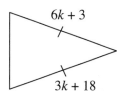

c.

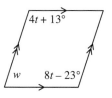

d.

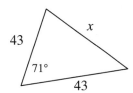

6-86.　**Multiple Choice:** What is the distance between the points (–2, –5) and (6, 3)?

a.　8　　　　　b.　$8\sqrt{2}$　　　　　c.　16　　　　　d.　64

Chapter 6 Closure   What have I learned?

Reflection and Synthesis

The activities below offer you a chance to reflect on what you have learned in this chapter. As you work, look for concepts that you feel very comfortable with, ideas that you would like to learn more about, and topics you need more help with. Look for connections between ideas as well as connections with material you learned previously.

①   TEAM BRAINSTORM

       With your team, brainstorm a list for each of the following two subjects. Be as detailed as you can. How long can you make your list? Challenge yourselves. Be prepared to share your team's ideas with the class.

Topics: What have you studied in this chapter? What ideas and words were important in what you learned? Remember to be as detailed as you can.

Connections: How are the topics, ideas, and words that you learned in previous courses connected to the new ideas in this chapter? Again, make your list as long as you can.

MAKING CONNECTIONS

The following is a list of the vocabulary used in this chapter. The words that appear in bold are new to this chapter. Make sure that you are familiar with all of these words and know what they mean. Refer to the glossary or index for any words that you do not yet understand.

AAS ≅	arrow diagrams	**ASA** ≅
conditional statement	congruent	conjecture
converse	corresponding	**diagonal**
flowchart	**HL** ≅	**SAS** ≅
similar	**SSS** ≅	

Make a concept map showing all of the connections you can find among the key words and ideas listed above. To show a connection between two words, draw a line between them and explain the connection, as shown in the example below. A word can be connected to any other word as long as there is a justified connection. For each key word or idea, provide a sketch of an example.

Your teacher may provide you with vocabulary cards to help you get started. If you use the cards to plan your concept map, be sure to either re-draw your concept map on your paper or glue the vocabulary cards to a poster with all of the connections explained for others to see and understand.

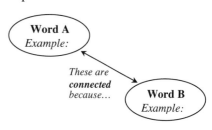

While you are making your map, your team may think of related words or ideas that are not listed above. Be sure to include these ideas on your concept map.

③ SUMMARIZING MY UNDERSTANDING

This section gives you an opportunity to show what you know about certain math topics or ideas. Your teacher will direct you how to do this.

④ WHAT HAVE I LEARNED?

This section will help you recognize those types of problems you feel comfortable with and those you need more help with. This section will appear at the end of every chapter to help you check your understanding. Even if your teacher does not assign this section, it is a good idea to try these problems and find out for yourself what you know and what you need to work on.

Solve each problem as completely as you can. The table at the end of this closure section has answers to these problems. It also tells you where you can find additional help and practice on similar problems.

CL 6-87. Write the converse of each statement and then determine whether or not the converse is true.

    a.   If two lines are parallel, then corresponding angles are equal.

    b.   In $\triangle ABC$, if the sum of $m\angle A$ and $m\angle B$ is $110°$, then $m\angle C = 70°$.

    c.   If alternate interior angles k and s are not equal, then the two lines cut by the transversal are not parallel.

    d.   If Johan throws coins in the fountain, then he loses his money.

CL 6-88. Determine whether or not the two triangles in each pair are congruent. If they are congruent, show your **reasoning** in a flowchart. If the triangles are not congruent or you cannot determine that they are, **justify** your conclusion.

    a.

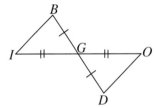

    b.

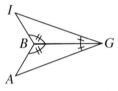

    c.

    d.

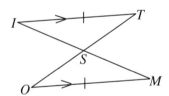

CL 6-89. For each part, determine which lines, if any, are parallel. Be sure to **justify** your decisions.

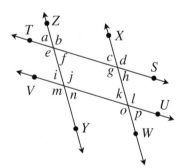

    a. $\angle e \cong \angle m$

    b. $\angle c \cong \angle o$

    c. $\angle d \cong \angle o$

    d. $\angle a \cong \angle m \cong \angle o$

    e. $\angle a \cong \angle k$

    f. $\angle k \cong \angle c \cong \angle f$

CL 6-90. Determine whether or not the two triangles in each pair below are similar. If so, write a flowchart to show your **reasoning**. If not, explain why not.

a.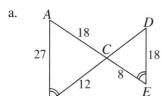

b.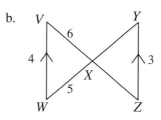

CL 6-91. For each diagram below, solve for the variable.

a.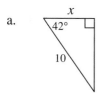

b.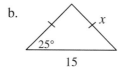

c.

CL 6-92. For each diagram, solve for the variable. Be sure to include the names of any relationships you used to get your solution.

a. b. c.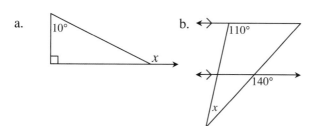

CL 6-93. Yee Ping thought about swimming across Redleaf Lake. She knows that she can swim about 1000 meters. She decided that she would feel more confident if she knew how far she would have to swim. To determine the length of the lake, she put posts at both ends of the lake (points A and B) and a third post on one side of the lake (point C). The distances between the posts are shown in the diagram at right. She measured the angle between the two posts and found that it was 150°. Use this information to determine the length of the lake. Do you think that Yee Ping will be able to swim between points A and B?

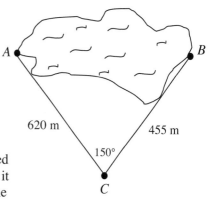

CL 6-94. Margarite has 9 pieces of copper pipe with which she plans to make 3 triangular frames. She has organized them into groups of three based on their coloring. The lengths of the pipes in each group are listed below.

        i.   23, 21, 4   *ii.*   2, 11, 10   *iii.*   31, 34, 3

    a.   Which groups, if any, will she actually be able to use to make a triangular frame if she is unable to cut any of the pipes? How do you know?

    b.   If possible, arrange the 9 pieces she has so that she can make 3 triangular frames. If so, how? If not, why not?

CL 6-95. At a story-telling class, Barbara took notes on the following story. However, all the parts of the story got mixed up. Help her make sense of the story by organizing the following details in a flowchart.

    a.   Maggie was happy she could play on the same team as Julie and Cheryl.

    b.   Julie was hoping to make the A team again this year as she grabbed her basketball and got on a bus in Bellingham.

    c.   Cheryl, having been named most valuable player in Port Townsend, started the drive to the statewide basketball camp.

    d.   Because of her skill in the first game, Maggie moved up to the A team.

    e.   At camp, Julie and Cheryl were placed on the A team.

    f.   Julie, Maggie and Cheryl met at a statewide basketball camp. Shortly after they met, they were placed on teams.

    g.   This year was Maggie's first year at camp, and was placed on the B team.

    h.   On the train to camp, Maggie thought about how surprising it was that her basketball coach chose her to attend camp.

CL 6-96. Check your answers using the table at the end of this section. Which problems do you feel confident about? Which problems were hard? Have you worked on problems like these in math classes you have taken before? Use the table to make a list of topics you need help on and a list of topics you need to practice more.

This course focuses on five different **Ways of Thinking**: investigating, examining, reasoning and justifying, visualizing, and choosing a strategy/tool. These are some of the ways in which you think while trying to make sense of a concept or attempting to solve a problem (even outside of math class). During this chapter, you have probably used each Way of Thinking multiple times without even realizing it!

Review each of the Ways of Thinking described in the closure sections of Chapters 1 through 5. Then choose three of these Ways of Thinking that you remember using while working in this chapter. Show and explain where and how you used each one. Describe why thinking in this way helped you solve a particular problem or understand something new. Be sure to include examples to demonstrate your thinking.

Answers and Support for Closure Activity #4
What Have I Learned?

Problem	Solutions	Need Help?	More Practice
CL 6-87.	a. True; If corresponding angles are equal, then the two lines are parallel	Lesson 6.1.4 Math Notes box	Problems 6-30, 6-31, 6-32, 6-38
	b. True; In triangle ABC, if angle C is 70°, then the sum of angles A and B is 110°.		
	c. True; If non-parallel lines form alternate interior angles, then those angles are not equal.		
	d. False; If Johan loses money, then he has thrown coins in the fountain.		

Problem	Solutions	Need Help?	More Practice
CL 6-88.	a. Congruent: 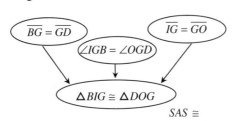	Lesson 6.1.3 Math Notes box	Problems 6-12, 6-21, 6-22, 6-37, 6-46, 6-51, 6-55, 6-65, 6-71
CL 6-88.	b. Congruent 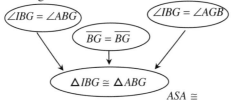	Lesson 6.1.3 Math Notes box	Problems 6-12, 6-21, 6-22, 6-37, 6-46, 6-51, 6-55, 6-65, 6-71

c. Not enough information.

d. Congruent:

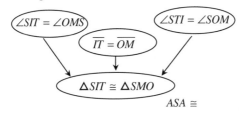

CL 6-89.	a. $\overline{ST}\,\|\,\overline{UV}$; corres. angles are equal.	Lessons 2.1.1, 2.1.4, and 2.2.1 Math Notes boxes, Problem 2-3	Problems 2-13, 2-16, 2-17, 2-23, 2-24, 2-25, 2-31, 2-32, 2-38, 2-49, 2-51, 2-62, 2-72, 2-111, 3-20, 3-31, 3-60, 3-70, 3-96, 4-6, 4-29, 4-44, 4-85, 4-93, 5-47, 6-30, 6-50
	b. Not enough information.		
	c. $\overline{ST}\,\|\,\overline{UV}$; alt. interior angles are equal.		
	d. $\overline{ZY}\,\|\,\overline{XW}$; corres. angles are equal.		
	e. Not enough information.		
	f. $\overline{ZY}\,\|\,\overline{XW}$; $\overline{ST}\,\|\,\overline{UV}$; corres. angles and alternate interior angles are equal.		

Problem	Solutions	Need Help?	More Practice
CL 6-90.	a. Similar: $\angle ABC = \angle DEC$ $\angle BCA = \angle ECD$ $\triangle ABC \sim \triangle DEC$ $AA \sim$ b. Similar; use of any 2 angles works. Ex: $\angle VWX = \angle ZYX$ $\angle WVX = \angle YZX$ $\triangle WXV \sim \triangle YXZ$ $AA \sim$	Lessons 2.1.4, 3.1.4, 3.2.1, 3.2.4, and 3.2.5 Math Notes boxes	Problems 3-64, 3-80, 3-83, 3-84, 3-86, 3-89, 4-7, 4-17, 4-38, 4-39, 4-83, 5-20, 5-49, 5-50, 5-57, 5-76, 5-86, 5-91, 6-5, 6-22, 6-46, 6-51, 6-61, 6-84
CL 6-91.	a. $x \approx 7.43$ b. $x \approx 8.28$ c. $\theta \approx 33.7°$	Lessons 4.1.4, 5.1.2, and 5.1.4 Math Notes boxes	Problems 4-12, 5-5, 5-7, 5-12, 5-16, 5-23, 5-89, 6-7, 6-14, 6-40, 6-43, 6-83
CL 6-92.	Reasons may vary. One possible set of reasons is listed for each: a. $x = 100°$ Triangle Angle Sum Theorem and supplementary angles b. $x = 30°$ Triangle Angle Sum Theorem, corresponding and supplementary angles c. $x = 14°$ corresponding angles	Lessons 2.1.1, 2.1.4, and 2.2.1 Math Notes boxes	Problems 2-17, 2-32, 2-38, 2-49, 2-51, 2-62, 2-72, 2-111, 3-20, 3-31, 3-60, 3-70, 3-96, 4-6, 4-29, 4-44, 4-85, 4-93, 5-47, 6-13, 6-24, 6-29, 6-35, 6-76, 6-82
CL 6-93.	The lake is about 1039 meters wide. This means Yee might have a problem.	Lesson 5.3.3 Math Notes box	Problems 5-100, 5-107, 5-109, 5-121, 6-14, 6-85

CL 6-94.	a. Sets *i* and *ii* work.
	b. It is possible
	    i. 4, 31, 34
	    ii. 3, 21, 23
	    iii. 2, 10, 11
		Lesson 2.3.2	Problems 2-107, 2-122, 3-60, 4-74, 6-18

Problem	Solutions	Need Help?	More Practice
CL 6-95.	One possible solution: 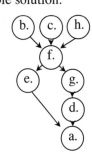	Lesson 3.2.4 Math Notes box	Problems 3-62, 3-56, 3-67, 3-87, 4-39, 5-49, 5-91, 6-8

CHAPTER 7   Proof and Quadrilaterals

This chapter opens with a set of explorations designed to introduce you to new geometric topics that you will explore further in Chapters 8 through 12. You will learn about the special properties of a circle, explore three-dimensional shapes, and use a hinged mirror to learn more about a rhombus.

Section 7.2 then builds upon your work from Chapters 3 through 6. Using congruent triangles, you will explore the relationships of the sides and diagonals of a parallelogram, kite, trapezoid, rectangle, and rhombus. As you explore new geometric properties, you will formalize your understanding of proof.

This chapter ends with an exploration of coordinate geometry.

In this chapter, you will learn:

➤ The relationships of the sides, angles, and diagonals of special quadrilaterals, such as parallelograms, rectangles, kites, and rhombi (plural of rhombus).

➤ How to write a convincing proof in a variety of formats, such as a flowchart or two-column proof.

➤ How to find the midpoint of a line segment.

➤ How to use algebraic tools to explore quadrilaterals on coordinate axes.

Guiding Questions

Think about these questions throughout this chapter:

What's the connection?

How can I prove it?

Is it convincing?

What tools can I use?

Chapter Outline

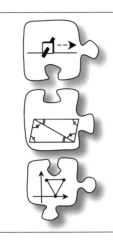

Section 7.1   This section contains four large **investigations** introducing you to geometry topics that will be explored further in Chapters 8 through 12.

Section 7.2   While **investigating** what congruent triangles can inform you about the sides, angles, and diagonals of a quadrilateral, you will develop an understanding of proof.

Section 7.3   This section begins a focus on coordinate geometry, the study of geometry on coordinate axes. During this section, you will use familiar algebraic tools (such as slope) to make and justify conclusions about shapes.

7.1.1 Does it roll smoothly?

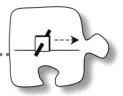

Properties of a Circle

In Chapters 1 through 6, you studied many different types of two-dimensional shapes, explored how they could be related, and developed tools to measure their lengths and areas. In Chapters 7 through 12, you will examine ways to extend these ideas to new shapes (such as polygons and circles) and will thoroughly **investigate** what we can learn about three-dimensional shapes.

To start, Section 7.1 contains four key **investigations** that will touch upon the big ideas of these chapters. As you explore these lessons, take note of what mathematical tools from Chapters 1 through 6 you are using and think about what new directions this course will take. Generate "What if…" questions that can be answered later once new tools are developed.

Since much of the focus of Chapters 7 through 12 is on the study of circles, this lesson will first explore the properties of a circle. What makes a circle special? Today you are going to answer that question and, at the same time, explore other shapes with surprisingly similar qualities.

7-1.   THE INVENTION OF THE WHEEL

One of the most important human inventions was the wheel. Many archeologists estimate that the wheel was probably first invented about 10,000 years ago in Asia. It was an important tool that enabled humans to transport very heavy objects long distances. Most people agree that impressive structures, such as the Egyptian pyramids, could not have been built without the help of wheels.

a.   One of the earliest types of "wheels" used were actually logs. Ancient civilizations laid multiple logs on the ground, parallel to each other, under a heavy item that needed to be moved. As long as the logs had the same thickness (called **diameter**) and the road was even, the heavy object had a smooth ride. What is special about a circle that allows it to be used in this way? In other words, why do circles enable this heavy object to roll smoothly?

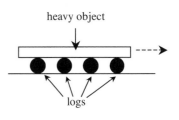

b.   What happens to a point on a wheel as it turns? For example, as the wheel at right rolls along the line, what is the path of point P? Imagine a piece of gum stuck to a tire as it rolls. On your paper, draw the motion of point P. If you need help, find a coin or other round object and test this situation.

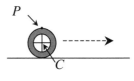

Problem continues on next page →

7-1.   *Problem continued from previous page.*

    c.   Now turn your attention to the center of the wheel (labeled *C* in the diagram above). As the wheel rolls along the line, what is the path of point *C*? Describe its motion. Why does that happen?

7-2.   DO CIRCLES MAKE THE BEST WHEELS?

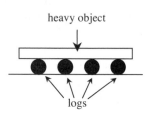

As you read in problem 7-1, ancient civilizations used circular logs to roll heavy objects. However, are circles the only shape they could have chosen? Are there any other shapes that could rotate between a flat road and a heavy object in a similar fashion?

Examine the shapes below. Would logs of any of these shapes be able to roll heavy objects in a similar fashion? Be prepared to defend your conclusion!

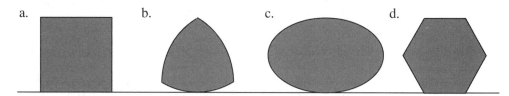

a.   b.   c.   d.

7-3.   Stanley says that he has a tricycle with square wheels and claims that it can ride as smoothly as a tricycle with circular wheels! Rosita does not believe him. Analyze this possibility with your team as you answer the questions below.

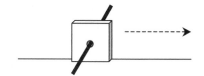

    a.   Is Stanley's claim possible? Describe what it would be like to ride a tricycle with square tires. What type of motion would the rider experience? Why does this happen?

    b.   When Rosita challenged him, Stanley confessed that he needed a special road so that the square wheels would be able to rotate smoothly and would keep Stanley at a constant height. What would his road need to look like? Draw an example on your paper.

    c.   How would Stanley need to change his road to be able to ride a tricycle with rectangular (but non-square) wheels? Draw an example on your paper.

    d.   Read the Math Notes box for this lesson to see a picture of Stanley and his tricycle. Then explain why the square wheel needed a modified road to ride smoothly, while the circular wheel did not. What is different between the two shapes?

7-4.   REULEAUX CURVES

Reuleaux curves (pronounced "roo **low**") are special because they
have a constant diameter. That means that as a Reuleaux curve
rotates, its height remains constant. Although the diagram at right
is an example of a Reuleaux curve based on an equilateral triangle,
these special curves can be based on any polygon with an odd
number of sides, including scalene triangles and pentagons.

a.   What happens to the center (point *C*) as the
Reuleaux wheel at right rolls?

b.   Since logs with a Reuleaux curve shape can
also smoothly roll heavy objects, why are
these shapes not used for bicycle wheels? In
other words, what is the difference between a
circle and a Reuleaux curve?

7-5.   A big focus of Chapters 7 through 12 is on circles. What did you
learn about circles today? Did you learn anything about other
shapes that was new or that surprised you? Write a Learning Log
entry explaining what you learned about the shapes of wheels. Title
this entry "Shapes of Wheels" and include today's date.

LOOKING **DEEPER**

Square Tires

MATH NOTES

As the picture at right
shows, square wheels are
possible if the road is
specially curved to accommodate
the change in the radius of the
wheel as it rotates. An example is
shown below.

*On a level
surface*

*On a curved
surface*

The picture above shows Stan Wagon riding his
special square-wheeled tricycle. This tricycle is
on display at Macalester College in St. Paul,
Minnesota. *Reprinted with permission.*

7-6.   **Examine** △*ABC* and △*DEF* at right.

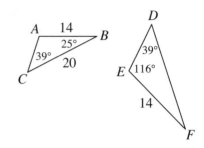

a.   Assume the triangles are not drawn to
        scale. Using the information provided
        in each diagram, write a mathematical
        statement describing the relationship
        between the two triangles. **Justify**
        your conclusion.

b.   Find *AC* and *DF* .

7-7.   Use the relationships in the diagram at right to find
        the values of each variable. Name which geometric
        relationships you used.

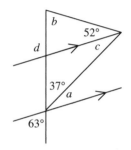

7-8.   A rectangle has one side of length 11 mm and a
        diagonal of 61 mm. Draw a diagram of this rectangle
        and find its width and area.

7-9.   Troy is thinking of a shape. He says that it has four sides and that no sides have equal
        length. He also says that no sides are parallel. What is the best name for his shape?

7-10.   Solve each system of equations below, if possible. If it is not possible, explain what
        the lack of an algebraic solution tells you about the graphs of the equations. Write
        each solution in the form (x, y). Show all work.

a. $y = -2x - 1$
        $y = \frac{1}{2}x - 16$

b. $y = x^2 + 1$
        $y = -x^2$

7.1.2 What can I build with a circle?

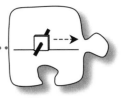

Building a Tetrahedron

In later chapters, you will learn more about polygons, circles, and 3-dimensional shapes. Later **investigations** will require that you remember key concepts you have already learned about triangles, parallel lines, and other angle relationships. Today you will have the opportunity to review some of the geometry you have learned while also beginning to think about what you will be studying in the future.

As you work with your team, consider the following focus questions:

Is there more than one way?

How can you be sure that is true?

What else can we try?

7-11.   IS THERE MORE TO THIS CIRCLE?

Circles can be folded to create many different shapes. Today, you will work with a circle and use properties of other shapes to develop a three-dimensional shape. Be sure to have **reasons** for each conclusion you make as you work. Each person in your team should start by obtaining a copy of a circle from your teacher and cutting it out.

a.   Fold the circle in half to create a crease that lies on a line of symmetry of the circle. Unfold the circle and then fold it in half again to create a new crease that is perpendicular to the first crease. Unfold your paper back to the full circle. How could you convince someone else that your creases are perpendicular? What is another name for the line segment represented by each crease?

b.   On the circle, label the endpoints of one diameter A and B. Fold the circle so that point A touches the center of the circle and create a new crease. Then label the endpoints of this crease C and D. What appears to be the relationship between $\overline{AB}$ and $\overline{CD}$? Discuss and **justify** with your team. Be ready to share your **reasons** with the class.

c.   Now fold the circle twice to form creases $\overline{BC}$ and $\overline{BD}$ and use scissors to cut out $\triangle BCD$. What type of triangle is $\triangle BCD$? How can you be sure? Be ready to convince the class.

7-12. ADDING DEPTH

Your equilateral triangle should now be flat (also called two-dimensional). **Two-dimensional** shapes have length and width, but not depth (or "thickness").

a. Label the vertices of $\triangle BCD$ if the labels were cut off. Then, with the unmarked side of the triangle facedown, fold and crease the triangle so that B touches the midpoint of $\overline{CD}$. Keep it in the folded position.

What does the resulting shape appear to be? What smaller shapes do you see inside the larger shape? **Justify** that your ideas are correct (for example, if you think that lines are parallel, you must provide evidence).

b. Open your shape again so that you have the large equilateral triangle in front of you. How does the length of a side of the large triangle compare to the length of the side of the small triangle formed by the crease? How many of the small triangles would fit inside the large triangle? In what ways are the small and large triangles related?

c. Repeat the fold in part (a) so that C touches the midpoint of $\overline{BD}$. Unfold the triangle and fold again so that D touches the midpoint of $\overline{BC}$. Create a three-dimensional shape by bringing points B, C, and D together. (A **three-dimensional** shape has length, width, and depth.) Use tape to hold your shape together.

d. Three-dimensional shapes formed with polygons have **faces** and **edges**, as well as **vertices**. Faces are the flat surfaces of the shape, while edges are the line segments formed when two faces meet. Vertices are the points where edges intersect. Discuss with your team how to use these words to describe your new shape. Then write a complete description. If you think you know the name of this shape, include it in your description.

7-13. Your team should now have 4 three-dimensional shapes (called **tetrahedra**). (If you are working in a smaller team, you should quickly fold more shapes so that you have a total of four.

a. Put four tetrahedra together to make an enlarged tetrahedron like the one pictured at right. Is the larger tetrahedron similar to the small tetrahedron? How can you tell?

b. To determine the edges and faces of the new shape, pretend that it is solid. How many edges does a tetrahedron have? Are all of the edges the same length? How does the length of an edge of the team shape compare with the length of an edge of one of the small shapes?

c. How many faces of the small tetrahedral would it take to cover the face of the large tetrahedron? Remember to count gaps as part of a face. Does the area of the tetrahedron change in the same way as the length?

Enlarged tetrahedron   Original

Geometry Connections

ETHODS AND MEANINGS

MATH NOTES

Parts of a Circle

A **circle** is the set of all points that are the same distance from a fixed central point, C. This text will use the notation ⊙C to name a circle with **center** at point C.

The distance from the center to the points on the circle is called the **radius** (usually denoted r), while the line segment drawn through the center of the circle with both endpoints on the circle is called a **diameter** (denoted d).

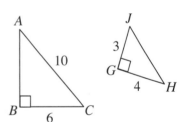

Notice that a diameter of a circle is always twice as long as the radius.

7-14.   What is the relationship of △ABC and △GHJ at right? **Justify** your conclusion.

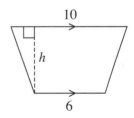

7-15.   The area of the trapezoid at right is 56 un².
What is h? Show all work.

7-16.   **Examine** the geometric relationships in each of the diagrams below. For each one, write and solve an equation to find the value of the variable. Name any geometric property or conjecture that you used.

a.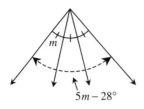

b.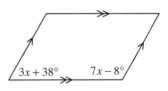

c.   △ABC below is equilateral.

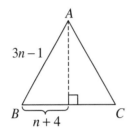

d. C is the center of the circle below. $AB = 11x - 1$ and $CD = 3x + 12$

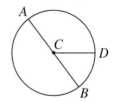

7-17.   In the Shape Factory, you created many shapes by rotating triangles about the midpoint of its sides. (Remember that the **midpoint** is the point exactly halfway between the endpoints of the line segment.) However, what if you rotate a trapezoid instead?

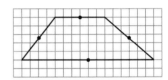

Carefully draw the trapezoid above on graph paper, along with the given midpoints. Then rotate the trapezoid 180° about one of the midpoints and **examine** the resulting shape formed by both trapezoids (the original and its image). Continue this process with each of the other midpoints, until you discover all the shapes that can be formed by a trapezoid and its image when rotated 180° about the midpoint of one of its sides.

7-18.   On graph paper, plot the points $A(-5, 7)$ and $B(3, 1)$.

a.   Find AB (the length of $\overline{AB}$).

b.   Locate the midpoint of $\overline{AB}$ and label it C. What are the coordinates of C?

c.   Find AC. Can you do this without using the Pythagorean Theorem?

7.1.3 What's the shortest distance?

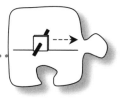

Shortest Distance Problems

Questions such as, "What length will result in the largest area?" or "When was the car traveling the slowest?" concern *optimization*. To **optimize** a quantity is to find the "best" possibility. Calculus is often used to solve optimization problems, but sometimes geometric tools offer surprisingly simple and elegant solutions.

7-19.   INTERIOR DESIGN

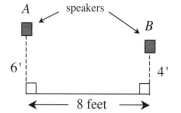

Laura needs your help. She needs to order expensive wire to connect her stereo to her built-in speakers and would like your help to save her money.

She plans to place her stereo somewhere on a cabinet that is 8 feet wide. Speaker A is located 6 feet above one end of the cabinet, while speaker B is located 4 feet above the other end. She will need wire to connect the stereo to speaker A, and additional wire to connect the stereo to speaker B.

Where should she place her stereo so that she needs the least amount of wire?

Your Task: Before you discuss this with your team, make your own guess. What does your intuition tell you? Then, using the Lesson 7.1.3 Resource Page, work with your team to determine where on the cabinet the stereo should be placed. How can you be sure that you found the best answer? In other words, how do you know that the amount of wire you found is the least amount possible?

Discussion Points

What is this problem about? What are you supposed to find?

What is a reasonable estimate of the total length of speaker wire?

What mathematical tools could be helpful to solve this problem?

Chapter 7: Proof and Quadrilaterals   343

7-20.   To help solve problem 7-19, first collect some data.

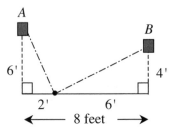

a.   Calculate the total length of wire needed if the stereo is placed 2 feet from the left edge of the cabinet (the edge below Speaker A), as shown in the diagram at right.

b.   Now calculate the total length of wire needed if the stereo is placed 3 feet from the same edge. Does this placement require more or less wire than that from part (a)?

c.   Continue testing placements for the stereo and create a table with your results. Where should the stereo be placed to minimize the amount of wire?

7-21.   This problem reminds Bradley of problem 3-93, *You Are Getting Sleepy…*, in which you and a partner created two triangles by standing and gazing into a mirror. He remembered that the only way two people could see each others' eyes in the mirror was when the triangles were similar. **Examine** your solution to problem 7-19. Are the two triangles created by the speaker wires similar? **Justify** your conclusion.

7-22.   Bradley enjoyed solving problem 7-19 so much that he decided to create other "shortest distance" problems. For each situation below, first predict where the shortest path lies using visualization and intuition. Then find a way to determine whether the path you chose is, in fact, the shortest.

a.   In this first puzzle, Bradley decided to test what would happen on the side of a cylinder, such as a soup can. On a can provided by your teacher, find points *A* and *B* labeled on the outside of the can. With your team, determine the shortest path from point *A* to point *B* along the surface of the can. (In other words, no part of your path can go inside the can.) Describe how you found your solution.

b.   What if the shape is a cube? Using a cube provided by your teacher, predict which path would be the shortest path from opposite corners of the cube (labeled points *C* and *D* in the diagram at right). Then test your prediction. Describe how you found the shortest path.

Geometry Connections

7-23. MAKING CONNECTIONS

As Bradley looked over his answer from
problem 7-19, he couldn't help but wonder
if there is a way to change this problem
into a straight-line problem like those in
problem 7-22.

a. On the Lesson 7.1.3 Resource Page, reflect
one of the speakers so that when the two
speakers are connected with a straight line,
the line passes through the horizontal
cabinet.

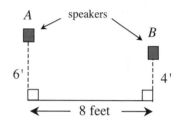

b. When the speakers from part (a) are connected
with a straight line, two triangles are formed.
How are the two triangles related? **Justify**
your conclusion.

c. Use the fact that the triangles are similar to find
where the stereo should be placed. Did your
answer match that from problem 7-19?

7-24. TAKE THE SHOT

While playing a game of pool,
"Montana Mike" needed to hit
the last remaining ball into
pocket A, as shown in the
diagram below. However, to
show off, he decided to make
the ball first hit at least one of
the rails of the table.

Your Task: On the Lesson 7.1.3
Resource Page provided by your
teacher, determine where Mike
could bounce the ball off a rail so
that it will land in pocket A. Work
with your team to find as many
possible locations as you can.
Can you find a way he could hit
the ball so that it would rebound
twice before entering pocket A?

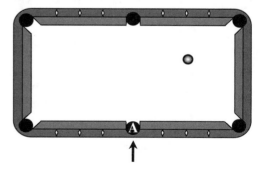

Be ready to share your solutions with the class.

7-25.   Look over your work from problems 7-19 to 7-24. What
mathematical ideas did you use? What connections, if any, did you
find? Can any other problems you have seen so far be solved using a
straight line? Describe the mathematical ideas you developed during
this lesson in your Learning Log. Title this entry "Shortest Distance"
and include today's date.

METHODS AND MEANINGS

Congruent Triangles → Congruent Corresponding Parts

MATH NOTES

As you learned in Chapter 3, if two shapes are congruent, then they
have exactly the same shape and the same size. This means that if you know
two triangles are congruent, you can state that corresponding parts are
congruent. This can be also stated with the arrow diagram:

$$\cong \triangle s \to \cong \text{ parts}$$

For example, if $\triangle ABC \cong \triangle PQR$, then it follows that $\angle A \cong \angle P$, $\angle B \cong \angle Q$,
and $\angle C \cong \angle R$. Also, $\overline{AB} \cong \overline{PQ}$, $\overline{AC} \cong \overline{PR}$, and $\overline{BC} \cong \overline{QR}$.

Review & Preview

7-26. $\triangle XYZ$ is reflected across $\overline{XZ}$, as shown at right.

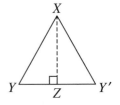

   a.   How can you **justify** that the points Y, Z, and Y'
all lie on a straight line?

   b.   What is the relationship between $\triangle XYZ$ and
$\triangle XY'Z$? Why?

   c.   Read the Math Notes box for this lesson. Then make all the statements you can
about the corresponding parts of these two triangles.

7-27.   Remember that a midpoint of a line segment is the point that divides the segment into
two segments of equal length. On graph paper, plot the points $P(0, 3)$ and $Q(0, 11)$.
Where is the midpoint M if $PM = MQ$? Explain how you found your answer.

7-28. Recall the three similarity shortcuts for triangles: SSS ~, SAS ~ and AA ~. For each pair of triangles below, decide whether the triangles are congruent and/or similar. **Justify** each conclusion.

a.

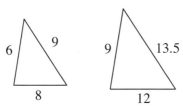

b.

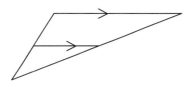

c.

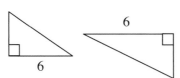

d.

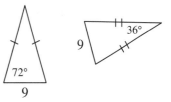

7-29. On graph paper, plot and connect the points $A(1, 1)$, $B(2, 3)$, $C(5, 3)$, and $D(4, 1)$ to form quadrilateral $ABCD$.

a. What is the best name for quadrilateral $ABCD$? **Justify** your answer.

b. Find and compare $m\angle DAB$ and $m\angle BCD$. What is their relationship?

c. Find the equations of diagonals $\overline{AC}$ and $\overline{BD}$. Are the diagonals perpendicular?

d. Find the point where diagonals $\overline{AC}$ and $\overline{BD}$ intersect.

7-30. Solve each system of equations below, if possible. If it is not possible, explain what having "no solution" tells you about the graphs of the equations. Write each solution in the form (x, y). Show all work.

a. $y = -\frac{1}{3}x + 7$

$y = -\frac{1}{3}x - 2$

b. $y = 2x + 3$

$y = x^2 - 2x + 3$

7-31. How long is the longest line segment that will fit inside a square of area 50 square units? Show all work.

7-32.   Graph and connect the points $G(-2, 2)$, $H(3, 2)$, $I(6, 6)$, and $J(1, 6)$ to form *GHIJ*.

a.   What specific type of shape is quadrilateral *GHIJ*? **Justify** your conclusion.

b.   Find the equations of the diagonals $\overline{GI}$ and $\overline{HJ}$.

c.   Compare the slopes of the diagonals. How do the diagonals of a rhombus appear to be related?

d.   Find J' if quadrilateral *GHIJ* is rotated 90° clockwise (↻) about the origin.

e.   Find the area of quadrilateral *GHIJ*.

7-33.   **Examine** the relationships in the diagrams below. For each one, write an equation and solve for the given variable(s). Show all work.

a.

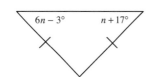

b.

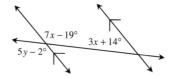

c.

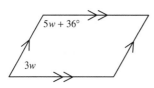

d.

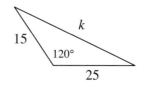

7-34.   Find the perimeter of the shape at right. Show all work.

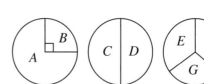

7-35.   Three spinners are shown at right. If each spinner is randomly spun and if spinners #2 and #3 are each equally divided, find the following probabilities.

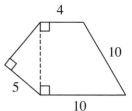

a.   P(spinning *A*, *C*, and *E*)

b.   P(spinning at least one vowel)

7.1.4 How can I create it?

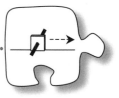

Using Symmetry to Study Polygons

In Chapter 1, you used a hinged mirror to study the special angles associated with regular polygons. In particular, you **investigated** what happens as the angle formed by the sides of the mirror is changed. Today, you will use a hinged mirror to determine if there is more than one way to build each regular polygon using the principals of symmetry. And what about other types of polygons? What can a hinged mirror help you understand about them?

As your work with your study team, keep these focus questions in mind:

Is there another way?

What types of symmetry can I find?

What does symmetry tell me about the polygon?

7-36.   THE HINGED MIRROR TEAM CHALLENGE

Obtain a hinged mirror, a piece of unlined colored paper, and a protractor from your teacher.

With your team, spend five minutes reviewing how to use the mirror to create regular polygons. (Remember that a **regular polygon** has equal sides and angles). Once everyone remembers how the hinged mirror works, select a team member to read the directions of the task below.

Your Task: Below are four challenges for your team. Each requires you to find a creative way to position the mirror in relation to the colored paper. You can tackle the challenges in any order, but you must work together as a team on each. Whenever you successfully create a shape, do not forget to measure the angle formed by the mirror, as well as draw a diagram on your paper of the core region in front of the mirror. If your team decides that a shape is impossible to create with the hinged mirror, explain why.

- Create a regular hexagon.

- Create an equilateral triangle at least <u>two</u> different ways.

- Create a rhombus that is <u>not</u> a square.

- Create a circle.

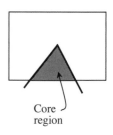

Core region

7-37.   ANALYSIS

How can symmetry help you to learn more about shapes? Discuss each question
below with the class.

a.   One way to create a regular hexagon with a hinged
       mirror is with six triangles, as shown in the diagram at
       right. (Note: the gray lines represent reflections of the
       bottom edges of the mirrors and the edge of the paper,
       while the core region is shaded.)

What is special about each of the triangles in the diagram? What is the
relationship between the triangles? Support your conclusions. Would it be
possible to create a regular hexagon with 12 triangles? Explain.

b.   If you have not done so already, create an equilateral triangle so that the core
       region in front of the mirror is a right triangle. Draw a diagram of the result
       that shows the different reflected triangles like the one above. What special
       type of right triangle is the core region? Can all regular polygons be created
       with a right triangle in a similar fashion?

c.   In problem 7-36, your team formed a rhombus that is not a square. On your
       paper, draw a diagram like the one above that shows how you did it. How can
       you be sure your resulting shape is a rhombus? Using what you know about the
       angle of the mirror, explain what must be true about the diagonals of a
       rhombus.

7-38.   Use what you learned today to answer the questions below.

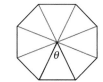

a.   **Examine** the regular octagon at right. What is
       the measure of angle θ ? Explain how you know.

b.   Quadrilateral *ABCD* at right is a
       rhombus. If $BD = 10$ units and
       $AC = 18$ units, then what is the
       perimeter of *ABCD*? Show all work.

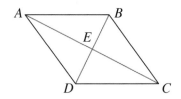

Review & Preview

7-39. Felipe set his hinged mirror so that its angle was 36° and the core region was isosceles, as shown at right.

a. How many sides did his resulting polygon have? Show how you know.

b. What is another name for this polygon?

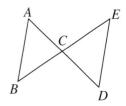

7-40. In problem 7-37 you learned that the diagonals of a rhombus are perpendicular bisectors. If *ABCD* is a rhombus with side length 15 mm and if *BD* = 24 mm, then find the length of the other diagonal, $\overline{AC}$. Draw a diagram and show all work.

7-41. Joanne claims that (2, 4) is the midpoint of the segment connecting the points (−3, 5) and (7, 3). Is she correct? Explain how you know.

7-42. If $\triangle ABC \cong \triangle DEC$, which of the statements below must be true? **Justify** your conclusion. Note: More than one statement may be true.

a. $\overline{AC} \cong \overline{DC}$ b. $m\angle B = m\angle D$ c. $\overline{AB} \, // \, \overline{DE}$

d. $AD = BE$ e. None of these are true.

7-43. On graph paper, graph the points *A*(2, 9), *B*(4, 3), and *C*(9, 6). Which point (*A* or *C*) is closer to point *B*? **Justify** your conclusion.

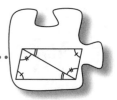

7.2.1 What can congruent triangles tell me?

• •

Special Quadrilaterals and Proof

In earlier chapters you studied the relationships between the sides and angles of a triangle, and solved problems involving congruent and similar triangles. Now you are going to expand your study of shapes to quadrilaterals. What can triangles tell you about parallelograms and other special quadrilaterals?

By the end of this lesson, you should be able to answer these questions:

> What are the relationships between the sides, angles,
> and diagonals of a parallelogram?

> How are congruent triangles useful?

7-44.   Carla is thinking about parallelograms, and wondering if there are as many special properties for parallelograms as there are for triangles. She remembers that is it possible to create a shape that looks like a parallelogram by rotating a triangle about the midpoint of one of its sides.

  a.   Carefully trace the triangle at right onto tracing paper. Be sure to copy the angle markings as well. Then rotate the triangle to make a shape that looks like a parallelogram.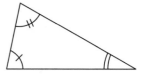

  b.   Is Carla's shape truly a parallelogram? Use the angles to convince your teammates that the opposites sides must be parallel. Then write a convincing argument.

  c.   What else can the congruent triangles tell you about a parallelogram? Look for any relationships you can find between the angles and sides of a parallelogram.

  d.   Does this work for all parallelograms? That is, does the diagonal of a parallelogram always split the shape into two congruent triangles? Draw the parallelogram at right on your paper. Knowing only that the opposite sides of a parallelogram are parallel, create a flowchart to show that the triangles are congruent.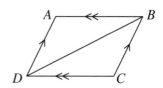

7-45. CHANGING A FLOWCHART INTO A PROOF

The flowchart you created for part (d) of problem 7-44 shows how you can conclude that if a quadrilateral is a parallelogram, then its each of its diagonals splits the quadrilateral into two congruent triangles.

However, to be convincing, the facts that you listed in your flowchart need to have **justification**. This shows the reader how you know the facts are true and helps to **prove** your conclusion.

Therefore, with the class or your team, decide how to add reasons to each statement (bubble) in your flowchart. You may need to add more bubbles to your flowchart to add **justification** and to make your proof more convincing.

7-46. Kip is confused. He put his two triangles from problem 7-44 together as shown at right, but he didn't get a parallelogram.

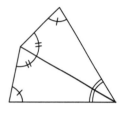

a.   What shape did he make? **Justify** your conclusion.

b.   What transformation(s) did Kip use to form his shape?

c.   What do the congruent triangles tell you about the angles of this shape?

7-47. KITES

Kip shared his findings about his kite with his teammate, Carla, who wants to learn more about the diagonals of a kite. Carla quickly sketched the kite at right onto her paper with a diagonal showing the two congruent triangles.

a.   **EXPLORE:** Trace this diagram onto tracing paper and carefully add the other diagonal. Then, with your team, consider how the diagonals may be related. Use tracing paper to help you explore the relationships between the diagonals. If you make an observation you think is true, move on to part (b) and write a conjecture.

b.   **CONJECTURE:** If you have not already done so, write a conjecture based on your observations in part (a).

Problem continues on next page →

7-47.   *Problem continued from previous page.*

     c.   **PROVE:** When she drew the second diagonal, Carla noticed that four new triangles appeared. "If any of these triangles are congruent, then they may be able to help us prove our conjecture from part (b)," she said.

        Examine △*ABC* below. Are △*ACD* and △*BCD* congruent? Create a flowchart proof like the one from problem 7-45 to **justify** your conclusion.

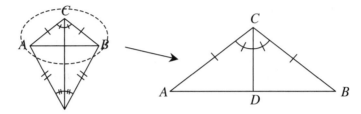

     d.   Now extend your proof from part (c) to prove your conjecture from part (b).

7-48.   Reflect on all of the interesting facts about parallelograms and kites you have proven during this lesson. Obtain a Theorem Toolkit (Lesson 7.2.1A Resource Page) from your teacher or from www.cpm.org. In it, record each **theorem** (proven conjecture) that you have proven about the sides, angles, and diagonals of a parallelogram. Do the same for a kite. Be sure your diagrams contain appropriate markings to represent equal parts.

Ⓜ️ETHODS AND MEANINGS

MATH NOTES

Reflexive Property of Equality

In problem 7-44, you used the fact that two triangles formed by the diagonal of a parallelogram share a side of the same length to help show that the triangles were congruent.

The **Reflexive Property of Equality** states that the measure of any side or angle is equal to itself. For example, in the parallelogram at right, $\overline{BD} \cong \overline{DB}$ because of the Reflexive Property.

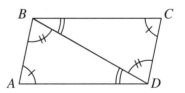

Geometry Connections

7-49.   Use the information given for each diagram below to solve for *x*. Show all work.

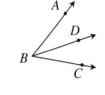

a. $\overrightarrow{BD}$ bisects $\measuredangle ABC$. (Remember that this means it divides the angle into two equal parts.) If $m\measuredangle ABD = 5x - 10°$ and $m\measuredangle ABC = 65°$, solve for *x*.

b.   Point *M* is a midpoint of $\overline{EF}$. If $EM = 4x - 2$ and $MF = 3x + 9$, solve for *x*.

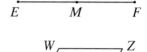

c.   *WXYZ* at right is a parallelogram. If $m\measuredangle W = 9x - 3°$ and $m\measuredangle Z = 3x + 15°$, solve for *x*.

7-50.   Jamal used a hinged mirror to create a regular polygon like he did in Lesson 7.1.4.

a.   If his hinged mirror formed a 72° angle and the core region in front of the mirror was isosceles, how many sides did his polygon have?

b.   Now Jamal has decided to create a regular polygon with 9 sides, called a **nonagon**. If his core region is again isosceles, what angle is formed by his mirror?

7-51.   Sandra wants to park her car so that she is the shortest distance possible from the entrances of both the Art Museum and the Public Library. Locate where on the street she should park so that her total distance directly to each building is the shortest.

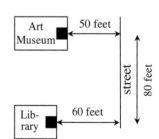

7-52.   Use the geometric relationships in the diagrams below to solve for *x*.

a.

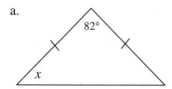

b.

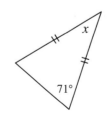

Chapter 7: Proof and Quadrilaterals   355

7-53.   Which pairs of triangles below are congruent and/or similar? For each part, explain
         how you know. Note: The diagrams are not necessarily drawn to scale.

a.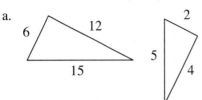

b.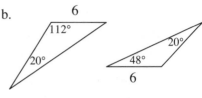

c.
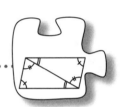

d.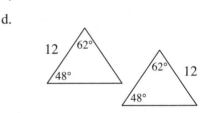

7.2.2 What is special about a rhombus?

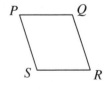

Properties of Rhombi

In Lesson 7.2.1, you learned that congruent triangles can be a useful tool to discover new
information about parallelograms and kites. But what about other quadrilaterals? Today you
will use congruent triangles to **investigate** and prove special properties of rhombi (the plural of
rhombus). At the same time, you will continue to develop your ability to make conjectures and
prove them convincingly.

7-54.   Audrey has a favorite quadrilateral – the rhombus. Even though a rhombus is
         defined as having four congruent sides, she suspects that the sides of a rhombus have
         other special properties.

    a.   **EXPLORE:** Draw a rhombus like the one at right
          on your paper. Mark the side lengths equal.

    b.   **CONJECTURE:** What else might be special
          about the sides of a rhombus? Write a
          conjecture.

    c.   **PROVE:** Audrey knows congruent triangles can help prove other properties
          about quadrilaterals. She starts by adding a diagonal $\overline{PR}$ to her diagram so that
          two triangles are formed. Add this diagonal to your diagram and prove that the
          triangles are congruent. Then use a flowchart with **reasons** to show your logic.
          Be prepared to share your flowchart with the class.

    d.   How can the triangles from part (c) help you prove your conjecture from part
          (b) above? Discuss with the class how to extend your flowchart to convince
          others. Be sure to **justify** any new statements with reasons.

7-55.   Now that you know the opposite sides of a rhombus are parallel, what else can you prove about a rhombus? Consider this as you answer the questions below.

a.   **EXPLORE:** Remember that in Lesson 7.1.4, you explored the shapes that could be formed with a hinged mirror. During this activity, you used symmetry to form a rhombus. Think about what you know about the reflected triangles in the diagram. What do you think is true about the diagonals $\overline{SQ}$ and $\overline{PR}$? What is special about $\overline{ST}$ and $\overline{QT}$? What about $\overline{PT}$ and $\overline{RT}$?

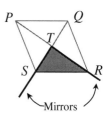

b.   **CONJECTURE:** Use your observations from part (a) to write a conjecture on the relationship of the diagonals of a rhombus.

c.   **PROVE:** Write a flowchart proof that proves your conjecture from part (b). Remember that to be convincing, you need to **justify** each statement with a reason. To help guide your discussion, consider the questions below.

- Which triangles should you use? Find two triangles that involve the lengths $\overline{ST}$, $\overline{QT}$, $\overline{PT}$ and $\overline{RT}$.

- How can you prove these triangles are congruent? Create a flowchart proof with **reasons** to prove these triangles must be congruent.

- How can you use the congruent triangles to prove your conjecture from part (b)? Extend your flowchart proof to include this **reasoning** and prove your conjecture.

7-56.   There are often many ways to prove a conjecture. You have rotated triangles to create parallelograms and used congruent parts of congruent triangles to **justify** that opposite sides are parallel. But is there another way?

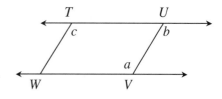

Ansel wants to prove the conjecture *"If a quadrilateral is a parallelogram, then opposite angles are congruent."* He started by drawing parallelogram *TUVW* at right. Copy and complete his flowchart. Make sure that each statement has a **reason**.

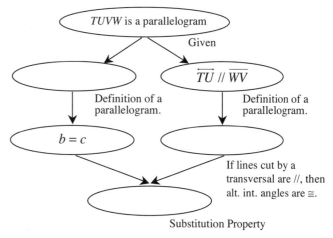

7-57.   Think about the new facts you have proven about
         rhombi during this lesson. On your Theorem Toolkit
         (Lesson 7.2.1A Resource Page), record each new
         theorem you have proven about the angles and
         diagonals of a rhombus. Include clearly labeled
         diagrams to illustrate your findings.

7-58.   Point *M* is the midpoint of $\overline{AB}$ and *B* is
         the midpoint of $\overline{AC}$. What are the values
         of *x* and *y*? Show all work and **reasoning**.

$$\overset{4x-1}{\underset{A}{\bullet}} \quad \overset{x+8}{\underset{M}{\bullet}} \quad \underset{B}{\bullet} \qquad \overset{5y+2}{\qquad} \underset{C}{\bullet}$$

7-59.   In the diagram at right, $\angle DCA$ is referred to as an
         exterior angle of $\triangle ABC$ because it lies outside the
         triangle and is formed by extending a side of the
         triangle.

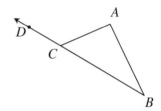

         a.   If $m\angle CAB = 46°$ and $m\angle ABC = 37°$, what is
               $m\angle DCA$? Show all work.

         b.   If $m\angle DCA = 135°$ and $m\angle ABC = 43°$, then what is $m\angle CAB$?

7-60.   On graph paper, graph quadrilateral *MNPQ* if *M*(–3, 6), *N*(2, 8), *P*(1, 5), and *Q*(–4, 3).

         a.   What shape is *MNPQ*? Show how you know.

         b.   Reflect *MNPQ* across the *x*-axis to find $M'N'P'Q'$. What are the coordinates
               of P'?

Geometry Connections

7-61.   Jester started to prove that the triangles at right are congruent. He is only told that point E is the midpoint of segments $\overline{AC}$ and $\overline{BD}$.

Copy and complete his flowchart below. Be sure that a **reason** is provided for every statement.

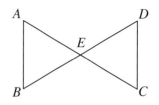

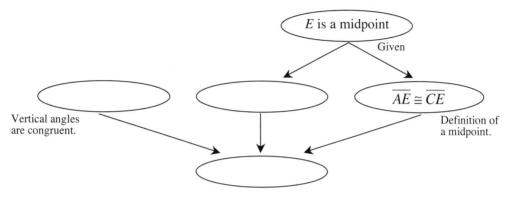

E is a midpoint

Given

Vertical angles are congruent.

$\overline{AE} \cong \overline{CE}$

Definition of a midpoint.

7-62.   On graph paper, graph and shade the solutions for the inequality below.

$$y < -\tfrac{2}{3}x + 5$$

7.2.3 What else can be proved?

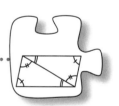

More Proof with Congruent Triangles

In Lessons 7.2.1 and 7.2.2, you used congruent triangles to learn more about parallelograms, kites, and rhombi. You now possess the tools to do the work of a geometrician: to discover and prove new properties about the sides and angles of shapes.

As you **investigate** these shapes, focus on proving your ideas. Remember to ask yourself and your teammates questions such as, "*Why does that work?*" and "*Is it always true?*" Decide whether your argument is convincing and work with your team to provide all of the necessary **justification**.

7-63.   Carla decided to turn her attention to
         rectangles. Knowing that a rectangle is
         defined as a quadrilateral with four
         right angles, she drew the diagram at
         right.

         After some exploration, she conjectured
         that all rectangles are also parallelograms.
         Help her prove that her rectangle *ABCD*
         must be a parallelogram. That is, prove
         that the opposite sides must be parallel.
         Then add this theorem to your Theorem
         Toolkit (your Lesson 7.2.1A Resource
         Page).

7-64.   For each diagram below, find the value of *x*, if possible. If the triangles are
         congruent, state which triangle congruence property was used. If the triangles are not
         congruent or if there is not enough information, state, "Cannot be determined."

a.   *ABC* below is a triangle.

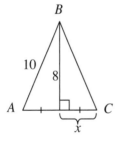

b.

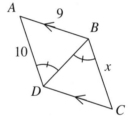

c.

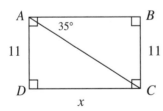

d. $\overline{AC}$ and $\overline{BD}$ are straight line
     segments.

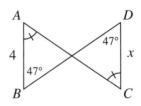

7-65.   With the class or your team, create a flowchart to prove your answer to part (b) of
         problem 7-64. That is, prove that $\overline{AD} \cong \overline{CB}$. Be sure to include a diagram for your
         proof and **reasons** for every statement. Make sure your argument is convincing and
         has no "holes."

METHODS AND MEANINGS

MATH NOTES

Definitions of Quadrilaterals

When proving properties of shapes, it is necessary to know exactly how a shape is defined. Below are the definitions of several quadrilaterals that you will study in this chapter and the chapters that follow.

Quadrilateral: A closed four-sided polygon.

Kite: A quadrilateral with two distinct pairs of consecutive congruent sides.

Trapezoid: A quadrilateral with at least one pair of parallel sides.

Parallelogram: A quadrilateral with two pairs of parallel sides.

Rhombus: A quadrilateral with four sides of equal length.

Rectangle: A quadrilateral with four right angles.

Square: A quadrilateral with four sides of equal length and four right angles.

7-66.   Identify if each pair of triangles below is congruent or not. Remember that the diagram may not be drawn to scale. **Justify** your conclusion.

a.

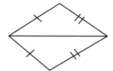

b.

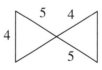

c.

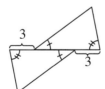

d.

7-67.   **Examine** the information provided in each diagram below.
Decide if each figure is possible or not. If the figure is not
possible, explain why.

a.

b.
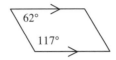

c.
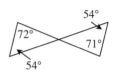

7-68.   Tromika wants to find the area of the isosceles triangle at right.

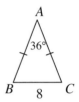

a.   She decided to start by drawing a height from vertex
A to side $\overline{BC}$ as shown below. Will the two smaller
triangles be congruent? In other words, is
$\triangle ABD \cong \triangle ACD$? Why or why not?

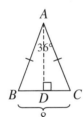

b.   What is $m\angle DAB$? BD?

c.   Find AD. Show how you got your answer.

d.   Find the area of $\triangle ABC$.

7-69.   On graph paper, graph quadrilateral $ABCD$ if $A(0, 0)$, $B(6, 0)$, $C(8, 6)$, and $D(2, 6)$.

a.   What is the best name for $ABCD$? **Justify** your answer.

b.   Find the equation of the lines containing each diagonal. That is, find the
equations of lines $\overleftrightarrow{AC}$ and $\overleftrightarrow{BD}$.

7-70.   For each diagram below, solve for x. Show all work.

a.

b.
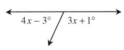

c.
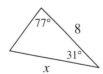

d.

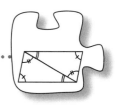

7.2.4 What else can I prove?

More Properties of Quadrilaterals

Today you will work with your team to apply what you have learned to other shapes. Remember to ask yourself and your teammates questions such as, *"Why does that work?"* and *"Is it always true?"* Decide whether your argument is convincing and work with your team to provide all of the necessary **justification**. By the end of this lesson, you should have a well-crafted mathematical argument proving something new about a familiar quadrilateral.

7-71.   WHAT ELSE CAN CONGRUENT TRIANGLES TELL US?

Your Task: For each situation below, determine how congruent triangles can tell you more information about the shape. Then prove your conjecture using a flowchart. Be sure to provide a **reason** for each statement. For example, stating "$m\angle A = m\angle B$" is not enough. You must give a convincing reason, such as *"Because vertical angles are equal"* or *"Because it is given in the diagram."* Use your triangle congruence properties to help prove that the triangles are congruent.

Later, your teacher will select one of these flowcharts for you to place on a poster. On your poster, include a diagram and all of your statements and reasons. Clearly state what you are proving and help the people who look at your poster understand your logic and **reasoning**.

a.   In Chapter 1, you used symmetry of an isosceles triangle to show that the base angles must be congruent. But what if you only know that the base angles are congruent? Does the triangle have to be isosceles?

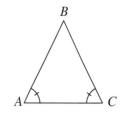

Assume that you know that the two base angles of $\triangle ABC$ are congruent. With your team, decide how to split $\triangle ABC$ into two triangles that you can show are congruent to show that $AB = CB$.

b.   What can congruent triangles tell us about the diagonals and angles of a rhombus? **Examine** the diagram of the rhombus at right. With your team, decide how to prove that the diagonals of a rhombus bisect the angles. That is, prove that $\angle ABD \cong \angle CBD$.

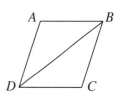

Problem continues on next page →

7-71. *Problem continued from previous page.*

c. What can congruent triangles tell us about the diagonals of a rectangle? **Examine** the rectangle at right. Using the fact that the opposite sides of a rectangle are parallel (which you proved in problem 7-63), prove that the diagonals of the rectangle are congruent. That is, prove that $AC = BD$.

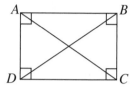

Ⓜ ETHODS AND MEANINGS

Diagonals of a Rhombus

A **rhombus** is defined as a quadrilateral with four sides of equal length. In addition, you proved in problem 7-55 that the diagonals of a rhombus are perpendicular bisectors of each other.

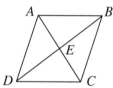

For example, in the rhombus at right, E is a midpoint of both $\overline{AC}$ and $\overline{DB}$. Therefore, $AE = CE$ and $DE = BE$. Also, $m\angle AEB = m\angle BEC = m\angle CED = m\angle DEA = 90°$.

In addition, you proved in problem 7-71 that the diagonals bisect the angles of the rhombus. For example, in the diagram above, $m\angle DAE = m\angle BAE$.

7-72. Use Tromika's method from problem 7-68 to find the area of an equilateral triangle with side length 12 units. Show all work.

7-73. The guidelines set forth by the National Kitchen & Bath Association recommends that the perimeter of the triangle connecting the refrigerator (F), stove, and sink of a kitchen be 26 feet or less. Lashayia is planning to renovate her kitchen and has chosen the design at right. (Note: All measurements are in feet.) Does her design conform to the National Kitchen and Bath Association's guidelines? Show how you got your answer.

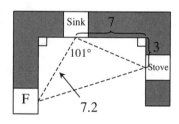

7-74.   For each figure below, determine if the two smaller triangles in each figure are congruent. If so, explain why and solve for x. If not, explain why not.

a.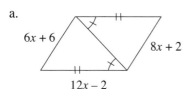

6x + 6

8x + 2

12x − 2

b.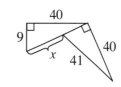

40

9

x

41

40

7-75.   The diagonals of a rhombus are 6 units and 8 units long. What is the area of the rhombus? Draw a diagram and show all reasoning.

7-76.   A hotel in Las Vegas is famous for its large-scale model of the Eiffel Tower. The model, built to scale, is 128 meters tall and 41 meters wide at its base. If the real tower is 324 meters tall, how wide is the base of the real Eiffel Tower?

7.2.5 How else can I write it?

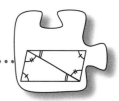

•••

Two-Column Proofs

Today you will continue to work with constructing a convincing argument, otherwise known as writing a proof. You will use what you know about flowchart proofs to write a convincing argument using another format, called a "two-column" proof.

7-77.   The following pairs of triangles are not necessarily congruent even though they appear to be. Use the information provided in the diagram to show why. **Justify** your statements.

a.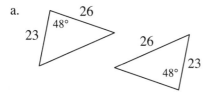

26

48°

23

26

23

48°

b.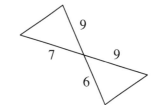

9

7   9

6

7-78.   Write a flowchart to prove that if E is the midpoint of $\overline{AD}$ and $\angle A$ and $\angle D$ are both right angles, then $\overline{AB} \cong \overline{DC}$.

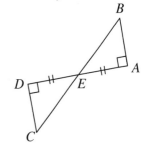

7-79.   Another way to organize a proof is called a **two-column proof**. Instead of using arrows to indicate the order of logical reasoning, this style of proof lists statements and reasons in a linear order.

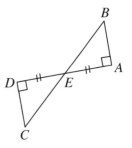

The proof from problem 7-78 has been converted to a two-column proof below. Copy and complete the proof on your paper using your statements and **reasons** from problem 7-78.

If: E is the midpoint of $\overline{AD}$ and $\angle A$ and $\angle D$ are both right angles,
Prove: $\overline{AB} \cong \overline{DC}$

Statements	Reasons (This statement is true because…)
E is the midpoint of $\overline{AD}$ and $\angle A$ and $\angle D$ are both right angles	Given
$\angle A \cong \angle D$	Angles with the same measure are congruent.
	Definition of a midpoint
$\angle DEC \cong \angle AEB$	

7-80.   Examine the posters of flowchart proofs from problem 7-71. Convert each flowchart proof to a two-column proof. Remember that one column must contain the statements of fact while the other must provide the **reason** (or **justification**) explaining why that fact must be true.

Geometry Connections

7-81.   So far in Section 7.2, you have proven many special
         properties of quadrilaterals and other shapes. When a
         conjecture is proven, it is called a **theorem**. For example,
         once you proved the relationship between the sides of a right
         triangle, you were able to refer to that relationship as the
         Pythagorean Theorem. Find your Theorem Toolkit (Lesson
         7.2.1A Resource Page) and make sure it contains all of the
         theorems you and your classmates have proven so far about
         various quadrilaterals. Be sure that your records include
         diagrams for each statement.

7-82.   Reflect on the new proof format you learned today. Compare it to
         the flowchart proof format that you have used earlier. What are the
         strengths and weaknesses of each style of proof? Which format is
         easier for you to use? Which is easier to read? Title this entry
         "Two-Column Proofs" and include today's date.

7-83.   Suppose you know that $\triangle TAP \cong \triangle DOG$ and that $TA = 14$, $AP = 18$, $TP = 21$, and
         $DG = 2y + 7$.

         a.   On your paper, draw a reasonable sketch of $\triangle TAP$ and $\triangle DOG$.

         b.   Find y. Show all work.

7-84. $\measuredangle a$, $\measuredangle b$, and $\measuredangle c$ are exterior angles of the triangle at right.
         Find $m\measuredangle a$, $m\measuredangle b$, and $m\measuredangle c$. Then find $m\measuredangle a + m\measuredangle b + m\measuredangle c$.

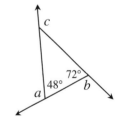

7-85.   Prove that if a pair of opposite sides of a
         quadrilateral are congruent and parallel, then the
         quadrilateral must be a parallelogram.

         For example, for the quadrilateral $ABCD$ at right,
         given that $\overline{AB} \parallel \overline{CD}$ and $\overline{AB} \cong \overline{CD}$, show that
         $\overline{BC} \parallel \overline{AD}$. Organize your **reasoning** in a flowchart.
         Then record your theorem in your Theorem Toolkit
         (your Lesson 7.2.1A Resource Page).

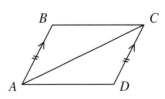

7-86.   Find the area and perimeter of the trapezoid
         at right.

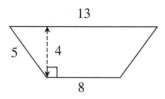

7-87.   For each pair of triangles below, determine if the
         triangles are congruent. If the triangles are congruent,

         •   complete the correspondence statement,

         •   state the congruence property,

         •   and record any other ideas you use that make your conclusion true.

         Otherwise, explain why you cannot conclude that the triangles are congruent. Note
         that the figures are not necessarily drawn to scale.

         a. $\triangle ABC \cong \triangle$ ____   b. $\triangle SQP \cong \triangle$ ____   c. $\triangle PLM \cong \triangle$ ____

         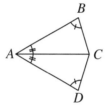   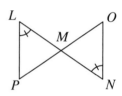

         d. $\triangle WXY \cong \triangle$ ____   e. $\triangle EDG \cong \triangle$ ____   f. $\triangle ABC \cong \triangle$ ____

         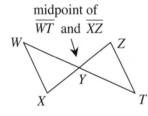   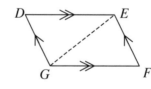   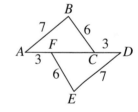

Geometry Connections

7.2.6 What can I prove?

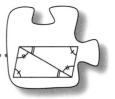

Explore-Conjecture-Prove

So far, congruent triangles have helped you to discover and prove many new facts about triangles and quadrilaterals. But what else can you discover and prove? Today your work will mirror the real work of professional mathematicians. You will **investigate** relationships, write a conjecture based on your observations, and then prove your conjecture.

7-88.   TRIANGLE MIDSEGMENT THEOREM

As Sergio was drawing shapes on his paper, he drew a line segment that connected the midpoints of two sides of a triangle. (This is called the **midsegment** of a triangle.) "I wonder what we can find out about this midsegment," he said to his team. **Examine** his drawing at right.

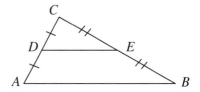

a.   **EXPLORE: Examine** the diagram of $\triangle ABC$, drawn to scale above. How do you think $\overline{DE}$ is related to $\overline{AB}$? How do their lengths seem to be related?

b.   **CONJECTURE:** Write a conjecture about the relationship between segments $\overline{DE}$ and $\overline{AB}$.

c.   **PROVE:** Sergio wants to prove that $AB = 2DE$. However, he does not see any congruent triangles in the diagram. How are the triangles in this diagram related? How do you know? Prove your conclusion with a flowchart.

d.   What is the common ratio between side lengths in the similar triangles? Use this to write a statement relating lengths DE and AB.

e.   Now Sergio wants to prove that $\overline{DE} \parallel \overline{AB}$. Use the similar triangles to find all the pairs of equal angles you can in the diagram. Then use your knowledge of angle relationships to make a statement about parallel segments.

7-89.   The work you did in problem 7-88 mirrors the work of many professional
         mathematicians. In the problem, Sergio **examined** a geometric shape and thought
         there might be something new to learn. You then helped him by finding possible
         relationships and writing a conjecture. Then, to find out if the conjecture was true
         for all triangles, you wrote a convincing argument (or proof). This process is
         summarized in the diagram below.

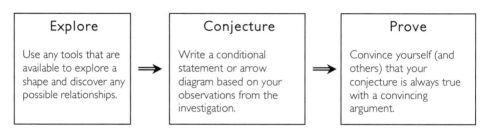

Explore		Conjecture		Prove
Use any tools that are available to explore a shape and discover any possible relationships.	⟹	Write a conditional statement or arrow diagram based on your observations from the investigation.	⟹	Convince yourself (and others) that your conjecture is always true with a convincing argument.

         Discuss this process with the class and describe when you have used this process
         before (either in this class or outside of class). Why do mathematicians rely on this
         process?

7-90.   RIGHT TRAPEZOIDS

         Consecutive angles of a polygon occur at opposite
         ends of a side of the polygon. What can you learn
         about a quadrilateral with two consecutive right
         angles?

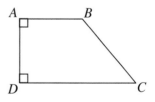

         a.   **EXPLORE: Examine** the quadrilateral at right
              with two consecutive right angles. What do you
              think is true of $\overline{AB}$ and $\overline{DC}$?

         b.   **CONJECTURE:** Write a conjecture about what type of quadrilateral has two
              consecutive right angles. Write your conjecture in conditional ("If…, then…")
              form.

         c.   **PROVE:** Prove that your conjecture from part (b) is true for all quadrilaterals
              with two consecutive right angles. Write your proof using the two-column
              format introduced in Lesson 7.2.4. (Hint: Look for angle relationships.)

         d.   The quadrilateral you worked with in this problem is called a **right trapezoid**.
              Are all quadrilaterals with two right angles a right trapezoid?

7-91.   ISOSCELES TRAPEZOIDS

An **isosceles trapezoid** is a trapezoid with a pair of
congruent base angles. What can you learn about
the sides of an isosceles trapezoid?

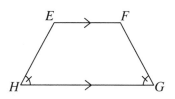

a.   **EXPLORE: Examine** *EFGH* at right. How
     do the side lengths appear to be related?

b.   **CONJECTURE:** Write a conjecture about
     side lengths in an isosceles trapezoid. Write
     your conjecture in conditional ("If…, then…")
     form.

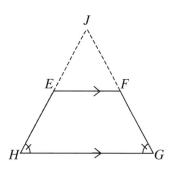

c.   **PROVE:** Now prove that your conjecture
     from part (b) is true for all isosceles
     trapezoids. Write your proof using the two-
     column format introduced in Lesson 7.2.5.
     To help you get started, the isosceles
     trapezoid is shown at right with its sides
     extended to form a triangle.

7-92.   Add the theorems you have proved in this lesson to your
         Theorem Toolkit (your Lesson 7.2.1A Resource Page). Be
         sure that your records include diagrams for each statement.

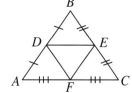

Ⓜ ETHODS AND MEANINGS

Triangle Midsegment Theorem

MATH NOTES

A **midsegment** of a triangle is a segment that connects the midpoints
of any two sides of a triangle. Every triangle has three midsegments,
as shown below.

A midsegment between two sides of a triangle is
half the length of and parallel to the third side of
the triangle. For example, in $\triangle ABC$ at right, $\overline{DE}$
is a midsegment, $\overline{DE} \parallel \overline{AC}$, and $DE = \frac{1}{2}AC$.

Review & Preview

7-93. One way a shape can be special is to have two congruent sides. For example, an isosceles triangle is special because it has a pair of sides that are the same length. Think about all the shapes you know and list the other special properties shapes can have. List as many as you can. Be ready to share your list with the class at the beginning of Lesson 7.3.1.

7-94. Carefully **examine** each diagram below and explain why the geometric figure cannot exist. Support your statements with **reasons**. If a line looks straight, assume that it is.

a.

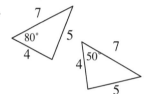

b.

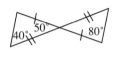

c.

d.

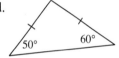

e.

f.

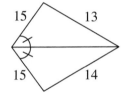

7-95. For each pair of numbers, find the number that is exactly halfway between them.

a.　9 and 15　　　　　　b.　3 and 27　　　　　　c.　10 and 21

7-96. Penn started the proof below to show that if $\overline{AD} \parallel \overline{EH}$ and $\overline{BF} \parallel \overline{CG}$, then $a = d$. Unfortunately, he did not provide reasons for his proof. Copy his proof and provide **justification** for each statement.

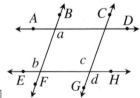

Statements	Reasons
1. $\overline{AD} \parallel \overline{EH}$ and $\overline{BF} \parallel \overline{CG}$	
2. $a = b$	
3. $b = c$	
4. $a = c$	
5. $c = d$	
6. $a = d$	

7-97.   After finding out that her kitchen does not conform
         to industry standards, Lashayia is back to the
         drawing board. (See problem 7-73). Where can she
         locate her sink along her top counter so that its
         distance from the stove and refrigerator is as small
         as possible? And will this location keep her
         perimeter below 26 feet? Show all work.

7.3.1 What makes a quadrilateral special?

· ·

Studying Quadrilaterals on a Coordinate Grid

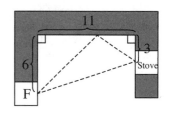

In Section 7.2 you **investigated** special types of quadrilaterals, such as parallelograms, kites,
and rhombi. Each of these quadrilaterals has special properties you have proved: parallel sides,
sides of equal length, equal opposite angles, bisected diagonals, etc.

But not all quadrilaterals have a special name. How can you tell if a quadrilateral belongs to
one of these types? And if a quadrilateral doesn't have a special name, can it still have special
properties? In Section 7.3 you will use both algebra and geometry to **investigate** quadrilaterals
defined on coordinate grids.

7-98.   PROPERTIES OF SHAPES

         Think about the special quadrilaterals you have
         studied in this chapter. Each shape has many
         properties that make it special. For example, a
         rhombus has two diagonals that are perpendicular.
         With the class, brainstorm the other types of
         properties that a shape can have. You may want to
         refer to your work from problem 7-93. Be ready to
         share your list with the class.

7-99.   Review some of the algebra **tools** you already have. On graph paper, draw $\overline{AB}$ given
         $A(0, 8)$ and $B(9, 2)$, and $\overline{CD}$ given $C(1, 3)$ and $D(9, 15)$.

         a.   Draw these two segments on a coordinate grid. Find the length of each
               segment.

         b.   Find the equation of $\overrightarrow{AB}$ and the equation of $\overrightarrow{CD}$. Write both equations in
               $y = mx + b$ form.

         c.   Is $\overline{AB} \, / / \, \overline{CD}$? Is $\overline{AB} \perp \overline{CD}$? **Justify** your answer.

         d.   Use algebra to find the coordinates of the point where $\overline{AB}$ and $\overline{CD}$ intersect.

7-100.   AM I SPECIAL?

Shayla just drew quadrilateral *SHAY*, shown at
right. The coordinates of its vertices are:

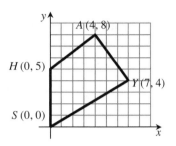

S(0, 0)   H(0, 5)   A(4, 8)   Y(7, 4)

a.   Shayla thinks her quadrilateral is a trapezoid.
Is she correct? Be prepared to **justify** your
answer to the class.

b.   Does Shayla's quadrilateral look like it is one of the other kinds of special
quadrilaterals you have studied? If so, which one?

c.   Even if Shayla's quadrilateral doesn't have a special name, it may still have
some special properties like the ones you listed in problem 7-98. Use algebra
and geometry tools to **investigate** Shayla's quadrilateral and see if it has any
special properties. If you find any special properties, be ready to **justify** your
claim that this property is present.

7-101.   THE MUST BE / COULD BE GAME

Mr. Quincey plays a game with his class. He says, "My quadrilateral has four right
angles." His students say, "Then it *must be* a rectangle" and "It *could be* a square."
For each description of a quadrilateral below, say what special type the quadrilateral
must be and/or what special type the quadrilateral *could be*. Look out: Some
descriptions may have no "must be"s, and some descriptions may have many "could
be"s!

a.   "My quadrilateral has four equal sides."

b.   "My quadrilateral has two pairs of
opposite parallel sides."

c.   "My quadrilateral has two consecutive
right angles."

d.   "My quadrilateral has two pairs of equal sides."

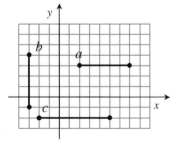

7-102.   The diagram at right shows three bold segments.
Find the coordinates of the midpoint of each
segment.

7-103.   **Examine** the diagram at right.

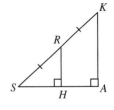

a.   Are the triangles in this diagram similar? **Justify** your answer.

b.   What is the relationship between the lengths of *HR* and *AK*? Between the lengths of *SH* and *SA*? Between the lengths of *SH* and *HA*?

c.   If *SK* = 20 units and *RH* = 8 units, what is *DA* ?

7-104.   For each pair of triangles below, determine if the triangles are congruent. If the triangles are congruent, state the congruence property that **justifies** your conclusion. If you cannot conclude that the triangles are congruent, explain why not.

a.   △*CAB* ≅ △_____

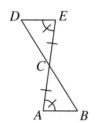

b.   △*CBD* ≅ △_____

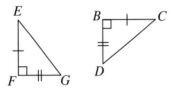

c.   △*LJI* ≅ △_____

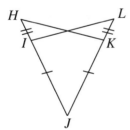

d.   △*PRQ* ≅ △_____

7-105.   Carolina compared her proof to that of Penn in problem 7-96.
          Like him, she wanted to prove that if $\overline{AD} \parallel \overline{EH}$ and $\overline{BF} \parallel \overline{CG}$,
          then $a = d$. Unfortunately, her statements were in a different
          order. **Examine** her proof below and help her decide if her
          statements are in a logical order in order to prove that $a = d$.

Statements	Reasons
1. $\overline{AD} \parallel \overline{EH}$ and $\overline{BF} \parallel \overline{CG}$	Given
2. $a = b$	If lines are parallel, alternate interior angles are equal.
3. $a = c$	Substitution
4. $b = c$	If lines are parallel, corresponding angles are equal.
5. $c = d$	Vertical angles are equal.
6. $a = d$	Substitution

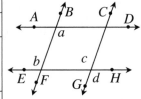

7-106.   Describe what the minimum information you would need to know about the shapes
          below in order to identify it correctly. For example, to know that a shape is a square,
          you must know that it has four sides of equal length and at least one right angle. Be
          as thorough as possible.

          a.   rhombus   b.   trapezoid

7.3.2 How can I find the midpoint?

Coordinate Geometry and Midpoints

In Lesson 7.3.1, you applied your existing algebraic tools to analyze geometric shapes on a
coordinate grid. What other algebraic processes can help us analyze shapes? And what else can
be learned about geometric shapes?

7-107.   Cassie wants to confirm her theorem on midsegments (from Lesson 7.2.6) using a coordinate grid. She started with $\triangle ABC$, with $A(0, 0)$, $B(2, 6)$, and $C(7, 0)$.

   a.   Graph $\triangle ABC$ on graph paper.

   b.   With your team, find the coordinates of P, the midpoint of $\overline{AB}$. Likewise, find the coordinates of Q, the midpoint of $\overline{BC}$.

   c.   Verify that the length of the midsegment, $\overline{PQ}$, is half the length of $\overline{AC}$. Also verify that $\overline{PQ}$ is parallel to $\overline{AC}$.

7-108.   As Cassie worked on problem 7-107, her teammate, Esther, had difficulty finding the midpoint of $\overline{BC}$. The study team decided to try to find another way to find the midpoint of a line segment.

   a.   As part of her team, Cassie wants you to draw $\overline{AM}$, with $A(3, 4)$ and $M(8, 11)$, on graph paper. Then extend the line segment to find a point B so that M is the midpoint of $\overline{AB}$. **Justify** your location of point B by drawing and writing numbers on the graph.

   b.   Esther thinks she understands how to find the midpoint on a graph. "I always look for the middle of the line segment. But what if the coordinates are not easy to graph?" she asks. With your team, find the midpoint of $\overline{KL}$ if $K(2, 125)$ and $L(98, 15)$. Be ready to share your method with the class.

   c.   Test your team's method by verifying that the midpoint between $(-5, 7)$ and $(9, 4)$ is $(2, 5.5)$.

7-109.   Randy has decided to study the triangle graphed at right.

   a.   Consider all the special properties this triangle can have. Without using any algebra tools, predict the best name for this triangle.

   b.   For your answer to part (a) to be correct, what is the minimum amount of information that must be true about $\triangle RND$?

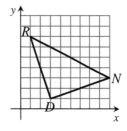

   c.   Use your algebra tools to verify each of the properties you listed in part (b). If you need, you may change your prediction of the shape of $\triangle RND$.

   d.   Randy wonders if there is anything special about the midpoint of $\overline{RN}$. Find the midpoint M, and then find the lengths of $\overline{RM}$, $\overline{DM}$, and $\overline{MN}$. What do you notice?

7-110.   Tomika remembers that the diagonals of a rhombus are perpendicular to each other.

a.   Graph on *ABCD* if *A*(1, 4), *B*(6, 6), *C*(4, 1), and *D*(–1, –1). Is *ABCD* a rhombus? Show how you know.

b.   Find the equation of the lines on which the diagonals lie. That is, find the equations of $\overrightarrow{AC}$ and $\overrightarrow{BD}$.

c.   Compare the slopes of $\overrightarrow{AC}$ and $\overrightarrow{BD}$. What do you notice?

7-111.   In your Learning Log, explain what a midpoint is and the method you prefer for finding midpoints of a line segment when given the coordinates of its endpoints. Include any diagram or example that helps explain why this method works. Title this entry "Finding a Midpoint" and include today's date.

MⒶTHODS AND MEANINGS

MATH NOTES

Coordinate Geometry

Coordinate geometry is the study of geometry on a coordinate grid. Using common algebraic and geometric tools, you can learn more about a shape such as the if it has a right angle or if two sides are the same length.

One useful tool is the Pythagorean Theorem. For example, the Pythagorean Theorem could be used to determine the length of side $\overline{AB}$ of *ABCD* at right. By drawing the slope triangle between points *A* and *B*, the length of $\overline{AB}$ can be found to be $\sqrt{2^2 + 5^2} = \sqrt{29}$ units.

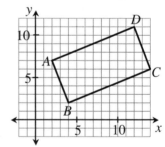

Similarly, slope can help analyze the relationships between the sides of a shape. If the slopes of two sides of a shape are equal, then those sides are **parallel**. For example, since the slope of $\overline{BC} = \frac{2}{5}$ and the slope of $\overline{AD} = \frac{2}{5}$, then $\overline{BC} \,/\!/\, \overline{AD}$.

Also, if the slopes of two sides of a shape are opposite reciprocals, then the sides are **perpendicular** (meaning they form a 90° angle). For example, since the slope of $\overline{BC} = \frac{2}{5}$ and the slope of $\overline{AB} = -\frac{5}{2}$, then $\overline{BC} \perp \overline{AB}$.

By using multiple algebraic and geometric tools, you can identify shapes. For example, further analysis of the sides and angles of *ABCD* above shows that $AB = DC$ and $BC = AD$. Furthermore, all four angles measure 90°. These facts together indicate that *ABCD* must be a rectangle.

7-112. Find another valid, logical order for the statements for Penn's proof from problem 7-96. Explain how you know that changing the order the way you did does not affect the logic.

7-113. Each of these numberlines shows a segment in bold. Find the midpoint of the segment in bold. (Note that the diagrams are *not* to scale.)

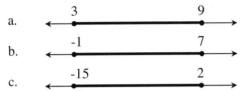

a.

b.

c.

7-114. **Examine** the diagram at right.

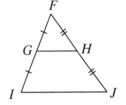

a. Are the triangles in this diagram similar? Explain.

b. Name all the pairs of congruent angles in this diagram you can.

c. Are $\overline{GH}$ and $\overline{IJ}$ parallel? Explain how you know.

d. If $GH = 4x - 3$ and $IJ = 3x + 14$, find x. Then find the length of $\overline{GH}$.

7-115. Consider $\triangle ABC$ with vertices $A(2, 3)$, $B(6, 3)$, and $C(6, 10)$.

a. Draw $\triangle ABC$ on graph paper. What kind of triangle is $\triangle ABC$?

b. Reflect $\triangle ABC$ across $\overline{AC}$. Where is B'? And what shape is $ABCB'$?

7-116. MUST BE / COULD BE

Here are some more challenges from Mr. Quincey. For each description of a quadrilateral below, say what special type the quadrilateral *must be* and/or what special type the quadrilateral *could be*. Look out: Some descriptions may have no "must be"s, and some descriptions may have many "could be"s!

a. "My quadrilateral has a pair of equal sides and a pair of parallel sides."

b. "The diagonals of my quadrilateral bisect each other."

What kind of quadrilateral is it?

Quadrilaterals on a Coordinate Plane

Today you will use algebra tools to **investigate** the properties of a quadrilateral and then will use those properties to identify the type of quadrilateral it is.

7-117.   MUST BE / COULD BE

Mr. Quincey has some new challenges for you! For each description below, decide what special type the quadrilateral *must be* and/or what special type the quadrilateral *could be*. Look out: Some descriptions may have no "must be"s, and some descriptions may have many "could be"s!

a.   "My quadrilateral has three right angles."

b.   "My quadrilateral has a pair of parallel sides."

c.   "My quadrilateral has two consecutive equal angles."

7-118.   THE SHAPE FACTORY

You just got a job in the Quadrilaterals Division of your uncle's Shape Factory. In the old days, customers called up your uncle and described the quadrilaterals they wanted over the phone: "I'd like a parallelogram with…". "But nowadays," your uncle says, "customers using computers have been emailing orders in lots of different ways." Your uncle needs your team to help analyze his most recent orders listed below to identify the quadrilaterals and help the shape-makers know what to produce.

Your Task: For each of the quadrilateral orders listed below,

- Create a diagram of the quadrilateral on graph paper.

- Decide if the quadrilateral ordered has a special name. To help the shape-makers, your name must be as specific as possible. (Don't just call a shape a rectangle when it's also a square!)

- Record and be ready to present a **justification** that the quadrilateral ordered must be the kind you say it is. It is not enough to say that a quadrilateral *looks* like it is of a certain type or *looks* like it has a certain property. Customers will want to be sure they get the type of quadrilateral they ordered!

Problem continues on next page →

Geometry Connections

7-118.   *Problem continued from previous page.*

Discussion Points

What special properties might a quadrilateral have?

What algebra tools could be useful?

What types of quadrilaterals might be ordered?

The orders:

a.   A quadrilateral formed by the intersection of these lines:

$$y = -\tfrac{3}{2}x + 3 \qquad y = \tfrac{3}{2}x - 3 \qquad y = -\tfrac{3}{2}x + 9 \qquad y = \tfrac{3}{2}x + 3$$

b.   A quadrilateral with vertices at these points:

$A(0, 2)$ $\qquad\qquad$ $B(1, 0)$ $\qquad\qquad$ $C(7, 3)$ $\qquad\qquad$ $D(4, 4)$

c.   A quadrilateral with vertices at these points:

$W(0, 5)$ $\qquad\qquad$ $X(2, 7)$ $\qquad\qquad$ $Y(5, 7)$ $\qquad\qquad$ $Z(5, 1)$

Ⓜ️ETHODS AND MEANINGS

MATH NOTES

Finding a Midpoint

A **midpoint** is a point that divides a line segment into two parts of equal length. For example, M is the midpoint of $\overline{AB}$ at right.

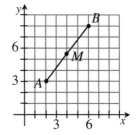

There are several ways to find the midpoint of a line segment if the coordinates of the endpoints are known. One way is to average the x-coordinates and to average the y-coordinates. Thus, if $A(2, 3)$ and $B(6, 8)$, then the x-coordinate of M is $\frac{2+6}{2} = 4$ and the y-coordinate is $\frac{3+8}{2} = 5.5$. So M is at $(4, 5.5)$.

7-119. Each problem below gives the endpoints of a segment. Find the coordinates of the midpoint of the segment. If you need help, consult the Math Notes box for this lesson.

   a. (5, 2) and (11, 14)   b. (3, 8) and (10, 4)

   c. (−3, 11) and (5, 6)   d. (−4, −1) and (8, 9)

7-120. Below are the equations of two lines and the coordinates of three points. For each line, determine which of the points, if any, lie on that line. (There may be more than one!)

   a. $y = \frac{1}{3}x + 15$ $X(0, 15)$ $Y(3, 16)$ $Z(7, 0)$

   b. $y - 16 = -4(x - 3)$

7-121. MUST BE / COULD BE

   Here are some more challenges from Mr. Quincey. For each description of a quadrilateral below, say what special type the quadrilateral *must be* and/or what special type the quadrilateral *could be*. Look out: Some descriptions may have no "must be"s, and some descriptions may have many "could be"s!

   a. "My quadrilateral has two right angles."

   b. "The diagonals of my quadrilateral are perpendicular."

7-122. The angle created by a hinged mirror when forming a regular polygon is called a **central angle**. For example, $\angle ABC$ in the diagram at right is the central angle of the regular hexagon.

   a. If the central angle of a regular polygon measures 18°, how many sides does the polygon have?

   b. Can a central angle measure 90°? 180°? 13°? For each angle measure, explain how you know.

7-123. On graph paper, graph lines $\overrightarrow{AB}$ and $\overrightarrow{CD}$ if $\overrightarrow{AB}$ can be represented by $y = -\frac{4}{3}x + 5$ and $\overrightarrow{CD}$ can be represented by $y = \frac{3}{4}x - 1$. Label their intersection E.

   a. What is the relationship between the lines? How do you know?

   b. If E is a midpoint of $\overline{CD}$, what type of quadrilateral could $ABCD$ be? Is there more than one possible type? Explain how you know.

Chapter 7 Closure What have I learned?

The activities below offer you a chance to reflect on what you have learned during this chapter. As you work, look for concepts that you feel very comfortable with, ideas that you would like to learn more about, and topics you need more help with. Look for **connections** between ideas as well as **connections** with material you learned previously.

① TEAM BRAINSTORM

With your team, brainstorm a list for each of the following three topics. Be as detailed as you can. How long can you make your list? Challenge yourselves. Be prepared to share your team's ideas with the class.

Topics: What have you studied in this chapter? What ideas and words were important in what you learned? Remember to be as detailed as you can.

Connections: How are the topics, ideas, and words that you learned in previous courses are **connected** to the new ideas in this chapter? Again, make your list as long as you can.

The following is a list of the vocabulary used in this chapter. The words that appear in bold are new to this chapter. Make sure that you are familiar with all of these words and know what they mean. Refer to the glossary or index for any words that you do not yet understand.

bisect	**central angle**	circle
congruent	conjecture	**coordinate geometry**
diagonal	**diameter**	**exterior angle**
kite	**midpoint**	**midsegment**
opposite	parallel	parallelogram
perpendicular	**proof**	**quadrilateral**
radius	rectangle	**Reflexive Property**
regular polygon	**rhombus**	square
theorem	**three-dimensional**	trapezoid
two-column proof	**two-dimensional**	

Make a concept map showing all of the **connections** you can find among the key words and ideas listed above. To show a **connection** between two words, draw a line between them and explain the **connection**, as shown in the example below. A word can be **connected** to any other word as long as there is a **justified connection**. For each key word or idea, provide a sketch of an example.

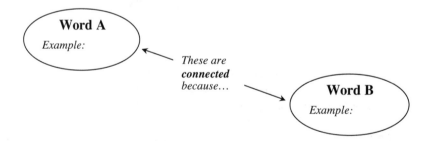

Your teacher may provide you with vocabulary cards to help you get started. If you use the cards to plan your concept map, be sure either to re-draw your concept map on your paper or to glue the vocabulary cards to a poster with all of the **connections** explained for others to see and understand.

While you are making your map, your team may think of related words or ideas that are not listed above. Be sure to include these ideas on your concept map.

③   SUMMARIZING MY UNDERSTANDING

This section gives you an opportunity to show what you know about certain math topics or ideas. Your teacher will give you directions for exactly how to do this. Your teacher may give you a "GO" page to work on.

To use the Quadrilateral "GO" (or Graphic Organizer), complete the diagram to show the relationships between special quadrilaterals. For each type of quadrilateral, draw a diagram and list the special properties (if any) that it has. Use words to explain how one type of quadrilateral is related to the others. The diagram is started below for you. You will need to add: **quadrilateral**, **kite**, **square**, **rectangle**, and **trapezoid**. Each quadrilateral should be connected to at least one other type of quadrilateral, but some can be related to more than one.

④   WHAT HAVE I LEARNED?

This section will help you evaluate which types of problems you have seen with which you feel comfortable and those with which you need more help. This section will appear at the end of every chapter to help you check your understanding. Even if your teacher does not assign this section, it is a good idea to try these problems and find out for yourself what you know and what you need to work on.

Solve each problem as completely as you can. The table at the end of this closure section has answers to these problems. It also tells you where you can find additional help and practice on problems like these.

CL 7-124. **Examine** the triangle pairs below, which are not necessarily drawn to scale. For each pair, determine:

- if they must be congruent, and state the congruence property (such as SAS ≅) and give a correct congruence statement (such as $\triangle PQR \cong \triangle STU$)

- if there is not enough information, and explain why.

- if they cannot be congruent, and explain why.

a.

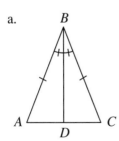

b.

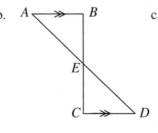

c.

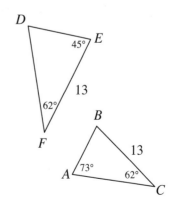

CL 7-125. Complete the following statements.

     a.   If $\triangle YSR \cong \triangle NVD$, then $\overline{DV} \cong$ __?__ and $m\measuredangle RYS =$ __?__

     b.   If $\overrightarrow{AB}$ bisects $\measuredangle DAC$, then __?__ $\cong$ __?__ ?

     c.   In $\triangle WQY$, if $\measuredangle WQY \cong \measuredangle QWY$, then __?__ $\cong$ __?__ .

     d.   If $ABCD$ is a parallelogram, and $m\measuredangle B = 148°$, then $m\measuredangle C =$ __?__ .

CL 7-126. Julius set his hinged mirror so that its angle was 72° and the core region was isosceles, as shown at right.

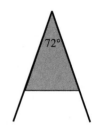

     a.   How many sides did his resulting polygon have? Show how you know.

     b.   What is another name for this polygon?

CL 7-127. Kelly started the proof below to show that if $\overline{TC} \cong \overline{TM}$ and $\overline{AT}$ bisects $\angle CTM$, then $\overline{CA} \cong \overline{MA}$. Copy and complete her proof.

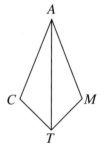

Statements	Reasons
1. $\overline{TC} \cong \overline{TM}$ and $\overline{AT}$ bisects $\angle CTM$	
2.	Definition of bisect
3. $\overline{AT} \cong \overline{AT}$	
4.	
5.	$\cong \Delta s \rightarrow \rightarrow$ parts

CL 7-128. $ABCD$ is a parallelogram. If $A(3, -4)$, $B(6, 2)$, $C(4, 6)$, then what are the possible locations of point D? Draw a graph and **justify** your answer.

CL 7-129. On graph paper, draw quadrilateral $MNPQ$ if $M(1, 7)$, $N(-2, 2)$, $P(3, -1)$, and $Q(6, 4)$.

     a.   Find the slopes of $\overline{MN}$ and $\overline{NP}$. What can you conclude about $\measuredangle MNP$?

     b.   What is the best name for $MNPQ$? **Justify** your answer.

     c.   Which diagonal is longer? Explain how you know your answer is correct.

     d.   Find the midpoint of $\overline{MN}$.

CL 7-130. **Examine** the geometric relationships in each of the diagrams below. For each one, write and solve an equation to find the value of the variable. Name any geometric property or conjecture that you used.

a.

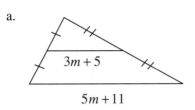

$3m + 5$

$5m + 11$

b. *PQRS* is a rhombus with perimeter = 28 units. *PR* = 8 units.

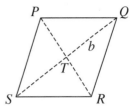

c.

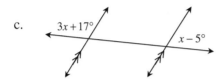

$3x + 17°$

$x - 5°$

d.

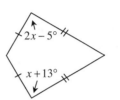

$2x - 5°$

$x + 13°$

CL 7-131. Given the information in the diagram at right, prove that $\triangle WXY \cong \triangle YZW$ using either a flowchart or a two-column proof.

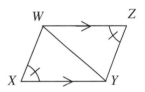

CL 7-132. Check your answers using the table at the end of the closure section. Which problems do you feel confident about? Which problems were hard? Use the table to make a list of topics you need help on and a list of topics you need to practice more.

⑤ HOW AM I THINKING?

This course focuses on five different **Ways of Thinking**: investigating, examining, reasoning & justifying, visualizing, and choosing a strategy/tool. These are some of the ways in which you think while trying to make sense of a concept or to solve a problem (even outside of math class). During this chapter, you have probably used each Way of Thinking multiple times without even realizing it!

Choose three of these Ways of Thinking that you remember using while working in this chapter. For each Way of Thinking that you choose, show and explain where you used it and how you used it. Describe why thinking in this way helped you solve a particular problem or understand something new. Be sure to include examples to demonstrate your thinking.

Answers and Support for Closure Activity #4
What Have I Learned?

Problem	Solution	Need Help?	More Practice
CL 7-124.	a. Congruent (SAS ≅), $\triangle ABD \cong \triangle CBD$ b. Not enough information (the triangles are similar (AA ~), but no side lengths are given to know if they are the same size.) c. Congruent (ASA ≅ or AAS ≅), $\triangle ABC \cong \triangle DEF$	Lessons 2.1.4, 2.2.1, and 6.1.3 Math Notes boxes	Problems 7-6, 7-14, 7-28, 7-42, 7-53, 7-66, 7-87, 7-104
CL 7-125.	a. $\overline{DV} \cong \overline{RS}$, $m\angle RYS = m\angle DNV$ b. $\angle DAB \cong \angle CAB$ c. $\overline{WY} \cong \overline{QY}$ d. $m\angle C = 32°$	Lessons 2.1.4, 3.1.4, and 7.1.3 Math Notes boxes, problems 7-49 and 7-71	Problems 7-26, 7-42, 7-53, 7-87
CL 7-126.	a. $360° \div 72° = 5$ sides b. regular pentagon	Lesson 7.1.4 Math Notes box, problems 7-37 and 7-38	Problems 7-39, 7-50, 7-122
CL 7-127.	<table><tr><th>Statements</th><th>Reasons</th></tr><tr><td>1. $\overline{TC} \cong \overline{TM}$ and $\overline{AT}$ bisects $\angle CTM$</td><td>Given</td></tr><tr><td>2. $\angle CTA \cong \angle MTA$</td><td>Definition of bisect</td></tr><tr><td>3. $\overline{AT} \cong \overline{AT}$</td><td>Reflexive Property</td></tr><tr><td>4. $\triangle CAT \cong \triangle MAT$</td><td>SAS ≅</td></tr><tr><td>5. $\overline{CA} \cong \overline{MA}$</td><td>≅ Δs → ≅ parts</td></tr></table>	Lessons 6.1.3, 7.1.3, and 7.2.1 Math Notes boxes, problems 7-45 and 7-79	Problems 7-61, 7-78, 7-85, 7-87, 7-96, 7-104, 7-105
CL 7-128.	Point D is at $(1, 0)$ or at $(5, -8)$.	Lessons 7.2.3 and 7.3.2 Math Notes boxes	Problems 7-29, 7-60, 7-69

Problem	Solution	Need Help?	More Practice

CL 7-129.
a. Slope of $\overline{MN} = \frac{5}{3}$ and $\overline{NP} = -\frac{3}{5}$, $\angle MNP$ is a right angle.

b. It is a square because all sides are equal and all angles are right angles.

c. The diagonals have equal length. Each is $\sqrt{68}$ units long.

d. $(-\frac{1}{2}, \frac{9}{2})$

Need Help? Lessons 3.2.4, 7.2.3, 7.3.2, and 7.3.3 Math Notes boxes, problem 7-40

More Practice: Problems 7-18, 7-27, 7-29, 7-32, 7-38, 7-40, 7-41, 7-43, 7-69, 7-99, 7-109, 7-110, 7-118, 7-119

CL 7-130.
a. $2(3m+5) = 5m+11$, $m = 1$

b. $b^2 + 4^2 = 7^2$, so $b = \sqrt{33} \approx 5.74$ units

c. $3x + 17° + x - 5° = 180°$, so $x = 42°$

d. $2x - 5° = x + 13°$, so $x = 18°$

Need Help? Lessons 2.1.4, 7.2.4, and 7.2.6 Math Notes boxes, problem 7-49

More Practice: Problems 7-16, 7-33, 7-49, 7-52, 7-70

CL 7-131.

$\overline{WZ} \parallel \overline{YX}$
Given

$\angle Z \cong \angle X$ $\angle ZWY \cong \angle XYW$ $\overline{WY} \cong \overline{YW}$
Given   If lines are //, then alt. int. angles are ≅.   Reflexive Property

$\triangle XYW \cong \triangle ZWY$
AAS ≅

Need Help? Lessons 3.2.4, 6.1.3, and 7.2.1 Math Notes boxes, problems 7-45 and 7-56

More Practice: Problems 7-61, 7-78, 7-85, 7-87, 7-96, 7-104, 7-105

CHAPTER 8

Polygons and Circles

In previous chapters, you have extensively studied triangles and quadrilaterals to learn more about their sides and angles. In this chapter, you will broaden your focus to include polygons with 5, 8, 10, and even 100 sides. You will develop a way to find the area and perimeter of a regular polygon and will study how the area and perimeter changes as the number of sides increases.

In Section 8.2, you will re-examine similar shapes to study what happens to the area and perimeter of a shape when the shape is enlarged or reduced.

Finally, in Section 8.3, you will connect your understanding of polygons with your knowledge of the area ratios of similar figures to find the area and circumference of circles of all sizes.

Guiding Questions

Think about these questions throughout this chapter:

How can I measure a polygon?

How does the area change?

Is there another method?

What if the polygon has infinite sides?

What's the connection?

In this chapter, you will learn:

> About special types of polygons, such as regular and non-convex polygons.

> How the measures of the interior and exterior angles of a regular polygon are related to the number of sides of the polygon.

> How the areas of similar figures are related.

> How to find the area and circumference of a circle and parts of circles and use this ability to solve problems in various contexts.

Chapter Outline

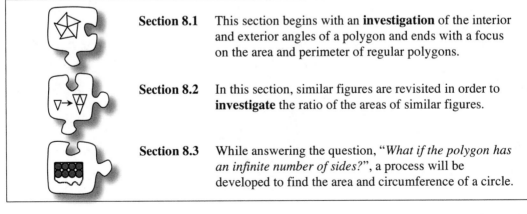

Section 8.1   This section begins with an **investigation** of the interior and exterior angles of a polygon and ends with a focus on the area and perimeter of regular polygons.

Section 8.2   In this section, similar figures are revisited in order to **investigate** the ratio of the areas of similar figures.

Section 8.3   While answering the question, "*What if the polygon has an infinite number of sides?*", a process will be developed to find the area and circumference of a circle.

8.1.1 How can I build it?

Pinwheels and Polygons

In previous chapters, you have studied triangles and quadrilaterals. In Chapter 8, you will broaden your focus to include all polygons and will study what triangles can tell us about shapes with 5, 8, or even 100 sides.

By the end of this lesson, you should be able to answer these questions:

How can you use the number of sides of a regular polygon to find the measure of the central angle?

What type of triangle is needed to form a regular polygon?

8-1. PINWHEELS AND POLYGONS

Inez loves pinwheels. One day in class, she noticed that if she put three congruent triangles together so that one set of corresponding angles are adjacent, she could make a shape that looks like a pinwheel.

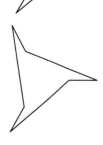

a. Can you determine any of the angles of her triangles? Explain how you found your answer.

b. The overall shape (outline) of Inez's pinwheel is shown at right. How many sides does it have? What is another name for this shape?

c. Inez's shape is an example of a **polygon** because it is a closed, two-dimensional figure made of straight line segments connected end-to-end. As you study polygons in this course, it is useful to use the names below because they identify how many sides a particular polygon has. Some of these words may be familiar, while others may be new. On your paper, draw an example of a *heptagon*.

Name of Polygon	Number of Sides	Name of Polygon	Number of Sides
Triangle	3	Octagon	8
Quadrilateral	4	Nonagon	9
Pentagon	5	Decagon	10
Hexagon	6	11-gon	11
Heptagon	7	*n*-gon	*n*

8-2.   Inez is very excited. She wants to know if
        you can build a pinwheel using *any* angle of
        her triangle. Obtain a Lesson 8.1.1 Resource
        Page from your teacher and cut out Inez's
        triangles. Then work with your team to build
        pinwheels and polygons by placing different
        corresponding angles together at the center.
        You will need to use the triangles from all
        four team members together to build one
        shape. Be ready to share your results with
        the class.

8-3.   Jorge likes Inez's pinwheels but wonders, "Will all triangles build a pinwheel or a
        polygon?"

        a. If you have not already done so, cut out the remaining triangles on the Lesson
            8.1.1 Resource Page. Work together to determine which congruent triangles
            can build a pinwheel (or polygon) when corresponding angles are placed
            together at the center. For each successful pinwheel, answer the questions
            below.

            • How many triangles did it take to build the pinwheel?

            • Calculate the measure of a **central angle** of the pinwheel. (Remember that
              a central angle is an angle of a triangle with a vertex at the center of the
              pinwheel.)

            • Is the shape familiar? Does it have a name? If so, what is it?

        b. Explain why one triangle may be able to create a pinwheel or polygon while
            another triangle cannot.

        c. Jorge has a triangle with angle measures 32°, 40°, and 108°. Will this triangle
            be able to form a pinwheel? Explain.

8-4.   Jasmine wants to create a pinwheel with equilateral triangles.

a.   How many equilateral triangles will she need? Explain how you know.

b.   What is the name for the polygon she created?

c.   Jasmine's shape is an example of a **convex polygon**, while Inez's shape, shown at right, is **non-convex**. Study the examples of convex and non-convex polygons below and then write a definition of a convex polygon on your paper.

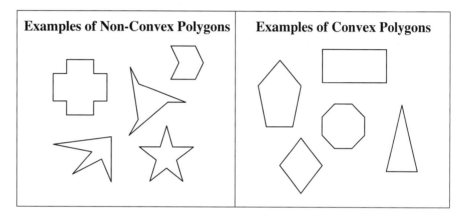

8-5.   When corresponding angles are placed together, why do some triangles form convex polygons while others result in non-convex polygons? Consider this as you answer the following questions.

a.   Carlisle wants to build a convex polygon using congruent triangles. He wants to select one of the triangles below to use. Which triangle(s) will build a convex polygon if multiple congruent triangles are placed together so that they share a common vertex and do not overlap? Explain how you know.

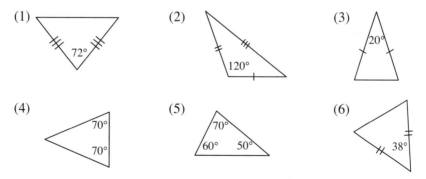

b.   For each triangle from part (a) that creates a convex polygon, how many sides would the polygon have? What name is most appropriate for the polygon?

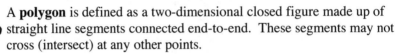

ETHODS AND MEANINGS

Convex and Non-Convex Polygons

A **polygon** is defined as a two-dimensional closed figure made up of straight line segments connected end-to-end. These segments may not cross (intersect) at any other points.

A polygon is referred to as a **regular polygon** if it is equilateral (all sides have the same length) and equiangular (all interior angles have equal measure). For example, the hexagon shown at right is a regular hexagon because all sides have the same length and each interior angle has the same measure.

A polygon is called **convex** if each pair of interior points can be connected by a segment without leaving the interior of the polygon. See the example of convex and non-convex shapes in problem 8-4.

8-6.   Solve for *x* in each diagram below.

a.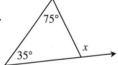

b.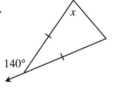

c.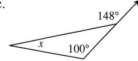

d.

8-7.   After solving for *x* in each of the diagrams in problem 8-6, Jerome thinks he sees a pattern. He notices that the measure of an exterior angle of a triangle is related to two of the angles of a triangle.

   a.   Do you see a pattern? To help find a pattern, study the results of problem 8-6.

Problem continues on next page →

8-7.   *Problem continued from previous page.*

b.   In the example at right, angles *a* and *b* are called **remote interior angles** of the given exterior angle because they are not adjacent to the exterior angle. Write a conjecture about the relationships between the remote interior and exterior angles of a triangle.

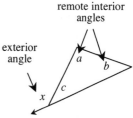

c.   Prove that the conjecture you wrote for part (b) is true for all triangles. Your proof can be written in any form, as long as it is convincing and provides **reasons** for all statements.

8-8.   **Examine** the geometric relationships in the diagram at right. Show all of the steps in your solutions for *x* and *y*.

8-9.   Steven has 100 congruent triangles that each has an angle measuring 15°. How many triangles would he need to use to make a pinwheel? Explain how you found your answer.

8-10.   Find the value of *x* in each diagram below, if possible. If the triangles are congruent, state which triangle congruence property was used. If the triangles are not congruent or if there is not enough information, state, "Cannot be determined."

a.

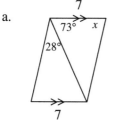

b.

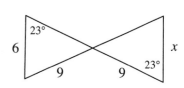

c.

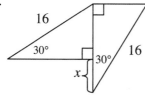

d.

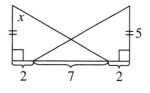

8-11.   Decide if the following statements are true or false. If a statement is false, provide a diagram of a counterexample.

a.   All squares are rectangles.

b.   All quadrilaterals are parallelograms.

c.   All rhombi are parallelograms.

d.   All squares are rhombi.

e.   The diagonals of a parallelogram bisect the angles.

8.1.2 What is its measure?

· ·

Interior Angles of a Polygon

In an earlier chapter you discovered that the sum of the interior angles of a triangle is always 180°. But what about the sum of the interior angles of other polygons, such as hexagons or decagons? Does it matter if the polygon is convex or not? Consider these questions today as you **investigate** the angles of a polygon.

8-12.   Copy the diagram of the regular pentagon at right onto your paper. Then, with your team, find the <u>sum</u> of the measures of the interior angles *as many ways as you can*. You may want to use the fact that the sum of the angles of a triangle is 180°. Be prepared to share your team's methods with the class.

8-13.   SUM OF THE INTERIOR ANGLES OF A POLYGON

In problem 8-12, you found the sum of the angles of a regular pentagon. But what about other polygons?

a.   Obtain a Lesson 8.1.2 Resource Page from your teacher. Then use one of the methods from problem 8-12 to find the sum of the interior angles of other polygons. Complete the table (also shown below) on the resource page.

Number of Sides of the Polygon	3	4	5	6	7	8	9	10	12
Sum of the Interior Angles of the Polygon	180°								

b.   Does the interior angle sum depend on whether the polygon is convex? Test this idea by drawing a few non-convex polygons (like the one at right) on your paper and determine if it matters whether the polygon is convex. Explain your findings.

c.   Find the sum of the interior angles of a 100-gon. Explain your **reasoning**.

d.   In your Learning Log, write an expression that represents the sum of the interior angles of an *n*-gon. Title this entry "Interior Angles of a Polygon" and include today's date.

8-14.   The pentagon at right has been dissected (broken up) into three triangles with the angles labeled as shown. Use the three triangles to prove that the sum of the interior angles of **any** pentagon is always 540°. If you need help, answer the questions below.

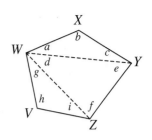

a.   What is the sum of the angles of a triangle? Use this fact to write three equations based on the triangles in the diagram.

b.   Add the three equations to create one long equation that represents the sum of all nine angles.

c.   Substitute the three-letter name for each angle of the pentagon for the lower case letters at each vertex of the pentagon. For example, $m\angle XYZ = c + e$.

8-15.   Use the angle relationships in each of the diagrams below to solve for the given variables. Show all work.

a.
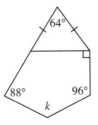

138°   106°
$m + 13°$
        $m - 9°$
120°
    133° m

b.
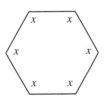

x x
x x
x x

c.
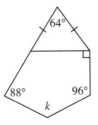

64°

88°   96°
    k

d.
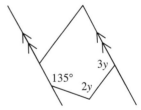

135° $3y$
    $2y$

METHODS AND MEANINGS

MATH NOTES

Special Quadrilateral Properties

In Chapter 7, you examined several special quadrilaterals and proved conjectures regarding many of their special properties. Review what you learned below.

Parallelogram: Opposite sides of a parallelogram are congruent and parallel. Opposite angles are congruent. Also, since the diagonals dissect the parallelogram into four congruent triangles, the diagonals bisect each other.

Parallelogram

Rhombus: Since a rhombus is a parallelogram, it has all of the properties of a parallelogram. In addition, its diagonals are perpendicular bisectors and bisect the angles of the rhombus.

Rhombus

Rectangle: Since a rectangle is a parallelogram, it has all of the properties of a parallelogram. In addition, its diagonals must be congruent.

Rectangle

Isosceles Trapezoid: The base angles (angles joined by a base) of an isosceles trapezoid are congruent.

Isosceles Trapezoid

Review & Preview

8-16. On graph paper, graph $\triangle ABC$ if $A(3, 0)$, $B(2, 7)$, and $C(6, 4)$.

    a. What is the best name for this triangle? **Justify** your answer using slope and/or lengths of sides.

    b. Find $m\angle A$. Explain how you found your answer.

8-17. The exterior angles of a quadrilateral are labeled a, b, c, and d in the diagram at right. Find the measures of a, b, c, and d and then find the sum of the exterior angles.

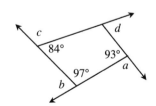

Geometry Connections

8-18.   Find the area and perimeter of the shape at right. Show all work.

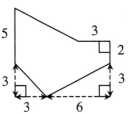

8-19.   Crystal is amazed! She graphed $\triangle ABC$ using the points $A(5, -1)$, $B(3, -7)$, and $C(6, -2)$. Then she rotated $\triangle ABC$ 90° counterclockwise (↺) about the origin to find $\triangle A'B'C'$. Meanwhile, her teammate took a different triangle ($\triangle TUV$) and rotated it 90° clockwise (↻) about the origin to find $\triangle T'U'V'$. Amazingly, $\triangle A'B'C'$ and $\triangle T'U'V'$ ended up using exactly the same points! Name the coordinates of the vertices of $\triangle TUV$.

8-20.   Suzette started to set up a proof to show that if $\overline{BC} \parallel \overline{EF}$, $\overline{AB} \parallel \overline{DE}$, and $AF = DC$, then $\overline{BC} \cong \overline{EF}$. **Examine** her work below. Then complete her missing statements and **reasons**.

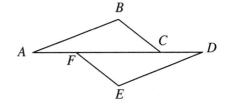

Statements	Reasons
1. $\overline{BC} \parallel \overline{EF}$, $\overline{AB} \parallel \overline{DE}$, and $AF = DC$	1.
2. $m\angle BCF = m\angle EFC$ and $m\angle EDF = m\angle CAB$	2.
3.	3. Reflexive Property
4. $AF + FC = CD + FC$	4. Additive Property of Equality (adding the same amount to both sides of an equation keeps the equation true)
5. $AC = DF$	5. Segment addition
6. $\triangle ABC \cong \triangle DEF$	6.
7.	7. $\cong \triangle s \rightarrow \cong$ parts

8-21.   **Multiple Choice:** Which equation below is **not** a correct statement based on the information in the diagram?

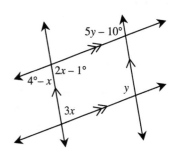

a. $3x + y = 180°$

b. $2x - 1° = 4° - x$

c. $2x - 1° = 5y - 10°$

d. $2x - 1° + 3x = 180°$

e.   None of these is correct

8.1.3 What if it is a regular polygon?

Angles of Regular Polygons

In Lesson 8.1.2 you discovered how to determine the sum of the interior angles of a polygon with any number of sides. But what more can you learn about a polygon? Today you will focus on the interior and exterior angles of regular polygons.

As you work today, keep the following focus questions in mind:

Does it matter if the polygon is regular?

Is there another way to find the answer?

What's the connection?

8-22. Diamonds, the most valuable naturally-occurring gem, have been popular for centuries because of their beauty, durability, and ability to reflect a spectrum of light. In 1919, a diamond cutter from Belgium, Marcel Tolkowsky, used his knowledge of geometry to design a new shape for a diamond, called the "round brilliant cut" (top view shown at right). He discovered that when diamonds are carefully cut with flat surfaces (called "facets" or "faces") in this design, the angles maximize the brilliance and reflective quality of the gem.

Notice that at the center of this design is a **regular octagon** with equal sides and equal interior angles. For a diamond cut in this design to achieve its maximum value, the octagon must be cut carefully and accurately. One miscalculation, and the value of the diamond can be cut in half!

a. Determine the measure of each interior angle of a regular octagon. Explain how you found your answer.

interior angle

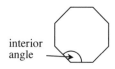

b. What about the interior angles of other regular polygons? Find the interior angles of a regular nonagon and a regular 100-gon.

c. Will the process you used for part (a) work for any regular polygon? Write an expression that will calculate the interior angle of an n-gon.

8-23.   Fern states, "If a triangle is equilateral, then all angles have equal measure and it must be a regular polygon." Does this logic work for polygons with more than three sides?

   a.   If all of the sides of a polygon (such as a quadrilateral) are equal, does that mean that the angles must be equal? If you can, draw a counterexample.

   b.   What if all the angles are equal? Does that force a polygon to be equilateral? Explain your thinking. Draw a counterexample on your paper if possible.

8-24.   Jeremy asks, "What about exterior angles? What can we learn about them?"

   a.   **Examine** the regular hexagon shown at right. Angle a is an example of an **exterior angle** because it is formed on the outside of the hexagon by extending one of its sides. Are all of the exterior angles of a regular polygon equal? Explain how you know.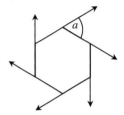

   b.   Find a. Be prepared to share how you found your answer.

   c.   This regular hexagon has six exterior angles, as shown in the diagram above. What is the sum of the exterior angles of a regular hexagon?

   d.   What about the exterior angles of other regular polygons? Explore this with your team. Have each team member choose a different shape from the list below to analyze. For each shape:

   • find the measure of one exterior angle of that shape

   • find the sum of the exterior angles.

   (1)   equilateral triangle   (2)   regular octagon

   (3)   regular decagon   (4)   regular dodecagon (12-gon)

   e.   Compare your results from part (d). As a team, write a conjecture about the exterior angles of polygons based on your observations. Be ready to share your conjecture with the rest of the class.

   f.   Is your conjecture from part (e) true for all polygons or for only regular polygons? Does it matter if the polygon is convex? Explore these questions using a dynamic geometric tool or obtain the Lesson 8.1.3 Resource Page and tracing paper from your teacher. Write a statement explaining your findings.

8-25.   Use your understanding of polygons to answer the questions below, if possible. If there is no solution, explain why not.

   a.   Gerardo drew a regular polygon that had exterior angles measuring 40°. How many sides did his polygon have? What is the name for this polygon?

   b.   A polygon has an interior angle sum of 2,520°. How many sides does it have?

   c.   A quadrilateral has four sides. What is the measure of each of its interior angles?

   d.   What is the measure of an interior angle of a regular 360-gon? Is there more than one way to find this answer?

8-26.   How can you find the interior angle of a regular polygon? What is the sum of the exterior angles of a polygon? Write a Learning Log entry about what you learned during this lesson. Title this entry "Interior and Exterior Angles of a Polygon" and include today's date.

8-27.   Find the area and perimeter of each shape below. Show all work.

   a.

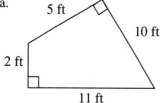

5 ft

10 ft

2 ft

11 ft

   b.

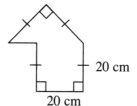

20 cm

20 cm

8-28.   In the figure at right, if $PQ = RS$ and $PR = SQ$, prove that $\angle P \cong \angle S$. Write your proof either in a flowchart or in two-column proof form.

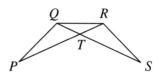

8-29.   Joey used 10 congruent triangles to create a regular decagon.

   a.   What kind of triangles is he using?

   b.   Find the three angle measures of one of the triangles. Explain how you know.

   c.   If the area of each triangle is 14.5 square inches, then what is the area of the regular decagon? Show all work.

Geometry Connections

8-30.   On graph paper, plot $A(2, 2)$ and $B(14, 10)$. If C is the midpoint of $\overline{AB}$, D is the midpoint of $\overline{AC}$, and E is the midpoint of $\overline{CD}$, find the coordinates of E.

8-31.   The arc at right is called a **quarter circle** because it is one-fourth of a circle.

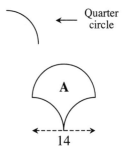

Quarter circle

    a.   Copy Region A at right onto your paper. If this region is formed using four quarter circles, can you find another shape that must have the same area as Region A? **Justify** your conclusion.

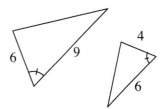

    b.   Find the area of Region A. Show all work.

8-32.   **Multiple Choice:** Which property below can be used to prove that the triangles at right are similar?

    a.   AA ~   b.   SAS ~

    c.   SSS ~   d.   HL ~

    e.   None of these

$8.1.4$ Is there another way?

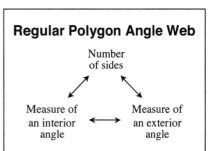

Regular Polygon Angle Connections

During Lessons 8.1.1 through 8.1.3, you have discovered many ways the number of sides of a regular polygon is related to the measures of the interior and exterior angles of the polygon. These relationships can be represented in the diagram at right.

How can these relationships be useful? And what is the most efficient way to go from one measurement to another? This lesson will explore these questions so that you will have a complete set of tools to analyze the angles of a regular polygon.

Regular Polygon Angle Web
Number of sides
Measure of an interior angle   ⟷   Measure of an exterior angle

8-33.   Which connections in the Polygon Angle Web do you
          already have? Which do you still need? Explore this
          as you answer the questions below.

   a.   If you know the number of sides of a regular
          polygon, how can you find the measure of an
          interior angle directly? Find the measurements
          of an interior angle of a 15-gon.

   b.   If you know the number of sides of a regular
          polygon, how can you find the measure of an exterior angle directly? Find the
          measurements of an exterior angle of a 10-gon.

   c.   What if you know that the measure of an interior angle of a regular polygon is
          162°? How many sides must the polygon have? Show all work.

   d.   If the measure of an exterior angle of a regular polygon is 15°, how many sides
          does it have? What is the measure of an interior angle? Show how you know.

8-34.   Suppose a regular polygon has an interior angle measuring 120°. Find the number of
          sides using *two* different **strategies**. Show all work. Which strategy was most
          efficient?

8-35.   Use your knowledge of polygons to answer the questions below, if possible.

   a.   How many sides does a polygon have if the sum of the measures of the interior
          angles is 1980°? 900°?

   b.   If the exterior angle of a regular polygon is 90°, how many sides does it have?
          What is another name for this shape?

   c.   Each interior angle of a regular pentagon has measure $2x + 4°$. What is x?
          Explain how you found your answer.

   d.   The measures of four of the exterior angles of a pentagon are 57°, 74°, 56°,
          and 66°. What is the measure of the remaining angle?

   e.   Find the sum of the interior angles of an 11-gon. Does it matter if it is regular
          or not?

8-36.   In a Learning Log entry, copy the Regular Polygon Angle Web that
          your class created. Explain what it represents and give an example
          of two of the connections. Title this entry "Regular Polygon Angle
          Web" and include today's date.

METHODS AND MEANINGS

MATH NOTES

Interior and Exterior Angles of a Polygon

The properties of interior and exterior angles in polygons, where n represents the number of sides in the polygon (n-gon), can be summarized as follows:

- The sum of the measures of the interior angles of an n-gon is $180(n-2)$.

- The measure of *each* angle in a regular n-gon is $\frac{180(n-2)}{n}$.

- The sum of the exterior angles of an n-gon is always $360°$.

Review & Preview

8-37.   Esteban used a hinged mirror to create an equilateral triangle, as shown in the diagram at right. If the area of the shaded region is 11.42 square inches, what is the area of the entire equilateral triangle? **Justify** your solution

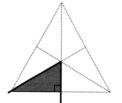

8-38.   Copy each shape below on your paper and state if the shape is convex or non-convex. You may want to compare each figure with the examples provided in problem 8-4.

a.   b.   c.   d.

8-39.   Find the area of each figure below. Show all work.

a. 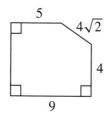   b.   c.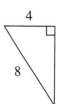

8-40.   Find the number of sides in a regular polygon if each interior angle has the following measures.

    a.   60°   b.   156°   c.   90°   d.   140°

8-41.   At right is a scale drawing of the floor plan for Nzinga's dollhouse. The actual dimensions of the dollhouse are 9 times the measurements provided in the floor plan at right.

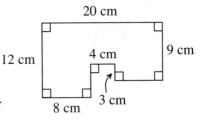

a.   Use the measurements provided in the diagram to find the area and perimeter of her floor plan.

b.   Draw a similar figure on your paper. Label the sides with the actual measurements of Nzinga's dollhouse. What is the perimeter and area of the floor of her actual dollhouse? Show all work.

c.   Find the ratio of the perimeters of the two figures. What do you notice?

d.   Find the ratio of the areas of the two figures. How does the ratio of the areas seem to be related to the zoom factor?

8-42.   **Multiple Choice:** A penny, nickel, and dime are all flipped once. What is the probability that at least one coin comes up heads?

a. $\frac{1}{3}$ b. $\frac{3}{8}$ c.   1   d. $\frac{7}{8}$

8.1.5 What's the area?

Finding the Area of Regular Polygons

In Lesson 8.1.4, you found the area of a regular hexagon. But what if you want to find the area of a regular pentagon or a regular decagon? Today you will explore these different polygons and generalize how to find the area of any regular polygon with n sides.

8-43.   USING MULTIPLE STRATEGIES

With your team, find the area of each shape below <u>twice</u>, each time using a distinctly different method or **strategy**. Make sure that your results from using different **strategies** are the same. Be sure that each member of your team understands each method.

a.   Square

b.   regular hexagon

8-44.   Create a poster or transparency that shows the two
          different methods that your team used to find the
          area of the regular hexagon in part (b) of problem
          8-43.

          Then, as you listen to other teams present, look for
          strategies that are different than yours. For each
          one, consider the questions below.

          • *Which geometric **tools** does this method use?*

          • *Would this method help find the area of other
            regular polygons (like a pentagon or 100-gon)?*

8-45.   Which method presented by teams in problem 8-44 seemed able to help find the area
          of other regular polygons? Discuss this with your team. Then find the area of the
          two regular polygons below. If your method does not work, switch to a different
          method. Assume *C* is the center of each polygon.

          a.   b.

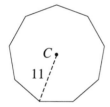   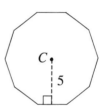

8-46.   So far, you have found the area of a regular hexagon, nonagon, and
          decagon. How can you calculate the area of *any* regular polygon?
          Write a Learning Log entry describing a general process for finding
          the area of a polygon with *n* sides.

8-47.   Beth needs to fertilize her flowerbed, which is in the
          shape of a regular pentagon. A bag of fertilizer states
          that it can fertilize up to 150 square feet, but Beth is
          not sure how many bags of fertilizer she should buy.

          Beth does know that each side of the pentagon is
          15 feet long. Copy the diagram of the regular
          pentagon below onto your paper. Find the area of the
          flowerbed and tell Beth how many bags of fertilizer to
          buy. Explain how you found your answer.

15 ft

8-48.   GO, ROWDY RODENTS!

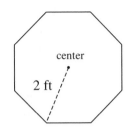

Recently, your school ordered a stained-glass window
with the design of the school's mascot, the rodent.
Your student body has decided that the shape of the
window will be a regular octagon, shown at right. To
fit in the space, the window must have a **radius** of 2
feet. That is, the distance from the center to each vertex
must be 2 feet.

a.   A major part of the cost of the window is the amount
of glass used to make it. The more glass used, the
more expensive the window. Your principal has
turned to your class to determine how much glass the
window will need. Copy the diagram onto your paper
and find its area. Explain how you found your answer.

b.   The edge of the window will have a polished brass
trim. Each foot of trim will cost $48.99. How much
will the trim cost? Show all work.

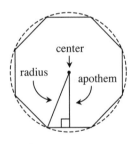

METHODS AND MEANINGS

Parts of a Regular Polygon

MATH NOTES

The **center** of a regular polygon is the center
of the smallest circle that completely encloses
the polygon.

A line segment that connects the center of a regular
polygon with a vertex is called a **radius**.

An **apothem** is the perpendicular line segment from
the center of a regular polygon to a side.

8-49.   The exterior angle of a regular polygon is 20°.

a.   What is the measure of an interior angle of this polygon? Show how you
know.

b.   How many sides does this polygon have? Show all work.

Geometry Connections

8-50.   Without using your calculator, find the exact values of *x* and *y* in each diagram below.

a.
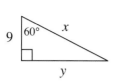

b.
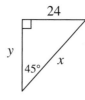

c.

8-51.   Find the coordinates of the point at which the diagonals of parallelogram *ABCD* intersect if *B*(−3, −17) and *D*(15, 59). Explain how you found your answer.

8-52.   Find the area of an equilateral triangle with side length 20 mm. Draw a diagram and show all work.

8-53.   For each equation below, solve for *w*, if possible. Show all work.

a. $5w^2 = 17$ b. $5w^2 - 3w - 17 = 0$ c. $2w^2 = -3$

8-54.   **Multiple Choice:** The triangles at right are congruent because of:

a.   SSA $\cong$ b.   HL $\cong$ c.   SAS $\cong$

d.   SSS $\cong$ e.   None of these

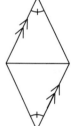

8-55.   Solve for *x* in each diagram below.

a.

b.
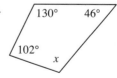

c.
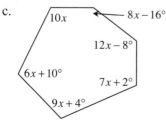

d.
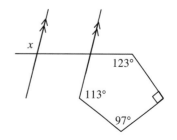

8-56.   What is another (more descriptive) name for each polygon described below?

    a.   A regular polygon with an exterior angle measuring 120°.

    b.   A quadrilateral with four equal angles.

    c.   A polygon with an interior angle sum of 1260°.

    d.   A quadrilateral with perpendicular diagonals.

8-57.   If $\triangle ABC$ is equilateral and if $A(0, 0)$ and $B(12, 0)$, then what do you know about the coordinates of vertex C?

8-58.   In the figure at right, $\overline{AB} \cong \overline{DC}$ and $\measuredangle ABC \cong \measuredangle DCB$.

    a.   Is $\overline{AC} \cong \overline{DB}$? Prove your answer.

    b.   Do the measures of $\measuredangle ABC$ and $\measuredangle DCB$ make any difference in your solution to part (a)? Explain why or why not.

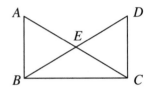

8-59.   On graph paper, graph the parabola $y = 2x^2 - x - 15$.

    a.   What are the roots (x-intercepts) of the parabola? Write your points in (x, y) form.

    b.   How would the graph of $y = -(2x^2 - x - 15)$ be the same or different? Can you tell without graphing?

8-60.   **Multiple Choice:** Approximate the length of $\overline{AB}$.

    a.   15.87   b.   21.84   c.   37.16

    d.   19.62   e.   None of these

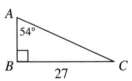

8.2.1 How does the area change?

Area Ratios of Similar Figures

Much of this course has focused on similarity. In Chapter 3, you **investigated** how to enlarge and reduce a shape to create a similar figure. You also have studied how to use proportional relationships to find the measures of sides of similar figures. Today you will study how the areas of similar figures are related. That is, as a shape is enlarged or reduced in size, how does the area change?

8-61.   MIGHTY MASCOT

To celebrate the victory of your school's championship girls' ice hockey team, the student body has decided to hang a giant flag with your school's mascot on the gym wall.

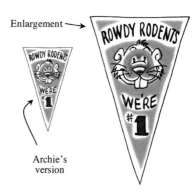

Enlargement →

Archie's version

To help design the flag, your friend Archie has created a scale version of the flag measuring 1 foot wide and 1.5 feet tall.

a.   The student body would like the final flag to be 3 feet tall. How wide will the final flag be? **Justify** your solution.

b.   If Archie used $2 worth of cloth to create his scale model, then how much will the cloth cost for the full-sized flag? Discuss this with your team. Explain your **reasoning**.

c.   Obtain the Lesson 8.2.1A Resource Page and scissors from your teacher. Carefully cut enough copies of Archie's scale version to fit into the large flag. How many did it take? Does this confirm your answer to part (b)? If not, what will the cloth cost for the flag?

d.

The student body is reconsidering the size of the flag. It is now considering enlarging the flag so that it is 3 or 4 times the width of Archie's model. How much would the cloth for a similar flag that is 3 times as wide as Archie's model cost? What if the flag is 4 times as wide?

To answer this question, first *estimate* how many of Archie's drawings would fit into each enlarged flag. Then obtain the Lesson 8.2.1B Resource Page (one for you and your team members to share) and confirm each answer by fitting Archie's scale version into the enlarged flags.

8-62.   Write down any observations or patterns you found while working on problem 8-61. For example, if the area of one shape is 100 times larger than the area of a similar shape, then what is the ratio of the corresponding sides (also called the **linear scale factor**)? And if the linear scale factor is r, then how many times larger is the area of the new shape?

8-63.   Use your pattern from problem 8-62 to answer the following questions.

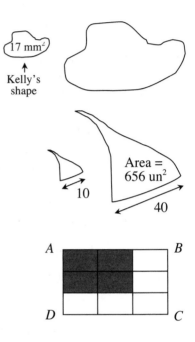

   a.   Kelly's shape at right has an area of 17 mm^2. If she enlarges the shape with a linear scale (zoom) factor of 5, what will be the area of the enlargement? Show how you got your answer.

   b.   **Examine** the two similar shapes at right. What is the linear scale factor? What is the area of the smaller figure?

   c.   Rectangle $ABCD$ at right is divided into nine smaller congruent rectangles. Is the shaded rectangle similar to $ABCD$? If so, what is the linear scale factor? And what is the ratio of the areas? If the shaded rectangle is not similar to $ABCD$, explain how you know.

   d.   While ordering carpet for his rectangular office, Trinh was told by the salesperson that a 16'-by-24' piece of carpet costs $200. Trinh then realized that he read his measurements wrong and that his office is actually 8'-by-12'. "Oh, that's no problem," said the salesperson. "That is half the size and will cost $100 instead." Is that fair? Decide what the price should be.

8-64.   If the side length of a hexagon triples, how does the area increase? First make a prediction using your pattern from problem 8-62. Then confirm your prediction by calculating and comparing the areas of the two hexagons shown at right.

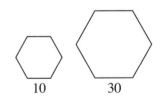

METHODS AND MEANINGS

Ratios of Similarity

Since Chapter 3, you have used the term **zoom factor** to refer to the ratio of corresponding dimensions of two similar figures. However, now that you will be using other ratios of similar figures (such as the ratio of the areas), this ratio needs a more descriptive name. From now on, this text will refer to the ratio of corresponding sides as the **linear scale factor**. The word "linear" is a reference to the fact that the ratio of the side lengths is a comparison of a single dimension of the shapes. Often, this value is represented with the letter r, for ratio.

For example, notice that the two triangles at right are similar because of AA ~. Since the corresponding sides of the new and original shape are 9 and 6, it can be stated that $r = \frac{9}{6} = \frac{3}{2}$.

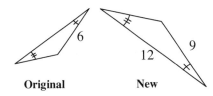

Original   New

8-65.   **Examine** the shape at right.

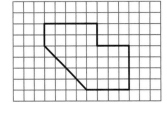

   a.   Find the area and perimeter of the shape.

   b.   On graph paper, enlarge the figure so that the linear scale factor is 3. Find the area and perimeter of the new shape.

   c.   What is the ratio of the perimeters of both shapes? What is the ratio of the areas?

8-66.   Sandip noticed that when he looked into a mirror that was lying on the ground 8 feet from him, he could see a clock on the wall. If Sandip's eyes are 64 inches off the ground, and if the mirror is 10 feet from the wall, how high above the floor is the clock? Include a diagram in your solution.

8-67.   Mr. Singer has a dining table in the shape of a regular hexagon. While he loves this design, he has trouble finding tablecloths to cover it. He has decided to make his own tablecloth!

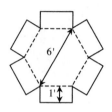

In order for his tablecloth to drape over each edge, he will add a rectangular piece along each side of the regular hexagon as shown in the diagram at right. Using the dimensions given in the diagram, find the total area of the cloth Mr. Singer will need.

8-68.   Your teacher has offered your class extra credit. She has created two spinners, shown at right. Your class gets to spin only one of the spinners. The number that the spinner lands on is the number of extra credit points each member of the class will get. Study both spinners carefully.

a.   Assuming that each spinner is divided into equal portions, which spinner do you think the class should choose to spin and why?

b.   What if the spot labeled "20" were changed to "100"? Would that make any difference?

8-69.   If the rectangles below have the same area, find x. Is there more than one answer? Show all work.

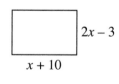

2x

$2x - 3$

$x + 10$

8-70.   **Multiple Choice:** A cable 100 feet long is attached 70 feet up the side of a building. If it is pulled taut (i.e., there is no slack) and staked to the ground as far away from the building as possible, approximately what angle does the cable make with the ground?

a.   39.99°   b.   44.43°   c.   45.57°   d.   12.22°

8.2.2 How does the area change?

Ratios of Similarity

Today you will continue **investigating** the ratios between similar figures. As you solve today's problems, look for connections between the ratios of similar figures and what you already know about area and perimeter.

8-71. TEAM PHOTO

Alice has a 4"-by-5" photo of your school's championship girls' ice hockey team. To celebrate their recent victory, your principal wants Alice to enlarge her photo for a display case near the main office.

a. When Alice went to the print shop, she was confronted with many choices of sizes: 7"-by-9", 8"-by-10", and 12"-by-16". She's afraid that if she picks the wrong size, part of the photo will be cut off. Which size should Alice pick and why?

b. The cost of the photo paper to print Alice's 4"-by-5" picture is $0.45. Assuming that the cost per square inch of photo paper remains constant, how much should it cost to print the enlarged photo? Explain how you found your answer.

c. Unbeknownst to her, the Vice-Principal also went out and ordered an enlargement of Alice's photo. However, the photo paper for his enlargement cost $7.20! What are the dimensions of his photo?

8-72.   So far, you have discovered and used the relationship between the areas of similar
         figures. How are the perimeters of similar figures related? Confirm your intuition
         by analyzing the pairs of similar shapes below. For each pair, calculate the areas and
         perimeters and complete a table like the one shown below. To help see patterns,
         reduce fractions to lowest terms or find the corresponding decimal values.

	Ratio of Sides	Perimeter	Ratio of Perimeters	Area	Ratio of Areas
small figure					
large figure					

a.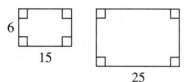

b.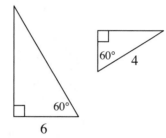

c.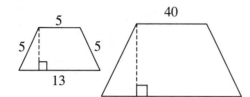

8-73.   While Jessie examines the two figures at right, she wonders if they
         are similar. Decide with your team if there is enough information
         to determine if the shapes are similar. **Justify** your conclusion.

8-74.   Your teacher enlarged the figure at right so that the
         area of the similar shape is 900 square cm. What is
         the perimeter of the enlarged figure? Be prepared
         to explain your method to the class.

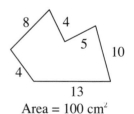

Area = 100 cm²

8-75.   Reflect on what you have learned during Lessons 8.2.1 and 8.2.2.
         Write a Learning Log entry that explains what you know about the
         areas and perimeters of similar figures. What connections can you
         make with other geometric concepts? Be sure to include an example.
         Title this entry "Area and Perimeter of Similar Figures" and include
         today's date.

8-76.　Assume Figure *A* and Figure *B*, at right, are similar.

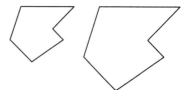

Figure A　　　Figure B

　　a.　If the ratio of similarity is $\frac{3}{4}$, then what is the ratio of the perimeters of *A* and *B*?

　　b.　If the perimeter of Figure A is *p* and the linear scale factor is *r*, what is the perimeter of Figure B?

　　c.　If the area of Figure A is *a* and the linear scale factor is *r*, what is the area of Figure B?

8-77.　Always a romantic, Marris decided to bake his girlfriend a cookie in the shape of a regular dodecagon (12-gon) for Valentine's Day.

　　a.　If the edge of the dodecagon is 6 cm, what is the area of the top of the cookie?

　　b.　His girlfriend decides to divide the cookie into 12 separate but congruent pieces. After 9 of the pieces have been eaten, what area of cookie is left?

8-78.　As her team was building triangles with linguini, Karen asked for help building a triangle with sides 5, 6, and 1. "I don't think that's possible," said her teammate, Kelly.

　　a.　Why is this triangle not possible?

　　b.　Change the lengths of one of the sides so that the triangle is possible.

8-79.   Callie started to prove that given the information in the
         diagram at right, then $\overline{AB} \cong \overline{CD}$. Copy her flowchart below
         on your paper and help her by **justifying** each statement.

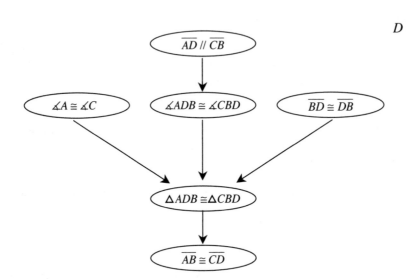

8-80.   For each pair of triangles below, decide if the triangles are congruent. If the triangles
         are congruent:

   - State which triangle congruence property proves that the triangles are
     congruent.

   - Write a congruence statement (such as $\triangle ABC \cong \triangle$____).

a.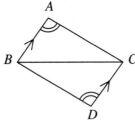

b.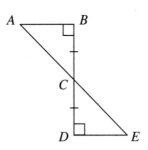

8-81.   **Multiple Choice:** What is the solution to the system
         of equations at right?

$$y = \tfrac{1}{2}x - 4$$
$$x - 4y = 12$$

   a.   (2, 0)   b.   (16, 4)

   c.   (−2, −5)   d.   (4, −2)

   e.   None of these

8.3.1 What if the polygon has infinite sides?

A Special Ratio

In Section 8.2, you developed a method to find the area and perimeter of a regular polygon with *n* sides. You carefully calculated the area of regular polygons with 5, 6, 8, and even 10 sides. But what if the regular polygon has an infinite number of sides? How can you predict its area?

As you **investigate** this question today, keep the following focus questions in mind:

What's the connection?

Do I see any patterns?

How are the shapes related?

8-82. POLYGONS WITH INFINITE SIDES

In order to predict the area and perimeter of a polygon with infinite sides, your team is going to work with other teams to generate data in order to find a pattern.

Your teacher will assign your team three of the regular polygons below. For each polygon, find the area and perimeter if the radius is 1 (as shown in the diagram of the regular pentagon at right). Leave your answer accurate to the nearest 0.01. Place your results into a class chart to help predict the area and perimeter of a polygon with an infinite number of sides.

a. equilateral triangle   b. regular octagon   c. regular 30-gon

d. square   e. regular nonagon   f. regular 60-gon

g. regular pentagon   h. regular decagon   i. regular 90-gon

j. regular hexagon   k. regular 15-gon   l. regular 180-gon

8-83. ANALYSIS OF DATA

With your team, analyze the chart created by the class.

a. What do you predict the area will be for a regular polygon with infinite sides? What do you predict its perimeter will be?

b. What is another name for a regular polygon with infinite sides?

c. Does the number 3.14… look familiar? If so, share what you know with your team. Be ready to share your idea with the class.

8-84.   Record the area and circumference of a circle with radius you're
         your Learning Log. Then, include a brief description of how you
         "discovered" π. Title this entry "Pi" and include today's date.

Ⓜ ETHODS AND MEANINGS

MATH NOTES

The Area of a Regular Polygon

 If a polygon is regular with n sides, it can be
subdivided into n congruent isosceles triangles.
One way to calculate the area of a regular polygon
is to multiply the area of one isosceles triangle by n.

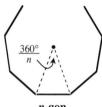

n-gon

To find the area of the isosceles triangle, it is helpful to
first find the measure of the polygon's central angle by
dividing 360° by n. The height of the isosceles triangle
divides the top vertex angle in half.

For example, suppose you want to find the area of a
regular decagon with side length 4 units. The central
angle is $\frac{360°}{10} = 36°$. Then the top angle of the shaded
right triangle at right would be $36° \div 2 = 18°$.

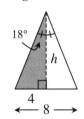

Use right triangle trigonometry to find the measurements of the right triangle,
then calculate its area. For the shaded triangle above, $\tan 18° = \frac{4}{h}$ and
$h \approx 12.311$. Use the height and the base to find the area of the isosceles
triangle: $\frac{1}{2}(8)(12.311) \approx 49.242$ sq. units. Then the area of the regular
decagon is approximately $10 \cdot 49.242 \approx 492.42$ sq. units. Use a similar
approach if you are given a different length of the triangle.

Review & Preview

8-85.   Find the area of the shaded region for the regular pentagon
         at right if the length of each side of the pentagon is 10
         units. Assume that point C is the center of the pentagon.

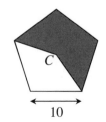

8-86.   For each triangle below, find the value of *x*, if possible. Name which triangle tool you used. If the triangle cannot exist, explain why.

a.
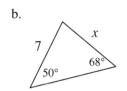

60°

x   28

b.

7   *x*

50°   68°

c.

19   17

x

d.   Area of the shaded region is 96 un².

x

←8→

8-87.   Find the measure of each interior angle of a regular 30-gon using **two different methods**.

8-88.   **Examine** the diagram at right. Assume that $\overline{BC} \cong \overline{DC}$ and $\angle A \cong \angle E$. Prove that $\overline{AB} \cong \overline{ED}$. Use the form of proof that you prefer (such as the flowchart or two-column proof format). Be sure to copy the diagram onto your paper and add any appropriate markings.

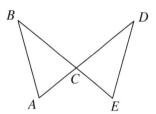

B   D

C

A   E

8-89.   On graph paper, plot the points $A(-3, -1)$ and $B(6, 11)$.

a.   Find the midpoint of $\overline{AB}$.

b.   Find the equation of the line that passes through points *A* and *B*.

c.   Find the distance between points *A* and *B*.

8-90.   **Multiple Choice:** What fraction of the circle at right is shaded?

a. $\frac{60}{360}$ b. $\frac{300}{360}$ c. $\frac{60}{180}$

d. $\frac{120}{180}$ e.   None of these

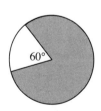

60°

8.3.2 What's the relationship?

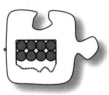

Area and Circumference of Circles

In Lesson 8.3.1, your class discovered that the area of a circle with radius 1 unit is π un^2 and that the circumference is 2π units. But what if the radius of the circle is 5 units or 13.6 units? Today, you will develop a method to find the area and circumference of circles when the radius is not 1. You will also explore parts of circles (called sectors and arcs) and learn about their measurements.

As you and your team work together, remember to ask each other questions such as:

Is there another way to solve it?

What's the relationship?

What is area? What is circumference?

8-91. **AREA AND CIRCUMFERENCE OF A CIRCLE**

Now that you know the area and circumference (perimeter) of a circle with radius 1, how can you find the area and circumference of a circle with any radius?

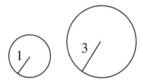

a. First **investigate** how the circles are related. **Examine** the circles at right. Since circles always have the same shape, what is the relationship between any two circles?

b. What is the ratio of the circumferences (perimeters)? What is the ratio of the areas? Explain.

c. If the area of a circle with radius of 1 is π square units, what is the area of a circle with radius 3 units? With radius 10 units? With radius r units?

d. Likewise, if the circumference (perimeter) of a circle is 2π units, what is the circumference of a circle with radius 3? With radius 7? With radius r?

8-92. Read the definitions of radius and diameter in the Math Notes box for this lesson. Then answer the questions below.

a. Find the area of a circle with radius 10 units.

b. Find the circumference of a circle with diameter 7 units.

c. If the area of a circle is 121π square units, what is its diameter?

d. If the circumference of a circle is 20π units, what is its area?

Geometry Connections

8-93. The giant sequoia trees in California are famous for their immense size and old age. Some of the trees are more than 2500 years old and tourists and naturalists often visit to admire their size and beauty. In some cases, you can even drive a car through the base of a tree!

One of these trees, the General Sherman tree in Sequoia National Park, is the largest living thing on the earth. The tree is so gigantic, in fact, that the base has a circumference of 102.6 feet! Assuming that the base of the tree is circular, how wide is the base of the tree? That is, what is its diameter? How does that diameter compare with the length and width of your classroom?

8-94. To celebrate their victory, the girls' ice-hockey team went out for pizza.

a. The goalie ate half of a pizza that had a diameter of 20 inches! What was the area of pizza that she ate? What was the length of crust that she ate? Leave your answers in exact form. That is, do not convert your answer to decimal form.

b. Sonya chose a slice from another pizza that had a diameter of 16 inches. If her slice had a central angle of 45°, what is the area of this slice? What is the length of its crust? Show how you got your answer.

c. As the evening drew to a close, Sonya noticed that there was only one slice of the goalie's pizza remaining. She measured the central angle and found out that it was 72°. What is the area of the remaining slice? What is the length of its crust? Show how you got your answer.

72°

d. A portion of a circle (like the crust of a slice of pizza) is called an **arc**. This is a set of connected points a fixed distance from a central point. The length of an arc is a part of the circle's circumference. If a circle has a radius of 6 cm, find the length of an arc with a central angle of 30°.

arc

e. A region that resembles a slice of pizza is called a **sector**. It is formed by two radii of a central angle and the arc between their endpoints on the circle. If a circle has radius 10 feet, find the area of a sector with a central angle of 20°.

sector

8-95.   Reflect on what you have learned today. How did you use
         similarity to find the areas and circumferences of circles? How are
         the radius and diameter of a circle related? Write a Learning Log
         entry about what you learned today. Title this entry "Area and
         Circumference of a Circle" and include today's date.

METHODS AND MEANINGS

Circle Facts

The area of a circle with radius $r = 1$ unit is π un^2.
(Remember that $\pi \approx 3.1415926...$)

Since all circles are similar, their areas increase by
a square of the zoom factor. That is, a circle with
radius 6 has an area that is 36 times the area of a
circle with radius 1. Thus, a circle with radius 6
has an area of 36π un^2, and a circle with radius r
has **area** $A = \pi r^2$ un^2.

Area $= \pi r^2$
Circumference $= 2\pi r = \pi d$

The **circumference** of a circle is its perimeter. It is the distance around a
circle. The circumference of a circle with radius $r = 1$ unit is 2π units. Since
the perimeter ratio is equal to the ratio of similarity, a circle with radius r has
circumference $C = 2\pi r$ units. Since the diameter of a circle is twice its radius,
another way to calculate the circumference is $C = \pi d$ units.

arc

A part of a circle is called an **arc**. This is a set of points a
fixed distance from a center and is defined by a central
angle. Since a circle does not include its interior region, an
arc is like the edge of a crust of a slice of pizza.

sector

A region that resembles a slice of pizza is called a **sector**. It is
formed by two radii of a central angle and the arc between their
endpoints on the circle.

Review & Preview

8-96.   The diagram at right shows a circle inscribed in a square.
         Find the area of the shaded region. Show all work.

10 units

8-97.   Reynaldo has a stack of blocks on his desk, as shown below at right.

   a.   If his stack is 2 blocks wide, 2 blocks long, and 2 blocks tall,
         how many blocks are in his stack?

   b.   What if his stack instead is 3 blocks wide, 3 blocks long, and 2 blocks tall?
         How many blocks are in this stack?

8-98.   Find the missing angle(s) in each problem below using the
         geometric relationships shown in the diagram at right. Be
         sure to write down the conjecture that **justifies** each
         calculation. Remember that each part is a separate problem.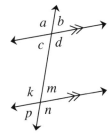

   a.   If $d = 110°$ and $k = 5x - 20°$, write an equation and
         solve for x.

   b.   If $b = 4x - 11°$ and $n = x + 26°$, write an equation and
         solve for x. Then find the measure of $\angle n$.

8-99.   An exterior angle of a regular polygon measures 18°.

   a.   How many sides does the polygon have?

   b.   If the length of a side of the polygon is 2, what is the area of the polygon?

8-100.   A regular hexagon with side length 4 has the same area as a square. What is the
         length of the side of the square? Explain how you know.

8-101.   **Multiple Choice:** Which type of quadrilateral below does not necessarily have
         diagonals that bisect each other?

   a.   square   b.   rectangle   c.   rhombus   d.   trapezoid

8.3.3 How can I use it?

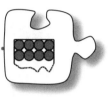

Circles in Context

In Lesson 8.3.1, you developed methods to find the area and circumference of a circle with radius *r*. During this lesson, you will work with your team to solve problems from different contexts involving circles and polygons.

As you and your team work together, remember to ask each other questions such as:

Is there another way to solve it?

What's the connection?

What is area? What is circumference?

8-102. While the earth's orbit (path) about the sun is slightly elliptical, it can be approximated by a circle with a radius of 93,000,000 miles.

a. How far does the earth travel in one orbit about the sun? That is, what is the approximate circumference of the earth's path?

b. Approximately how fast is the earth traveling in its orbit in space? Calculate your answer in miles per hour.

8-103. A certain car's windshield wiper clears a portion of a sector as shown shaded at right. If the angle the wiper pivots during each swing is 120°, find the area of the windshield that is wiped during each swing.

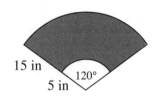

8-104.   THE GRAZING GOAT

Zoe the goat is tied by a rope to one corner of a 15 meter-by-25 meter rectangular barn in the middle of a large, grassy field. Over what area of the field can Zoe graze if the rope is:

a.   10 meters long?

b.   20 meters long?

c.   30 meters long?

d.   Zoe is happiest when she has at least 400 m^2 to graze. What possible lengths of rope could be used?

8-105.   THE COOKIE CUTTER

A cookie baker has an automatic mixer that turns out a sheet of dough in the shape of a square 12" wide. His cookie cutter cuts 3" diameter circular cookies as shown at right. The supervisor complained that too much dough was being wasted and ordered the baker to find out what size cookie would have the least amount of waste.

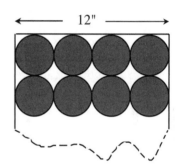

Your Task:

•   Analyze this situation and determine how much cookie dough is "wasted" when 3" cookies are cut. Then have each team member find the amount of dough wasted when a cookie of a different diameter is used. Compare your results.

•   Write a note to the supervisor explaining your results. **Justify** your conclusion.

METHODS AND MEANINGS

Arc Length and Area of a Sector

MATH NOTES

The ratio of the area of a sector to the area of a circle with the same radius equals the ratio of its central angle to 360°. For example, for the sector in circle C at right, the area of the entire circle is $\pi(8)^2 = 64\pi$ square units. Since the central angle is 50°, then the area of the sector can be found with the proportional equation:

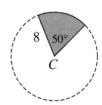

$$\frac{50°}{360°} = \frac{\text{area of sector}}{64\pi}$$

To solve, multiply both sides of the equation by 64π. Thus, the area of the sector is $\frac{50°}{360°}(64\pi) = \frac{80\pi}{9} \approx 27.93$ square units.

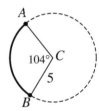

The length of an arc can be found using a similar process. The ratio of the length of an arc to the circumference of a circle with the same radius equals the ratio of its central angle to 360°. To find the length of $\overarc{AB}$ at right, first find the circumference of the entire circle, which is $2\pi(5) = 10\pi$ units. Then:

$$\frac{104°}{360°} = \frac{\text{arc length}}{10\pi}$$

Multiplying both sides of the equation by 10π, the arc length is $\frac{104°}{360°}(10\pi) = \frac{26\pi}{9} \approx 9.08$ units.

Review & Preview

8-106. Your teacher has constructed a spinner like the one at right. He has informed you that the class gets one spin. If the spinner lands on the shaded region, you will have a quiz tomorrow. What is the probability that you will have a quiz tomorrow? Explain how you know.

8-107. Use what you know about the area and circumference of circles to answer the questions below. Show all work. Leave answers in terms of π.

    a. If the radius of a circle is 14 units, what is its circumference? What is its area?

    b. If a circle has diameter 10 units, what is its circumference? What is its area?

    c. If a circle has circumference 100π units, what is its diameter? What is its radius?

8-108. Larry started to set up a proof to show that if $\overline{AB} \perp \overline{DE}$ and $\overline{DE}$ is a diameter of $\odot C$, then $\overline{AF} \cong \overline{FB}$. **Examine** his work below. Then complete his missing statements and **reasons**.

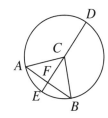

Statements	Reasons
1. $\overline{AB} \perp \overline{DE}$ and $\overline{DE}$ is a diameter of $\odot C$.	1.
2. $\angle AFC$ and $\angle BFC$ are right angles.	2.
3. $FC = FC$	3.
4. $\overline{AC} = \overline{BC}$	4. Definition of a Circle (radii must be equal)
5.	5. HL $\cong$
6. $\overline{AF} \cong \overline{FB}$	6.

8-109. Match each regular polygon named on the left with a statement about its qualities listed on the right.

    a. regular hexagon   (1) Central angle of 36°

    b. regular decagon   (2) Exterior angle measure of 90°

    c. equilateral triangle   (3) Interior angle measure of 120°

    d. square   (4) Exterior angle measure of 120°

8-110. **Examine** the graph of $f(x)$ at right. Use the graph to find the following values.

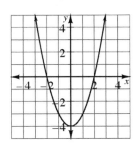

    a. $f(1)$ b. $f(0)$

    c. x if $f(x) = 4$ d. x if $f(x) = 0$

8-111.   **Multiple Choice:** How many cubes with edge length 1 unit would fit in a cube with edge length 3 units?

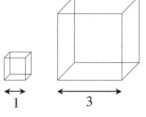

a.   3   b.   9

c.   10   d.   27

e.   None of these

8-112.   The city of Denver wants you to help build a dog park. The design of the park is a rectangle with two semicircular ends. (Note: A semicircle is half a circle.)

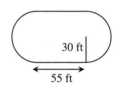

a.   The entire park needs to be covered with grass. If grass is sold by the square foot, how much grass should you order?

b.   The park also needs a fence for its perimeter. A sturdy chain-linked fence costs about $8 per foot. How much will a fence for the entire park cost?

c.   The local design board has rejected the plan because it was too small. "Big dogs need lots of room to run," the president of the board said. Therefore, you need to increase the size of the park with a zoom factor of 2. What is the area of the new design? What is the perimeter?

8-113.   For each diagram below, write and solve an equation to find x.

a.

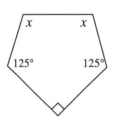

b.

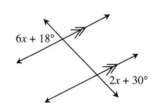

8-114. $\overline{BE}$ is the midsegment of $\triangle ACD$, shown at right.

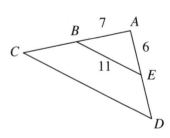

a.   Find the perimeter of $\triangle ACD$.

b.   If the area of $\triangle ABE$ is 54 cm², what is the area of $\triangle ACD$?

8-115.   Christie has tied a string that is 24 cm long into a closed loop,
          like the one at right.

          a.   She decided to form an equilateral triangle with her
                string. What is the area of the triangle?

          b.   She then forms a square with the same loop of string. What is the area of the
                square? Is it more or less than the equilateral triangle she created in part (a)?

          c.   If she forms a regular hexagon with her string, what would be its area?
                Compare this area with the areas of the square and equilateral triangle from
                parts (a) and (b).

          d.   What shape should Christie form to enclose the greatest area?

8-116.   The **Isoperimetric Theorem** states that of all closed figures on a flat surface with
          the same perimeter, the circle has the greatest area. Use this fact to answer the
          questions below.

          a.   What is the greatest area that can be enclosed by a loop of string that is 24 cm
                long?

          b.   What is the greatest area that can be enclosed by a loop of string that is 18π
                cm long?

8-117.   **Multiple Choice:** The diagram at right is not
          drawn to scale. If $\triangle ABC \sim \triangle KLM$, find KM.

          a.   6   b.   12

          c.   15   d.   21

          e.   None of these

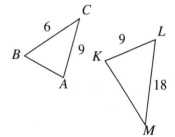

Chapter 8 Closure   What have I learned?

Reflection and Synthesis

The activities below offer you a chance to reflect on what you have learned during this chapter. As you work, look for concepts that you feel very comfortable with, ideas that you would like to learn more about, and topics you need more help with. Look for **connections** between ideas as well as **connections** with material you learned previously.

① TEAM BRAINSTORM

With your team, brainstorm a list for each of the following three topics. Be as detailed as you can. How long can you make your list? Challenge yourselves. Be prepared to share your team's ideas with the class.

Topics:	What have you studied in this chapter? What ideas and words were important in what you learned? Remember to be as detailed as you can.
Problem Solving:	What did you do to solve problems? What different strategies did you use?
Connections:	How are the topics, ideas, and words that you learned in previous courses are **connected** to the new ideas in this chapter? Again, make your list as long as you can.

MAKING CONNECTIONS

The following is a list of the vocabulary used in this chapter. The words that appear in bold are new to this chapter. Make sure that you are familiar with all of these words and know what they mean. Refer to the glossary or index for any words that you do not yet understand.

apothem	**arc**	area
central angle	**circumference**	**convex**
diameter	exterior angle	**interior angle**
linear scale factor	**non-convex**	perimeter
pi (π)	polygon	radius
regular polygon	**remote interior angle**	**sector**
similar	zoom factor	

Make a concept map showing all of the **connections** you can find among the key words and ideas listed above. To show a **connection** between two words, draw a line between them and explain the **connection**, as shown in the example below. A word can be **connected** to any other word as long as there is a **justified connection**. For each key word or idea, provide a sketch of an example.

Your teacher may provide you with vocabulary cards to help you get started. If you use the cards to plan your concept map, be sure either to re-draw your concept map on your paper or to glue the vocabulary cards to a poster with all of the **connections** explained for others to see and understand.

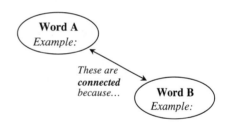

While you are making your map, your team may think of related words or ideas that are not listed above. Be sure to include these ideas on your concept map.

③ SUMMARIZING MY UNDERSTANDING

This section gives you an opportunity to show what you know about certain math topics or ideas. Your teacher will give you directions for exactly how to do this.

WHAT HAVE I LEARNED?

This section will help you evaluate which types of problems you have seen with which you feel comfortable and those with which you need more help. This section will appear at the end of every chapter to help you check your understanding. Even if your teacher does not assign this section, it is a good idea to try these problems and find out for yourself what you know and what you need to work on.

Solve each problem as completely as you can. The table at the end of this closure section has answers to these problems. It also tells you where you can find additional help and practice on problems like these.

CL 8-118.      Mrs. Frank loves the clock in her classroom because it has the school colors, green and purple. The shape of the clock is a regular dodecagon with a radius of 14 cm. Centered on the clock's face is a green circle of radius 9 cm. If the region outside the circle is purple, which color has more area?

CL 8-119. Graph the quadrilateral *ABCD* if *A*(–2, 6), *B*(2, 3), *C*(2, –2), and *D*(–2, 1).

   a.   What's the best name for this quadrilateral? **Justify** your conclusion.

   b.   Find the area of *ABCD*.

   c.   Find the slope of the diagonals, $\overline{AC}$ and $\overline{BD}$. How are the slopes related?

   d.   Find the point of intersection of the diagonals. What is the relationship between this point and diagonal $\overline{AC}$?

CL 8-120. **Examine** the diagram at right. If *M* is the midpoint of $\overline{KQ}$ and if $\angle P \cong \angle L$, prove that $\overline{KL} \cong \overline{QP}$.

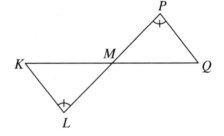

CL 8-121. A running track design is composed of two half circles connected by two straight line segments. Garrett is jogging on the inner lane (with radius *r*) while Devin is jogging on the outer (with radius *R*). If *r* = 30 meters and *R* = 33 meters, how much longer does Devin have to run to complete one lap?

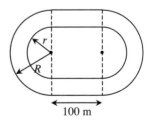

Geometry Connections

CL 8-122. Use the relationships in the diagrams below to solve for the given variable. **Justify** your solution with a definition or theorem.

a.

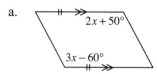

b. The perimeter of the quadrilateral below is 202 units.

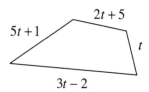

c. *CARD* is a rhombus.

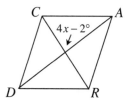

d.

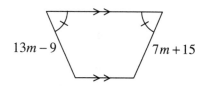

CL 8-123. Answer the following questions about polygons. If there is not enough information or the problem is impossible, explain why.

a. Find the sum of the interior angles of a dodecagon.

b. Find the number of sides of a regular polygon if its central angle measures 35°.

c. If the sum of the interior angles of a regular polygon is 900°, how many sides does the polygon have?

d. If the exterior angle of a regular polygon is 15°, find its central angle.

e. Find the exterior angle of a polygon with 10 sides.

CL 8-124. Check your answers using the table at the end of the closure section. Which problems do you feel confident about? Which problems were hard? Use the table to make a list of topics you need help on and a list of topics you need to practice more.

⑤   HOW AM I THINKING?

This course focuses on five
different **Ways of Thinking**:
investigating, examining, reasoning
and justifying, visualizing, and
choosing a strategy/tool. These are
some of the ways in which you
think while trying to make sense of
a concept or to solve a problem
(even outside of math class).
During this chapter, you have
probably used each Way of
Thinking multiple times without
even realizing it!

Choose three of these Ways of Thinking that you remember using while working in
this chapter. For each Way of Thinking that you choose, show and explain where
you used it and how you used it. Describe why thinking in this way helped you
solve a particular problem or understand something new. Be sure to include
examples to demonstrate your thinking.

Answers and Support for Closure Activity #4
What Have I Learned?

Problem	Solution	Need Help?	More Practice
CL 8-118.	Area of green = $81\pi \approx 254.5$ cm^2; area of purple = $588 - 81\pi \approx 333.5$ cm^2, so the area of purple is greater.	Lessons 5.1.2, 8.1.4, 8.1.5, 8.3.1, and 8.3.2 Math Notes boxes	Problems 8-45, 8-47, 8-48, 8-64, 8-67, 8-77, 8-82, 8-85, 8-92, 8-112
CL 8-119.	a. Rhombus. It is a quadrilateral with four equal sides. b. 20 square units c. The slopes are –2 and $\frac{1}{2}$. They are opposite reciprocals. d. The point of intersection is (0, 2). It is the midpoint of the diagonal.	Lessons 2.2.4, 7.2.3, 7.3.2, and 7.3.3 Math Notes boxes	Problems 7-29, 7-32, 7-69, 7-99, 7-107, 7-109, 7-110, 8-51, 8-89

Problem	Solution	Need Help?	More Practice

CL 8-120.

```
          ( M is a midpoint of KQ )
                     │
                   Given
                     ↓
  ( ∠P ≅ ∠L )   ( KM ≅ QM )   ( ∠KML ≅ ∠QMP )
     Given       Definition      Vertical
                 of a            angles are
                 midpoint        congruent
        ↘           ↓           ↙
          ( △KLM ≅ △QPM )
                AAS ≅
```

Need Help? Lessons 3.2.4, 6.1.3, and 7.1.3 Math Notes boxes, problems 7-56 and 7-79

More Practice: Problems 7-61, 7-78, 7-85, 7-87, 7-96, 7-104, 7-105, 8-20, 8-28, 8-58, 8-79, 8-88

CL 8-121. Devin must run 6π meters farther than Garrett on each lap.

Need Help? Lessons 8.3.1 and 8.3.2 Math Notes boxes

More Practice: Problems 8-92, 8-93, 8-102, 8-107, 8-112

CL 8-122.

a. $x = 110°$ (Opposite angles in a parallelogram are equal.)

b. $t = 18$

c. $x = 23°$ (Diagonals of a rhombus are perpendicular.)

d. $m = 4$ (Nonparallel sides of an isosceles trapezoid are congruent.)

Need Help? Lessons 1.1.3, 2.1.4, 7.2.4, and 8.1.2 Math Notes boxes

More Practice: Problems 7-16, 7-33, 7-40, 7-49, 7-52, 7-70, 8-6, 8-15, 8-21, 8-55, 8-113

CL 8-123.

a. 1800°

b. Impossible. In a regular polygon, the central angle must be a factor of 360°.

c. 7 sides

d. 15°

e. 36°

Need Help? Lessons 7.1.4, 8.1.1, and 8.1.4 Math Notes boxes, problems 8-1, 8-13, and 8-14

More Practice: Problems 8-15, 8-25, 8-29, 8-33, 8-34, 8-35, 8-40, 8-49, 8-55, 8-56, 8-87, 8-99, 8-109

SOLIDS AND CONSTRUCTIONS

9

CHAPTER 9   Solids and Constructions

In your study of geometry so far, you have focused your attention on two-dimensional shapes. You have **investigated** the special properties of triangles, parallelograms, regular polygons and circles, and have developed tools to help you describe and analyze those shapes. For example, you have tools to find an interior angle of a regular hexagon, to calculate the length of the hypotenuse of a right triangle, and to measure the perimeter of a triangle or the area of a circle.

In Section 9.1, you will turn your focus to three-dimensional shapes (called **solids**), such as cubes and cylinders. You will learn several ways to represent three-dimensional solids and develop methods to measure their volume and surface area.

Then, in Section 9.2, you will learn how to use special tools to construct accurate diagrams of two-dimensional shapes and geometric relationships. During this **investigation**, you will revisit many of the geometric conjectures and theorems that you have developed so far.

Guiding Questions

Think about these questions throughout this chapter:

How does it change?

How can I represent it?

How can I construct it?

What's the connection?

Is there another way?

In this chapter, you will learn:

> ➤ How to find the surface area and volume of three-dimensional solids, such as prisms and cylinders.

> ➤ How to represent a three-dimensional solid with a mat plan, a net, and side and top views.

> ➤ How the volume changes when a three-dimensional solid is enlarged proportionally.

> ➤ How to construct familiar geometric shapes (such as a rhombus and a regular hexagon) using construction tools such as tracing paper, a compass and straightedge, and a dynamic geometry tool.

Chapter Outline

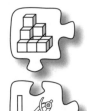

Section 9.1   This section is devoted to the study of three-dimensional solids and their measurement. You will also learn to use a variety of methods to represent the shapes of solids.

Section 9.2   This section will introduce you to the study of constructing geometric shapes and relationships. For example, you will learn how to construct a perpendicular bisector using only a compass and a straightedge.

9.1.1 How can I build it?

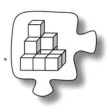

Three-Dimensional Solids

With your knowledge of polygons and circles, you are able to create and explore new, interesting shapes and make elaborate designs such as the one shown in the stained glass window at right. However, in the physical world, the objects we encounter every day are three-dimensional. In other words, physical objects cannot exist entirely on a flat surface, such as a tabletop.

To understand the shapes that you encounter daily, you will need to learn more about how three-dimensional shapes (called **solids**) can be created, described, and measured.

As you work with your team today, be especially careful to explain to your teammates how you "see" each solid. Remember that spatial **visualization** takes time and effort, so be patient with your teammates and help everyone understand how each solid is built.

Reprinted with permission by Rob Mielke, Blue Feather Stained Glass Designs.

9-1. Using blocks provided by your teacher, work with your team to build the three-dimensional solid at right. Assume that blocks cannot hover in midair. That is, if a block is on the second level, assume that it has a block below it to prop it up.

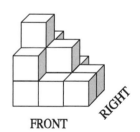

a. Is there more than one arrangement of blocks that could look like the solid drawn at right? Why or why not?

b. To avoid confusion, a **mat plan** can be used to show how the blocks are arranged in the solid. The number in each square represents the number of the blocks stacked in that location if you are looking from above. For example, in the lower right-hand corner, the solid is only 1 block tall, so there is a "1" in the corresponding corner of its mat plan. .

Mat Plan

Verify that the solid your team built matches the solid represented in the mat plan above.

c. What is the **volume** of the solid? That is, if each block represents a "cubic unit," how many blocks (cubic units) make up this solid?

9-2.   Another way to represent a three-
dimensional solid is by its **side**
and **top views**.

For example, the solid from
problem 9-1 can also be represented
by a top, front, and right-hand view,
as shown at right. Each view shows
all of the blocks that are visible
when looking directly at the solid
from that direction.

Examine the diagram of
blocks below. On graph
paper, draw the front, right,
and top views of this solid.
Assume that there are no
hidden blocks.

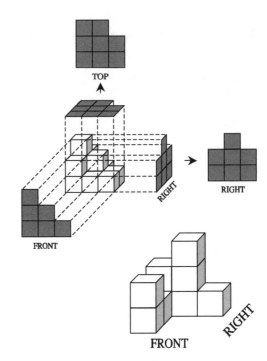

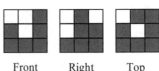

9-3.   For each of the mat plans below:

• Build the three-dimensional solid with the blocks provided by your teacher.

• Find the volume of the solid in cubic units.

• Draw the front, right, and top views of the solid on a piece of graph paper.

a.

0	3	0
2	3	1
0	2	0

RIGHT

FRONT

b.

0	2	1
0	3	0
3	2	1

RIGHT

FRONT

c.

1	1	3
2	1	2
0	0	1

RIGHT

FRONT

9-4.   Meagan built a shape with blocks and then drew
the views shown at right.

a.   Build Meagan's shape using blocks
provided by your teacher. Use as few
blocks as possible.

b.   What is the volume of Meagan's shape?

Front   Right   Top

Geometry Connections

9-5.   Draw a mat plan for each of the following solids. There may be more than one
         possible answer! Then find the possible volumes of each.

a.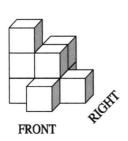

FRONT   RIGHT

b.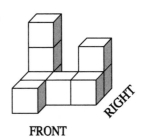

FRONT   RIGHT

9-6.   During this lesson, you have found the volume of several three-
         dimensional solids. However, what *is* volume? What does it
         measure? Write a Learning Log entry describing volume. Add at
         least one example. Title this entry "Volume of a Three-
         Dimensional Shape" and include today's date.

9-7.   **Examine** the solid at right.

a.   On your paper, draw a possible mat plan
         for this solid.

b.   Find the volume of this solid.

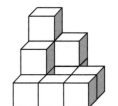

9-8.   Assume that two figures, *A* and *B*, are similar.

a.   If the linear scale factor is $\frac{2}{5}$, then what is the ratio of the areas of *A* and *B*?

b.   If the ratio of the perimeters of *A* and *B* is 14:1, what is the ratio of the areas?

c.   If the area of *A* is 81 times that of *B*, what is the ratio of the perimeters?

9-9.   Find the area of a regular decagon with perimeter 100 units. Show all work.

9-10.   The diagram at right shows a circle inscribed in a square.
         Find the area of the shaded region if the side length of the
         square is 6 meters.

9-11.   Solve each system of equations below. Write your solution in the form (x, y). Check your solution.

 a. $3x - y = 14$
      $x = 2y + 8$

 b. $x = 2y + 2$
      $x = -y - 10$

 c. $16x - y = -4$
      $2x + y = 13$

9-12.   **Multiple Choice:** What information would you need to know about the diagram at right in order to prove that $\triangle ABD \cong \triangle CBD$ by SAS $\cong$?

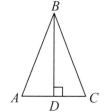

 a. $\overline{AD} \cong \overline{CD}$ b. $\overline{AB} \cong \overline{CB}$

 c. $\measuredangle A \cong \measuredangle C$ d. $\measuredangle ABD \cong \measuredangle CBD$

 e.   None of these

9.1.2   How can I measure it?

Volume and Surface Area of Prisms

Today you will continue to study three-dimensional solids and will practice representing a solid using a mat plan and its side and top views. You will also learn a new way to represent a three-dimensional object, called a **net**. As you work today, you will learn about a special set of solids called **prisms** and will study how to find the surface area and volume of a prism.

9-13.   The front, top, and right-hand views of Heidi's solid are shown at right.

 a.   Build Heidi's solid using blocks provided by your teacher. Use the smallest number of blocks possible. What is the volume of her solid?

      Front   Top   Right

 b.   Draw a mat plan for Heidi's solid. Be sure to indicate where the front and right sides are located.

 c.   Oh no! Heidi accidentally dropped her entire solid into a bucket of paint! What is the **surface area** of her solid? That is, what is the area that is now covered in paint?

9-14.   So far, you have studied three ways to represent a solid: a three-dimensional drawing, a mat plan, and its side and top views.

Another way to represent a three-dimensional solid is with a **net**, such as the one shown at right. When folded, a net will form the three-dimensional solid it represents.

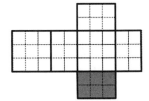

a.   With your team, predict what the three-dimensional solid formed by this net will look like. Assume the shaded squares make up the base (or bottom) of the solid.

b.   Obtain a Lesson 9.1.2 Resource Page and scissors from your teacher and cut out the net. Fold along the solid lines to create the three-dimensional solid. Did the result confirm your prediction from part (a)?

c.   Now build the shape with blocks and complete the mat plan at right for this solid.

FRONT

d.   What is the volume of this solid? How did you get your answer?

e.   What is the surface area of the solid? How did you find your answer? Be prepared to share any shortcuts with the class.

9-15.   Paul built a tower by stacking six identical layers of the shape at right on top of each other.

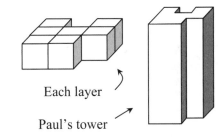

a.   What is the volume of his tower? How can you tell without building the shape?

Each layer

b.   What is the surface area of his tower?

Paul's tower

c.   Paul's tower is an example of a **prism** because it is a solid and two of its faces (called **bases**) are congruent and parallel. A prism must also have sides that connect the bases (called **lateral faces**). Each lateral face must be a parallelogram.

For each of the prisms below, find the volume and surface area.

(1)   (2)   (3)

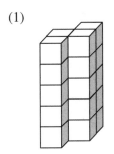   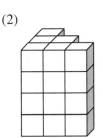   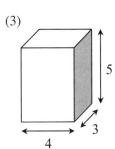

5

4

3

9-16.   Heidi created several more solids, represented below. Find the volume of each one.

a.

0	3	5
22	10	25
18	15	8
16	12	0

RIGHT

FRONT

b.

c.

9-17.   Pilar built a tower by stacking identical layers on top of each other. If her tower used a total of 312 blocks and if the bottom layer has 13 blocks, how tall is her tower? Explain how you know.

9-18.   What is the relationship between the area of the base of a prism, its height, and its volume? In a Learning Log entry, summarize how to find the volume of a solid. Be sure to include an example. Title this entry "Finding Volume" and include today's date.

(M)ETHODS AND MEANINGS

Polyhedra and Prisms

MATH NOTES

A closed three-dimensional solid that has flat, polygonal faces is called a **polyhedron**. The plural of polyhedron is **polyhedra**. "Poly" is the Greek root for "many," and "hedra" is the Greek root for "faces."

A **prism** is a special type of polyhedron. It must have two congruent, parallel **bases** that are polygons. Also, its **lateral faces** (the faces connecting the bases) are parallelograms formed by connecting the corresponding vertices of the two bases. Note that lateral faces may be any type of parallelogram, such as rectangles, rhombi, or squares.

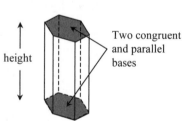

height

Two congruent and parallel bases

9-19. Mr. Wallis is designing a home. He found the plan for his dream house on the Internet and printed it out on paper.

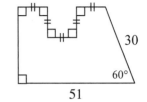

a. The design of the home is shown at right. If all measurements are in millimeters, find the area of the diagram.

b. Mr. Wallis took his home design to the copier and enlarged it 400%. What is the area of the diagram now? Show how you know.

9-20. At right is the solid from problem 9-7.

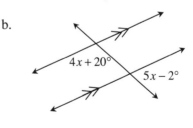

a. On graph paper, draw the front, right, and top views.

b. Find the total surface area of the solid.

9-21. Review what you know about the angles of polygons below.

a. If the exterior angle of a polygon is 29°, what is the interior angle?

b. If the interior angle of a polygon is 170°, can it be a regular polygon? Why or why not?

c. Find the sum of the interior angles of a regular 29-gon.

9-22. For each geometric relationship represented below, write and solve an equation for x. Show all work.

a.

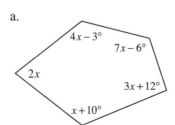

b.

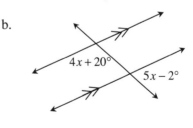

9-23.   On graph paper, graph $\triangle ABC$ if $A(-3, -4)$, $B(-1, -6)$, and $C(-5, -8)$.

   a.   What is AB (the length of $\overline{AB}$)?

   b.   Reflect $\triangle ABC$ across the x-axis to form $\triangle A'B'C'$. What are the coordinates of B'?

   c.   Rotate $\triangle A'B'C'$ 90° clockwise (↻) about the origin to form $\triangle A''B''C''$. What are the coordinates of C''?

   d.   Translate $\triangle ABC$ so that $(x, y) \rightarrow (x+5, y+1)$. What are the new coordinates of point A?

9-24.   **Multiple Choice:** Find the perimeter of the sector at right.

   a. 12π units   b. 3π units   c. $6 + 3\pi$ units

   d. $12 + \pi$ units   e.   None of these

9.1.3 What if the bases are not rectangles?

Prisms and Cylinders

In Lessons 9.1.1 and 9.1.2, you **investigated** volume, surface area, and special three-dimensional solids called prisms. Today you will explore different ways to find the volume and surface area of a prism and a related solid called a cylinder. You will also consider what happens to the volume of a prism or cylinder if it slants to one side or if it is enlarged proportionally.

9-25.   **Examine** the three-dimensional solid at right.

   a.   On graph paper, draw a net that, when folded, will create this solid.

   b.   Compare your net with those of your teammates. Is there more than one possible net? Why or why not?

   c.   Find the surface area and volume of this solid.

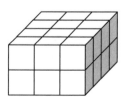

9-26.   SPECIAL PRISMS

The prism in problem 9-25 is an example of a **rectangular prism**, because its bases are rectangular. Similarly, the prism at right is called a **triangular prism** because the two congruent bases are triangular.

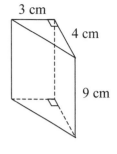

a.   Carefully draw the prism at right onto your paper. One way to do this is to draw the two triangular bases first and then to connect the corresponding vertices of the bases. Notice that hidden edges are represented with dashed lines.

b.   Find the surface area of the triangular prism. Remember that the surface area includes the areas of <u>all</u> surfaces – the sides and the bases. Carefully organize your work and verify your solution with your teammates.

c.   Find the volume of the triangular prism. Be prepared to share your team's method with the class.

d.   Does your method for finding surface area and volume work on other prisms? For example, what if the bases are hexagonal, like the one shown at right? Work with your team to find the surface area and volume of this hexagonal prism. Assume that the bases are regular hexagons with side length 4 inches.

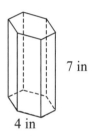

9-27.   CYLINDERS

Carter wonders, "*What if the bases are circular?*"

a.   Copy the **cylinder** at right onto your paper. Discuss with your team how to find its surface area and volume if the radius of the base is 5 units and the height of the cylinder is 8 units.

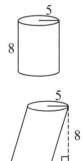

b.   Now Carter wants to figure out what happens to the volume of a cylinder when it slants as shown in the diagram at right. When the lateral faces of a prism or cylinder are not perpendicular to its base, the solid is referred to as an **oblique** cylinder or prism.

With your team, discuss whether the volume of the cylinder will increase, decrease or stay the same when the prism or cylinder is slanted. Assume that the radius and height of the cylinder do not change. When you agree, explain your answer on your paper. Be sure to provide **reasons** for your statements.

9-28.   Hernando needs to replace the hot water tank at his
         house. He estimates that his family needs a tank that
         can heat at least 75 gallons of water. His local water
         tank supplier has a cylindrical model that has a
         diameter of 2 feet and a height of 3 feet. If 1 gallon of
         water is approximately 0.1337 cubic feet, determine if
         the supplier's tank will provide enough water.

Ⓜ️ETHODS AND MEANINGS

Volume and Total Surface Area of a Solid

Volume measures the size of a three-dimensional space
enclosed within an object. It is expressed as the number
of 1x1x1 cubes (or parts of cubes) that fit inside a solid.

For example, the solid shown at right has a volume of 6 cubic
units.

Since volume reflects the number of cubes that fit within
a solid, it is measured in **cubic units**. For example, if the
dimensions of a solid are measured in feet, then the
volume would be measured in cubic feet (a cube with
dimensions 1' x 1' x 1').

On the other hand, the **total surface area** of a solid is the area of all of the
external faces of the solid. For example, the total surface area of the solid
above is 24 square units.

9-29.   In the diagram at right, $\overline{DE}$ is a midsegment of $\triangle ABC$. If
         the area of $\triangle ABC$ is 96 square units, what is the area of
         $\triangle ADE$?

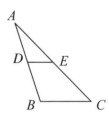

9-30.   Draw a rectangular prism as neatly as possible on your paper.
         If the width is 9 cm, the height is 14 cm, and the depth is 7 cm,
         find the surface area and volume.

9-31.   Are $\triangle EHF$ and $\triangle FGE$ congruent? If so, explain how you
         know. If not, explain why not.

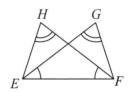

Geometry Connections

9-32.   Remember that the absolute value of a number is its positive value. For example, $|-5| = 5$ and $|5| = 5$. Use this understanding to solve the equations below, if possible. If there is no solution, explain how you know.

a. $|x| = 6$ b. $|x| = -2$ c. $|x + 7| = 10$

9-33.   Cindy's cylindrical paint bucket has a diameter of 12 inches and a height of 14.5 inches. If 1 gallon ≈ 231 in^3, how many gallons does her paint bucket hold?

9-34.   **Multiple Choice:** Which ratio below is the best approximation of the ratio between the circumference of a circle and its diameter?

a.   2   b.   3

c.   4   d.   6

$9.1.4$ How does the volume change?

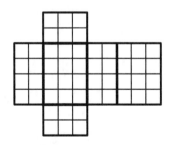

Volumes of Similar Solids

As you continue your study of three-dimensional solids, you will explore how the volume of a solid changes as the solid is enlarged proportionally.

9-35.   HOW DOES THE VOLUME CHANGE?

In Lesson 9.1.3, you began a study of the surface area and volume of similar solids. Today, you will continue that **investigation** in order to generalize about the ratios of similar solids.

a.   Describe the solid formed by the net at right. What are its dimensions (its length, width, and height)?

b.   Have each team member select a different enlargement ratio from the list below. On graph paper, carefully draw the net of a similar solid using your enlargement ratio. Then cut out your net and build the solid (so that the gridlines end up on the outside the solid) using scissors and tape.

(1)   1   (2)   2   (3)   3   (4)   4

Problem continues on next page $\rightarrow$

9-35.   *Problem continued from previous page.*

c.   Find the volume of your solid and compare it to the volume of the original solid. What is the ratio of these volumes? Share the results with your teammates so that each person can complete a table like the one below.

Linear Scale Factor	Original Volume	New Volume	Ratio of Volumes
1			
2			
3			
4			
r			

d.   How does the volume change when a three-dimensional solid is enlarged or reduced to create a similar solid? For example, if a solid's length, width, and depth are enlarged by a zoom factor of 10, then how many times bigger does the volume get? What if the solid is enlarged by a zoom factor of r? Explain.

9-36.   **Examine** the 1x1x3 solid at right.

a.   Build this solid with blocks provided by your teacher.

b.   If this shape is enlarged by a linear scale factor of 2, how wide will the new shape be? How tall? How deep?

c.   How many of the 1x1x3 solids would you need to build the enlargement described in part (b) above? Use blocks to prove your answer.

d.   What if the 1x1x3 solid is enlarged with a linear scale factor of 3? How many times larger would the volume of the new solid be? Explain how you found your answer.

9-37.      At the movies, Maurice counted the number of kernels of popcorn that filled his tub and found that it had 316 kernels. He decided that next time, he will get an enlarged tub that is similar, but has a linear scale factor of 1.5. How many kernels of popcorn should the enlarged tub hold?

9-38.   In your Learning Log, explain how the volume changes when a solid is enlarged proportionally. That is, if a three-dimensional object is enlarged by a zoom factor of 2, by what factor does the volume increase? Title this entry "Volumes of Similar Solids" and include today's date.

9-39.　Koy is inflating a spherical balloon for her brother's birthday party. She has used three full breaths so far and her balloon is only half the width she needs. Assuming that she puts the same amount of air into the balloon with each breath, how many more breaths does she need to finish the task? Explain how you know.

9-40.　Draw a cylinder on your paper. Assume the radius of the cylinder is 6 inches and the height is 9 inches.

a.　What is the surface area of the cylinder? What is the volume?

b.　If the cylinder is enlarged with a linear scale factor of 3, what is the volume of the enlarged cylinder? How do you know?

9-41.　While Katarina was practicing her figure skating, she wondered how far she had traveled. She was skating a "figure 8," which means she starts between two circles and then travels on the boundary of each circle, completing the shape of an "8." If both circles have a radius of 5 feet, how far does she travel when skating one "figure 8"?

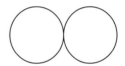

9-42.　For each triangle below, solve for x, if possible. If no solution is possible, explain why.

a.
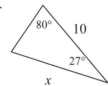

b.

c.
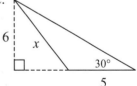

9-43.　The graph of the inequality $y > 2x - 3$ is shown at right. On graph paper, graph the inequality $y \le 2x - 3$. Explain what you changed about the graph.

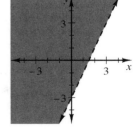

9-44.　**Multiple Choice:** The point $A(-2, 5)$ is rotated 90° counter-clockwise (↺). What are the new coordinates of point A?

a.　(2, 5)　　　　b.　(5, –2)　　　　c.　(2, –5)　　　　d.　(–5, –2)

9.1.5 How does the volume change?

Ratios of Similarity

Today, work with your team to analyze the following problems. As you work, think about whether the problem involves volume or area. Also think carefully about how similar solids are related to each other.

9-45. A statue to honor Benjamin Franklin will be placed outside the entry to the Liberty Bell exhibit hall. The designers decide that a smaller, similar version will be placed on a table inside the building. The dimensions of the life-size statue will be four times those of the smaller statue. Planners expect to need 1.5 pints of paint to coat the small statue. They also know that the small statue will weigh 14 pounds.

a. How many pints of paint will be needed to paint the life-size statue?

b. If the small statue is made of the same material as the enlarged statue, then its weight will change just as the volume changes as the statue is enlarged. How much will the life-size statue weigh?

9-46. The Blackbird Oil Company is considering the purchase of 20 new jumbo oil storage tanks. The standard model holds 12,000 gallons. Its dimensions are $\frac{4}{5}$ the size of the similarly shaped jumbo model, that is, the ratio of the dimensions is 4:5.

a. How much more storage capacity would the purchase of the twenty jumbo models give Blackbird Oil?

b. If jumbo tanks cost 50% more than standard tanks, which tank is a better buy?

9-47.   In problem 7-13, your class constructed a large tetrahedron like the one at right. Assume the dimensions of the shaded tetrahedron at right are half of the dimensions of the similar enlarged tetrahedron.

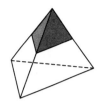

a.   If the volume of the large tetrahedron is 138 in^3, find the volume of the small shaded tetrahedron.

b.   Each face of a tetrahedron is an equilateral triangle. If the small shaded tetrahedron has an edge length of 16 cm, find the total surface area of the both tetrahedra.

c.   Your class tried to construct a tetrahedron using four smaller congruent tetrahedra. However, the result left a gap in the center, as shown in the diagram at right. If the volume of each small shaded tetrahedron is 50 in^3, what is the volume of the gap? Explain how you know.

METHODS AND MEANINGS

The $r : r^2 : r^3$ Ratios of Similarity

When a two-dimensional figure is enlarged proportionally, its perimeter and area also grow. If the linear scale factor is r, then the perimeter of the figure is enlarged by a factor of r while the area of the figure is enlarged by a factor of r^2. Examine what happens when the square at right is enlarged by a linear scale factor of 3.

$P = 4$ un $P = 4 \cdot 3 = 12$ un
$A = 1$ un^2 $A = 1 \cdot 3^2 = 9$ un^2

When a solid is enlarged proportionally, its surface area and volume also grow. If it is enlarged by a linear scale factor of r, then the surface area grows by a factor of r^2 and the volume grows by a factor of r^3. The example at right shows what happens to a solid when it is enlarged by a linear scale factor of 2.

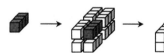

Original solid   Width, height,   **Result**:
SA = 14 un^2 and depth are   SA = 56 un^2
V = 3 un^3 doubled   V = 24 un^3

Thus, if a solid is enlarged proportionally by a linear scale factor of r, then:

New edge length = $r \cdot$ (corresponding edge length of original solid)

New surface area = $r^2 \cdot$ (original surface area)

New volume = $r^3 \cdot$ (original volume)

9-48.   Consider the two similar solids at right.

a.   What is the linear scale factor?

b.   Find the surface area of each solid. What
is the ratio of the surface areas? How is
this ratio related to the linear scale factor?

c.   Now find the volumes of each solid. How
are the volumes related? Compare this to
the linear scale factor and record your
observations.

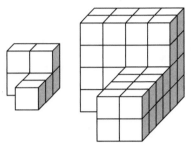

9-49.   Elliot has a modern fish tank that is in the shape of an
oblique prism, shown at right. If the slant of the prism
makes a 60° angle, find the volume of water the tank
can hold. Assume all measurements are in inches.

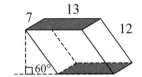

9-50.   Decide if the following statements are true or false. If they are true, explain how you
know. If they are false, provide a counterexample.

a.   If a quadrilateral has two sides that are parallel and two sides that are
congruent, then the quadrilateral must be a parallelogram.

b.   If the interior angles of a polygon add up to 360°, then the polygon must be a
quadrilateral.

c.   If a quadrilateral has 3 right angles, then the quadrilateral must be a rectangle.

d.   If the diagonals of a quadrilateral bisect each other, then
the quadrilateral must be a rhombus.

9-51.   Write and solve an equation based on the geometric
relationship shown at right.

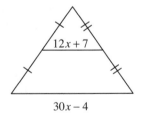

9-52.   Solve each equation below. Check your solution.

a. $20 - 6(5 + 2x) = 10 - 2x$

b. $2x^2 - 9x - 5 = 0$

c. $\frac{3}{5x-1} = \frac{1}{x+1}$

d. $|2x - 1| = 5$

9-53.   **Multiple Choice:** For $\angle ABE \cong \angle BEF$ in the
diagram at right, what must be true?

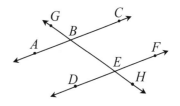

a. $\angle ABE \cong \angle BED$ b. $\angle ABE \cong \angle GBC$

c. $\overline{AC} \parallel \overline{GH}$ d. $\overline{AC} \parallel \overline{DF}$

e.   None of these.

9.2.1 How can I construct it?

Introduction to Construction

So far in this course, you have used tools such as rulers, tracing
paper, templates, protractors, and even computers to draw
geometric relationships and shapes. But how did ancient
mathematicians accurately construct shapes such as squares or
equilateral triangles without these types of tools?

Today you will start by exploring how to construct several
geometric relationships and figures with tracing paper. You
will then **investigate** how to construct geometric shapes with
tools called a compass and a straightedge, much like the
ancient Greeks did over 2000 years ago. As you study these
forms of **construction**, you will not only learn about new
geometric tools, but also gain a deeper understanding of some
of the special geometric relationships and shapes you have
studied so far in this course.

9-54.   CONSTRUCTING WITH TRACING PAPER

To start this focus on construction, you will begin with a familiar tool: tracing paper.
Obtain several sheets of tracing paper and a straightedge from your teacher.
Note: A straightedge is *not* a ruler. It does not have any markings or measurements
on it. A 3x5 index card makes a good straightedge.

a.   Starting with a smooth, square piece of tracing paper, find a way to create
parallel lines (or creases). Make sure the lines are *exactly* parallel. Be ready to
share with the class how you accomplished this.

b.   On a new piece of tracing paper, trace line segment
$\overline{AB}$ at right. Use your straightedge for accuracy. Can
you fold the tracing paper so that the resulting crease
not only finds the midpoint of $\overline{AB}$ but also is
perpendicular to $\overline{AB}$? (This is called a
perpendicular bisector.) Again, be ready to share
your method.

Problem continues on next page →

9-54.   *Problem continued from previous page.*

    c.   On the perpendicular bisector from part (b) above, choose a point C and then connect $\overline{AC}$ and $\overline{BC}$ to form $\triangle ABC$. What type of triangle did you construct? Use your geometry knowledge to **justify** your answer.

    d.   In part (b), you figured out how to use tracing paper and a straightedge to construct a line that bisects another line. How can you construct an angle bisector?

    On a piece of tracing paper, trace $\angle ABC$ at right. Construct an angle bisector. That is, find $\overrightarrow{BD}$ so that $\angle ABD \cong \angle CBD$.

9-55.   CONSTRUCTING WITH A COMPASS AND A STRAIGHTEDGE

Producing a geometric shape with a compass and a straightedge is another form of **construction**. Obtain a Lesson 9.2.1 Resource Page from your teacher (or download from www.cpm.org) and explore what types of shapes you can construct using these tools.

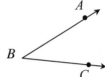

    a.   Find point C on the resource page. Use your compass to construct two circles with different radii that have a center at point C. (Note: Circles that have the same center are called **concentric** circles.)

    b.   With tracing paper, copying a line segment means just putting the tracing paper over the line and tracing it. But how can you copy a line using only a compass and a straightedge?

    On the resource page, find $\overline{AB}$. Next to $\overline{AB}$, use your straightedge to draw a new line. With your team, decide how to use the compass to mark off two points (C and D) so that $\overline{AB} \cong \overline{CD}$. Be ready to share your method with the class.

    c.   Now construct a new line segment, labeled $\overline{EF}$, that is twice as long as $\overline{AB}$. How can you be sure that $\overline{EF}$ is twice as long as $\overline{AB}$?

9-56.   In problem 9-55, you learned how to use a compass and a straightedge to copy a line segment. But how can you use these tools to copy an angle? On your Lesson 9.2.1 Resource Page, find $\angle X$. With your team, discuss how you can construct a new angle ($\angle Y$) that is congruent to $\angle X$. If you need help, use parts (a) through (c) below to guide you.

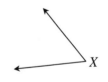

    a.   On the resource page, draw a ray with endpoint Y.

    b.   With your compass point at X, draw an arc that intersects both sides of $\angle X$. Now draw an arc with the same radius and with center Y.

Problem continues on next page →

Geometry Connections

9-56.   *Problem continued from previous page.*

    c.   How can you use your compass to measure the "width" of ∡X ? Discuss this with your teammates and then determine how to complete ∡Y . Be ready to share your method with the class.

9-57.   REGULAR HEXAGON

As Shui was completing her homework, she noticed that a regular hexagon has a special quality: when dissected into congruent triangles, the hexagon contains triangles that are all equilateral! "I bet I can use this fact to help me construct a regular hexagon," she told her team.

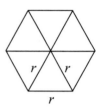

    a.   On the Lesson 9.2.1 Resource Page, construct a circle with radius *r* and center *H*.

    b.   Mark one point on the circle to be a starting vertex. Since each side of the hexagon has length *r*, the radius of the circle, carefully use the compass to mark off the other vertices of the hexagon on the circle. Then connect the vertices to create the regular hexagon.

    c.   When all vertices of a polygon lie on the same circle, the polygon is called **inscribed**. For example, the hexagon you constructed in part (b) is inscribed in ⊙*H* . After consulting with your teammates, construct an equilateral triangle that is also inscribed in ⊙*H* . You may want to use colored markers or pencils to help distinguish between the hexagon and the triangle.

9-58.   **Examine** the diagram of *ABCD* at right.

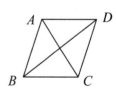

    a.   If opposite sides of the quadrilateral are parallel and all sides are congruent, what type of quadrilateral is *ABCD*?

    b.   List what you know about the diagonals of *ABCD*.

    c.   Find the area of *ABCD* if *BC* = 8 and *m∡ABC* = 60° .

9-59.   For a rectangular prism with base dimensions 3 cm and 5 cm and height 8 cm:

          a.   Sketch the prism on your paper.

          b.   Find the volume of the prism.

          c.   Find the surface area of the prism.

9-60.   Use the relationships given in the diagram at
          right to write and solve an equation for *x*.
          Show all work.

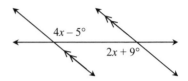

9-61.   For each pair of triangles below, determine if the triangles are congruent. If they are
          congruent, state the congruence property that assures their congruence and write a
          congruence statement (such as $\triangle ABC \cong \triangle$ _____).

          a.   b.   c.

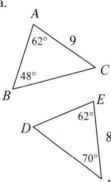

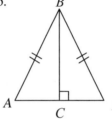

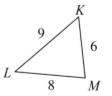

9-62.   Write the equation represented by the table below.

IN (*x*)	−4	−3	−2	−1	0	1	2	3	4
OUT (*y*)	−26	−20	−14	−8	−2	4	10	16	22

9-63.   **Multiple Choice:** Which net below will not produce a closed cube?

          a.   b.   c.   d.

                              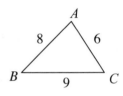

9.2.2 How can I construct it?

Constructing Bisectors

During Lesson 9.2.1, you studied how to construct geometric relationships such as congruent line segments using tools that include a compass and tracing paper. But what other geometric relationships and shapes can we construct using these tools? Today, as you **investigate** new ways to construct familiar geometric figures, look for connections to previous course material.

9-64.   INTERSECTING CIRCLES

As Ventura was doodling with his compass, he drew the diagram at right.

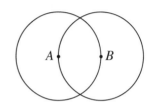

a.   Explain why ⊙A and ⊙B must have the same radius.

b.   On the Lesson 9.2.2 Resource Page provided by your teacher, construct two intersecting circles so that each passes through the other's center. Label the centers A and B.

c.   On your construction, locate the two points where the circles intersect each other. Label these points C and D. Then construct quadrilateral ACBD. What type of quadrilateral is ACBD? **Justify** your answer.

d.   Use what you know about the diagonals of ACBD to describe the relationship of $\overline{AB}$ and $\overline{CD}$. Make as many statements as you can.

9-65.   In problem 9-64, you constructed a rhombus and a perpendicular bisector.

a.   In your own words, describe how this process works. That is, given any line segment, how can you find its midpoint? How can you find a line perpendicular to it? Be sure to **justify** your statements.

b.   Test that your directions in part (a) work for line $\overline{KM}$ on the Lesson 9.2.2 Resource Page. In other words, construct a perpendicular bisector of $\overline{KM}$. Label the midpoint of $\overline{KM}$ point N.

c.   Return to your work from part (b) and use it to construct a 45°- 45°- 90° triangle. Prove that your triangle must be isosceles.

9-66.   In problem 9-64, you used the fact that the diagonals of a rhombus are perpendicular bisectors of each other to develop a construction. In fact, most constructions are rooted in the properties of many of the geometric shapes you have studied so far. A rhombus can help us with another important construction.

   a.   **Examine** the rhombus *ABCD* at right. What is the relationship between ∡*ABC* and $\overline{BD}$?

   b.   Since the diagonals of a rhombus bisect the angles, use this relationship to construct an angle bisector of ∡*R* on the resource page. That is, construct a rhombus so that *R* is one of its vertices. Use only a compass, a straightedge, and a pencil.

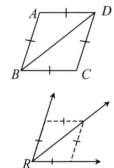

9-67.   CONSTRUCTION CHALLENGE

   On the Lesson 9.2.2 Resource Page, locate $\overline{PQ}$, $\overline{ST}$, and ∡*V*. In the space provided, use the construction **strategies** you have developed so far to construct a triangle with legs congruent to $\overline{PQ}$ and $\overline{ST}$, with an angle congruent to ∡*V* in between. Be sure you know how to do this two ways: with a compass and a straightedge and with tracing paper.

Ⓜ ETHODS AND MEANINGS

Rhombus Facts

Review what you have previously learned about a rhombus below.

A **rhombus** is a quadrilateral with four equal sides. All rhombi (the plural of rhombus) are parallelograms.

The diagonals of a rhombus are perpendicular bisectors of each other. That is, they intersect each other at their midpoints and form right angles at that point. Examine these relationships in the diagram at right.

9-68.   Unlike a straightedge, a ruler has measurement markings. With a ruler, it is fairly simple to construct a line segment of length 6 cm or a line segment with length 3 inches. But how can we construct a line segment of $\sqrt{2} \approx 1.414213562...$ centimeters? Consider this as you answer the questions below.

   a.   With a ruler, construct a line segment of 1 cm.

   b.   What about 1.4 cm? Adjust your line segment from part (a) so that its length is 1.4 cm. Did your line get longer or shorter?

   c.   Now change the line segment so that its length is 1.41 cm. How did it change?

   d.   Karen wants to continue this process until her line segment is exactly $\sqrt{2} \approx 1.414213562...$ centimeters long. What do you think will happen?

9-69.   The floor plan of Marina's local drug store is shown at right. While shopping one day, Marina tied her dog, Mutt, to the building at point F. If Mutt's leash is 4 meters long and all measurements in the diagram are in meters, what is the area that Mutt can roam? Draw a diagram and show all work.

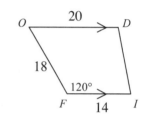

9-70.   Find the area of the Marina's drugstore (*FIDO*) in problem 9-69. Show all work.

9-71.   Given the information in the diagram at right, prove that $\angle C \cong \angle D$. Write your proof using any format studied so far.

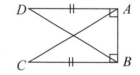

9-72.   Which has greater measure: an exterior angle of an equilateral triangle or an interior angle of a regular heptagon (7-gon)? Show all work.

9-73.   **Multiple Choice:** A solid with a volume of 26 in^3 was enlarged to create a similar solid with a volume of 702 in^3. What is the linear scale factor between the two solids?

   a.   1   b.   2   c.   3   d.   4

9.2.3 How do I construct it?

• •

More Exploration with Constructions

So far, several geometric relationships and properties have helped you develop constructions using a compass and a straightedge. For example, constructing a rhombus helped you construct an angle bisector. Constructing intersecting circles helped you construct a perpendicular bisector. What other relationships can help you develop constructions?

Today you will review some of your triangle knowledge to **investigate** how to construct congruent triangles and special triangles, such as 30°- 60°- 90° triangles. You'll also find a way to construct a line segment with a seemingly impossible length!

In Lesson 9.2.2, you developed a method to construct a rhombus within a given angle. This not only allowed you to construct an angle bisector, but it also helped you construct parallel lines, since the opposite sides of a rhombus are parallel. Today you will explore how to use your parallel line conjectures to construct a line parallel to a given line through a point not on the line.

9-74.   Find △*ABC* on the Lesson 9.2.3 Resource Page provided by your teacher.

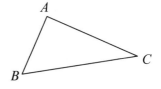

   a.   Using a compass and a straightedge, construct △*DEF* so that △*DEF* ≅ △*ABC*. Share construction ideas with your teammates.

   b.   Are there other ways to copy a triangle with a compass and a straightedge? Brainstorm as many ways as you can with your team. Be ready to share your ideas with the class.

9-75.   Consider what you know about all 30°- 60°- 90° triangles.

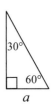

   a.   Using the information in the triangle at right, how long is the hypotenuse? Explain how you know.

   b.   Negin (pronounced "Nay-GEEN") wants to use this relationship to construct a 30°- 60°- 90° triangle. On the Lesson 9.2.3 Resource Page, locate her work so far. She has constructed perpendicular lines and has constructed one side ($\overline{MN}$). Complete her construction so that her triangle has angles 30°, 60°, and 90°.

Geometry Connections

9-76.   CONSTRUCTING AN IRRATIONAL LENGTH

Revisit your work from problem 9-68 from homework.

a.   Explain why you cannot construct a line segment of exactly length
$\sqrt{2} \approx 1.414213562...$ with a ruler.

b.   The number $\sqrt{2}$ is known as an **irrational number** because it cannot be
represented as a fraction of two integers. One way to spot an irrational number
is by looking at its decimal form. Irrational numbers have decimal numbers
that never repeat and "go on forever" (meaning they never terminate, like
$\frac{1}{2} = 0.5$ or $\frac{3}{8} = 0.375$ do).

Negin thinks that a right triangle may be able to help her construct a line
segment of length $\sqrt{2}$ units. First find two lengths of the legs of a right
triangle that will have a hypotenuse of $\sqrt{2}$ units. Then, on your Lesson 9.2.3
Resource Page, construct a right triangle with these dimensions.

9-77.   CONSTRUCTING PARALLEL LINES

So far, you have used geometric concepts such as triangle congruence and the
special properties of a rhombus to create constructions. How can angle relationships
formed by parallel lines help with construction? Consider this as you answer the
questions below.

a.   **Examine** the diagram at right. If $l \parallel m$, what
do you know about $\measuredangle a$ and $\measuredangle b$? $\measuredangle a$ and
$\measuredangle c$? **Justify** your answer.

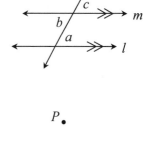

b.   Negin thinks that angle relationships can help her
construct a line parallel to another line through a
given point not on the line. On the Lesson 9.2.3
Resource Page, find line l and point P. Help
Negin construct a line parallel to l through point
P by first constructing a transversal through point
P that intersects line l.

c.   If you have not already done so, complete Negin's construction by copying an
angle formed by the transversal and line l. Explain how you used alternate
interior angles or corresponding angles.

9-78. Negin started to construct a parallel line with tracing paper but got off-task. She started by tracing line *l* and point *P* on her tracing paper, as shown in the diagram at right. While experimenting, she folded the tracing paper so that line *l* passed through point *P*. After creating a crease, she unfolded the tracing paper and then folded it again at a different place so that line *l* still passed through point *P*. She continued this process until she had over 20 creases on her tracing paper!

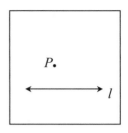

a. With your team, predict what shape the creases created.

b. With your own tracing paper, recreate Negin's experiment. First draw a line *l* and a point *P* that is not on the line, as shown above. Then fold the tracing paper as described above so that each fold causes line *l* to pass through point *P* at a different point on line *l*. What shape emerged?

MᴇTHODS AND Mᴇᴀɴɪɴɢs

Constructing a Perpendicular Bisector

MATH NOTES

A perpendicular bisector of a given segment can be constructed using tracing paper or using a compass and a straightedge.

With tracing paper: To construct a perpendicular bisector with tracing paper, first copy the line segment onto the tracing paper. Then fold the tracing paper so that the endpoints coincide (so that they lie on top of each other). When the paper is unfolded, the resulting crease is the perpendicular bisector of the line segment.

With a compass and a straightedge: One way to construct a perpendicular bisector with a compass and a straightedge is to construct a circle at each endpoint of the line segment with a radius equal to the length of the line segment. Then use the straightedge to draw a line through the two points where the circles intersect. This line will be the perpendicular bisector of the line segment.

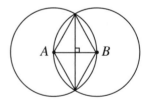

Geometry Connections

9-79.   **Examine** the mat plan of a three-dimensional solid at right.

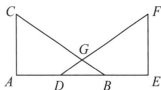

a.   On your paper, draw the front, right, and top views of this solid.

b.   Find the volume of the solid.

c.   If the length of each edge of the solid is divided by 2, what will the new volume be? Show how you got your answer.

9-80.   **Examine** the diagram at right. Given that $\triangle ABC \cong \triangle EDF$, prove that $\triangle DBG$ is isosceles. Use any format of proof that you prefer.

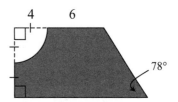

9-81.   The Portland Zoo is building a new children's petting zoo. One of the designs being considered is shown at right (the shaded portion). If the measurements are in meters, find:

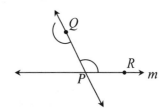

a.   The area of the petting zoo.

b.   The length of fence needed to enclose the petting zoo area.

9-82.   Sylvia has 14 coins, all nickels and quarters. If the value of the coins is $2.90, how many of each type of coin does she have? Explain your method.

9-83.   West High School has a math building in the shape of a regular polygon. When Mrs. Woods measured an interior angle of the polygon (which was in her classroom), she got 135°. How many sides does the math building have? Show how you got your answer.

9-84.   **Multiple Choice:** Jamila has started to construct a line parallel to line m through point Q at right. Which of the possible **strategies** below make the most sense to help her find the line parallel to m through point Q?

a.   Measure $\angle QPR$ with a protractor.

b.   Use the compass to measure the arc centered at P, then place the point of the compass where the arc centered at Q meets $\overline{QP}$, and mark that measure off on the arc.

c.   Construct $\overline{QR}$.

d.   Measure PR with a ruler.

9.2.4 What more can I construct?

Finding a Centroid

So far in this section, you have developed a basic library of constructions that can help create many of the geometric shapes and relationships you have studied in Chapters 1 through 8. For example, you can construct a rhombus, an isosceles triangle, a right triangle, a regular hexagon, and an equilateral triangle.

As you continue your **investigation** of geometric constructions today, keep in mind the following focus questions:

> What geometric principles or properties can I use?
>
> Why does it work?
>
> Is there another way?

9-85.   TEAM CHALLENGE

Albert has a neat trick. Given any triangle, he can place it on the tip of his pencil and it balances on his first try! The whole class wonders, "How does he do it?"

Your Task: Construct a triangle and find its point of balance. This point, called a **centroid**, is special not only because it is the center of balance, but also because it is where the **medians** of the triangle meet. Read more about medians of a triangle in the Math Notes box for this lesson and then follow the directions below.

a.   After reading about medians and centroids in the Math Notes box for this lesson, draw a large triangle on a piece of unlined paper provided by your teacher. (Note: Your team will work together on one triangle.)

b.   Working together, carefully construct the three medians and locate the centroid of the triangle.

c.   Once your team is convinced that your centroid is accurate, glue the paper to a piece of cardstock or cardboard provided by your teacher. Carefully cut out the triangle and demonstrate that your centroid is, in fact, the center of balance of your triangle! Good luck!

Geometry Connections

9-86.   CONSTRUCTING OTHER GEOMETRIC SHAPES

On a plain, unlined piece of paper, use a compass and a
straightedge to construct a kite. Remember that a kite is defined
as a quadrilateral with two pairs of adjacent, congruent sides. Be
prepared to explain to the class how you constructed your kite.

9-87.   In Lesson 9.2.3, you figured out how to construct a triangle
congruent to a given triangle. However, what if you are not
given the triangle and are instead given only its side lengths?

1 unit

a.   On a plain, unlined piece of paper, use a compass and a straightedge to
construct a triangle that has side lengths 3, 4, and 5 units. Note that the length
of 1 unit is provided above.

b.   What kind of triangle did you construct? **Justify** your conclusion.

9-88.   Albert wants to construct a triangle with side lengths
2, 3, and 6 units. On a plain, unlined piece of paper,
use a compass and a straightedge to construct
Albert's triangle. Use the unit length provided
below. Explain to him what happened.

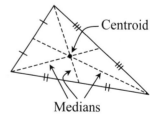

1 unit

METHODS AND MEANINGS

Centroid and Medians of a Triangle

MATH NOTES

A line segment connecting a vertex of a
triangle to the midpoint of the side
opposite the vertex is called a **median**.

Since a triangle has three vertices, it has three
medians. An example of a triangle with its
three medians is provided at right.

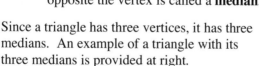

Centroid

Medians

The point at which the three medians intersect is called a **centroid**. The
centroid is also the center of balance of a triangle.

Since the three medians intersect at a single point, this point is called a **point
of concurrency**. You will learn about other points of concurrency in a later
chapter.

9-89. Find the volumes of the solids below.

a. cylinder

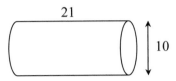

b. regular octagonal prism

9-90. Jillian is trying to construct a square. She has started by constructing two perpendicular lines, as shown at right. If she wants each side of the square to have length k, as defined at right, describe how she should finish her construction.

9-91. Without using a calculator, find the sum of the interior angles of a 1,002-gon. Show all work.

9-92. York County, Maine, is roughly triangular in shape. To help calculate its area, Sergio has decided to use a triangle, as shown at right. According to his map, the border with New Hampshire is 165 miles long, while the coastline along the Atlantic Ocean is approximately 100 miles long. If the angle at the tip of Maine is 43°, as shown in the diagram, what is the area of York County?

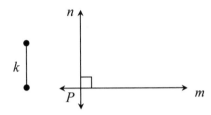

9-93. Copy the following words and their lines of reflection onto your paper. Then use your **visualization** skills to help draw the reflected images.

a. **REFLECT**

b. **PRISM**

9-94. **Multiple Choice:** Solve this problem without a calculator:

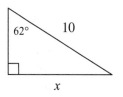

Examine the triangle at right. Find the approximate value of *x*. Use the values in the trigonometric table below as needed.

a. 4.69   b. 5.32

c. 8.83   d. 18.81

e. None of these

θ	$\cos\theta$	$\sin\theta$	$\tan\theta$
28°	0.883	0.469	0.532
62°	0.469	0.883	1.881

Chapter 9 Closure   What have I learned?

Reflection and Synthesis

The activities below offer you a chance to reflect on what you have learned during this chapter. As you work, look for concepts that you feel very comfortable with, ideas that you would like to learn more about, and topics you need more help with. Look for **connections** between ideas as well as **connections** with material you learned previously.

① TEAM BRAINSTORM

With your team, brainstorm a list for each of the following three topics. Be as detailed as you can. How long can you make your list? Challenge yourselves. Be prepared to share your team's ideas with the class.

Topics: What have you studied in this chapter? What ideas and words were important in what you learned? Remember to be as detailed as you can.

Problem Solving: What did you do to solve problems? What different **strategies** did you use?

Connections: How are the topics, ideas, and words that you learned in previous courses are **connected** to the new ideas in this chapter? Again, make your list as long as you can.

The following is a list of the vocabulary used in this chapter. The words that appear in bold are new to this chapter. Make sure that you are familiar with all of these words and know what they mean. Refer to the glossary or index for any words that you do not yet understand.

base	bisect	**centroid**
circle	**compass**	**concentric circles**
construction	cylinder	**inscribed**
irrational number	**lateral face**	line segment
linear scale factor	**mat plan**	**median**
net	**oblique**	perimeter
perpendicular bisector	polygon	**polyhedra**
prism	ratio	rhombus
similar	**solid**	**straightedge**
surface area	three-dimensional	**volume**

Make a concept map showing all of the **connections** you can find among the key words and ideas listed above. To show a **connection** between two words, draw a line between them and explain the **connection**, as shown in the example below. A word can be **connected** to any other word as long as there is a **justified connection**. For each key word or idea, provide a sketch of an example.

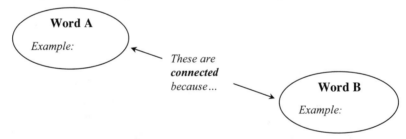

Your teacher may provide you with vocabulary cards to help you get started. If you use the cards to plan your concept map, be sure either to re-draw your concept map on your paper or to glue the vocabulary cards to a poster with all of the **connections** explained for others to see and understand.

While you are making your map, your team may think of related words or ideas that are not listed above. Be sure to include these ideas on your concept map.

③   SUMMARIZING MY UNDERSTANDING

This section gives you an opportunity to show what you know about certain math topics or ideas. Your teacher will give you directions for exactly how to do this. Your teacher may give you a "GO" page to work on. "GO" stands for "Graphic Organizer," a tool you can use to organize your thoughts and communicate your ideas clearly.

④   WHAT HAVE I LEARNED?

This section will help you evaluate which types of problems you have seen with which you feel comfortable and those with which you need more help. This section will appear at the end of every chapter to help you check your understanding. Even if your teacher does not assign this section, it is a good idea to try these problems and find out for yourself what you know and what you need to work on.

Solve each problem as completely as you can. The table at the end of this closure section has answers to these problems. It also tells you where you can find additional help and practice on problems like these.

CL 9-95.   On her paper, Kaye has a line with points A and B on it. Explain how she can use a compass to find a point C so that B is a midpoint of $\overline{AC}$. If you have access to a compass, try this yourself.

CL 9-96.   Assume that the solid at right has no hidden cubes.

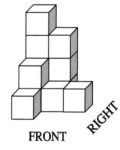

a.   On graph paper, draw the front, right, and top views of this solid.

b.   Find the volume and surface area of the cube.

c.   Which net(s) below would have the same volume as the solid at right when it is folded to create a box?

FRONT   RIGHT

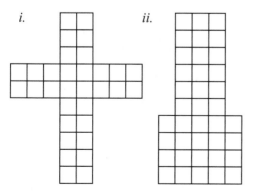

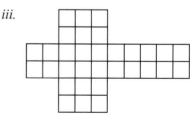

CL 9-97. The solid from problem CL 9-96 is redrawn at right.

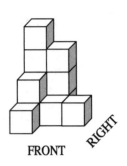

a. If this solid were enlarged by a linear scale factor of 4, what would the volume and surface area of the new solid be?

b. Enrique enlarged the solid at right so that its volume was 1500 cubic units. What was his linear scale factor? **Justify** your answer.

FRONT   RIGHT

CL 9-98. After constructing a $\triangle ABC$, Pricilla decided to try a little experiment. She chose a point V outside of $\triangle ABC$ and then constructed rays $\overrightarrow{VA}$, $\overrightarrow{VB}$, and $\overrightarrow{VC}$. Her result is shown at right. Copy this diagram onto your paper.

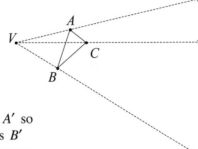

a. Pricilla then used a compass to mark point A' so that $VA = AA'$. She also constructed points B' and C' using the same method. For the diagram on your paper, locate A', B', and C'.

b. Now connect $\triangle A'B'C'$. What do you notice? What appears to be the relationship between $\triangle ABC$ and $\triangle A'B'C'$? Explain what happened.

c. If the area of $\triangle ABC$ is 19 cm^2 and its perimeter is 15 cm, find the area and perimeter of $\triangle A'B'C'$.

CL 9-99. Find the volume and surface area of the prism at right if the base is a regular octagon with side length 14 mm and the height of the prism is 28 mm.

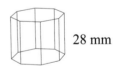

28 mm

CL 9-100. Answer the questions about the angles of polygons below, if possible. If it is not possible, explain how you know it is not possible.

a. Find the sum of the interior angles of a 28-gon.

b. If the exterior angle of a regular polygon is 42°, how many sides does the polygon have?

c. Find the measure of each interior angle of a pentagon.

d. Find the measure of each interior angle of a regular decagon.

Geometry Connections

CL 9-101. Fill in the blanks in each statement below with one of the quadrilaterals listed at right so that the statement is <u>true</u>. Use each quadrilateral name only once.

List:

Kite

Rectangle

Rhombus

Trapezoid

    a.   If a shape is a square, then it must also be a _____.

    b.   The diagonals of a _____ must be perpendicular to each other.

    c.   If the quadrilateral has only one line of symmetry, then it could be a _____.

    d.   If a quadrilateral has only two sides that are congruent, then the shape could be a _____.

CL 9-102. Copy quadrilateral *DART*, shown at right, onto your paper. If $\overline{DR}$ bisects $\angle ADT$ and if $\angle A \cong \angle T$, prove that $\overline{DA} \cong \overline{DT}$.

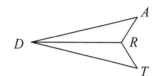

CL 9-103. After Myong's cylindrical birthday cake was sliced, she received the slice at right. If her birthday cake originally had a diameter of 14 inches and a height of 6 inches, find the volume of her slice of cake.

CL 9-104. Check your answers using the table at the end of the closure section. Which problems do you feel confident about? Which problems were hard? Use the table to make a list of topics you need help on and a list of topics you need to practice more.

⑤   HOW AM I THINKING?

This course focuses on five different **Ways of Thinking**: investigating, examining, reasoning and justifying, visualizing, and choosing a strategy/tool. These are some of the ways in which you think while trying to make sense of a concept or to solve a problem (even outside of math class). During this chapter, you have probably used each Way of Thinking multiple times without even realizing it!

Choose three of these Ways of Thinking that you remember using while working in this chapter. For each Way of Thinking that you choose, show and explain where you used it and how you used it. Describe why thinking in this way helped you solve a particular problem or understand something new. Be sure to include examples to demonstrate your thinking.

Answers and Support for Closure Activity #4
What Have I Learned?

Problem	Solution	Need Help?	More Practice
CL 9-95.	She should match the length of $\overline{AB}$ with her compass. Then, with the point of the compass at B, she should mark a point on the line on the side of point B opposite point A. Then she should label that point C.	Lesson 7.3.3 Math Notes box, problem 9-55	Problems 7-108, 8-30, 9-64, 9-65, 9-67, 9-74, 9-75, 9-76, 9-86, 9-87, 9-90
CL 9-96.	a. Front Right Top b. $V = 12$ un^3, $SA = 42$ un^2 c. All three nets will form a box with volume 12 un^3.	Lesson 9.1.3 Math Notes box, problems 9-1, 9-2, and 9-14	Problems 9-3, 9-4, 9-5, 9-7, 9-13, 9-15, 9-16, 9-20, 9-25, 9-35, 9-63, 9-79
CL 9-97.	a. $V = 12(4)^3 = 768$ un^3, $SA = 42(4)^2 = 672$ un^2 b. Linear scale factor = 5	Lesson 9.1.5 Math Notes box, problems 9-1, 9-35 and 9-36	Problems 9-37, 9-39, 9-40, 9-45, 9-46, 9-47, 9-48, 9-73, 9-79
CL 9-98.	a. b. $\triangle ABC$ was enlarged (or dilated) to create a similar triangle with a linear scale factor of 2. c. $A = 19(2)^2 = 76$ un^2; $P = 15(2) = 30$ un	Lessons 3.1.1, 7.2.6, and 9.1.5 Math Notes boxes	Problems 7-114, 7-103, 8-63, 8-65, 8-74, 8-76, 8-114, 9-8, 9-19, 9-29
CL 9-99.	Area of base ≈ 946.37 un^2 Volume $\approx 26,498.41$ un^3 Surface Area ≈ 5028.74 un^2	Lessons 8.3.1, 9.1.2, and 9.1.3 Math Notes boxes, problem 9-15	Problems 9- 16, 9-17, 9-26, 9-27, 9-28, 9-33, 9-40, 9-59, 9-89

Problem	Solution	Need Help?	More Practice
CL 9-100.	a. 4680° b. Not possible because 42° does not divide evenly into 360°. c. Not possible because it is not stated that the pentagon is regular. d. 144°	Lessons 7.1.4, 8.1.1, and 8.1.4 Math Notes boxes, problem 8-1	Problems 8-15, 8-25, 8-29, 8-33, 8-34, 8-35, 8-40, 8-49, 8-55, 8-56, 8-87, 8-99, 8-109, 9-21, 9-50, 9-72, 9-83, 9-91
CL 9-101.	a. Rectangle b. Rhombus c. Kite d. Trapezoid	Lessons 7.2.3, 8.1.2, and 9.2.2 Math Notes boxes	Problems 7-101, 7-106, 7-116, 7-117, 7-121, 8-11, 8-56, 9-50
CL 9-102.	$\overline{DR}$ bisects $\angle ADT$ Given $\angle A \cong \angle T$ (Given) $\angle ADR \cong \angle TDR$ (Definition of angle bisector) $\overline{DR} \cong \overline{DR}$ (Reflexive Property) $\triangle ADR \cong \triangle TDR$ AAS $\cong$ $\overline{DA} \cong \overline{DT}$ $\cong \triangle s \rightarrow \cong$ parts	Lessons 3.2.4, 6.1.3, 7.1.3, and 7.2.1 Math Notes boxes, problems 7-56 and 7-79	Problems 7-61, 7-78, 7-85, 7-87, 7-96, 7-104, 7-105, 8-20, 8-28, 8-58, 8-79, 8-88, 9-12, 9-31, 9-71, 9-80
CL 9-103.	$V \approx 97.49$ cubic inches	Lessons 8.3.2, 8.3.3, 9.1.2, and 9.1.3 Math Notes boxes	Problems 8-94, 8-96, 8-103, 8-104, 8-106, 9-16, 9-17, 9-26, 9-27, 9-28, 9-33, 9-40, 9-59, 9-89

10

CIRCLES AND EXPECTED VALUE

CHAPTER 10 Circles and Expected Value

In Chapter 8, you developed a method for finding the area and circumference of a circle, while in Chapter 9 you constructed many shapes and relationships using circles. In Section 10.1, you will explore the relationships between angles, arcs, and chords in a circle.

The focus turns to probability and games of chance in Section 10.2. As you analyze the probabilities of different outcomes on spinners, you will develop an understanding of expected value, which is a method to predict the outcome of a random event.

Circles will be revisited in Section 10.3 when you find the equation of a circle using the Pythagorean Theorem.

Guiding Questions

Think about these questions throughout this chapter:

What's the relationship?

How can I solve it?

What's the connection?

Is there another method?

How can I predict it?

In this chapter, you will learn:

➢ How to use the relationships between angles, arcs, and line segments within a circle to solve problems.

➢ How to find a circle inscribed in (and circumscribed about) a triangle.

➢ How to find the expected outcome of a game of chance.

➢ How to find the equation of a circle.

Chapter Outline

 Section 10.1   The relationships between angles, arcs, and line segments in a circle will be **investigated** to develop "circle tools" that can help solve problems involving circles.

 Section 10.2   After data is collected, the design of a spinner will be revealed and analyzed. Then the concept of expected value will be developed as a tool to predict the outcome of a random generator.

 Section 10.3   The Pythagorean Theorem will help to find the equation of a circle when it is graphed on coordinate axes.

10.1.1 What's the diameter?

••

Introduction to Chords

In Chapter 8, you learned that the diameter of a circle is the distance across the center of the circle. This length can be easily measured if the entire circle is in front of you and the center is marked, or if you know the radius of the circle. However, what if you only have part of a circle (called an **arc**)? Or what if the circle is so large that it is not practical to measure its diameter using standard measurement tools (such as finding the diameter of the Earth's equator)?

Today you will consider a situation that demonstrates the need to learn more about the parts of a circle and the relationships between them.

10-1.   THE WORLD'S WIDEST TREE

The baobab tree is a species of tree found in Africa and Australia. It is often referred to as the "world's widest tree" because it has been known to be up to 45 feet in diameter!

While digging at an archeological site, Rafi found a fragment of a fossilized baobab tree that appears to be wider than any tree on record! However, since he does not have the remains of the entire tree, he cannot simply measure across the tree to find its diameter. He needs your help to determine the radius of this ancient tree. Assume that the shape of the tree's cross-section is a circle.

Tree fragment

a.   Obtain the Lesson 10.1.1 Resource Page from your teacher. On it, locate $\overset{\frown}{AB}$, which represents the curvature of the tree fragment. Trace this arc as neatly as possible on tracing paper. Then decide with your team how to fold the tracing paper to find the center of the tree. (Note: This will take more than one fold.) Be ready to share with the class how you found the center.

b.   In part (a), you located the center of a circle. Use a ruler to measure the radius of that circle. If 1 cm represents 10 feet of tree, find the approximate radius and diameter of the tree. Does the tree appear to be larger than 45 feet in diameter?

10-2.   PARTS OF A CIRCLE, Part One

A line segment that connects the endpoints of an arc is
called a **chord**. Thus, $\overline{AB}$ in the diagram at right is an
example of a chord.

a.   One way to find the center of a circle when given an
arc is to fold it so that the two parts of the arc
coincide (lie on top of each other).

If you fold $\overset{\frown}{AB}$ so that A lies on B, what is the
relationship between the resulting crease and the
chord $\overline{AB}$? Explain how you know.

b.   The tree fragment in problem 10-1 was an arc between points A and B.
However, the missing part of the tree formed another larger arc of the tree.
With your team, find the larger arc formed by the circle and points A and B
above. Then propose a way to name the larger arc to distinguish it from $\overset{\frown}{AB}$.

c.   In problem 10-1, the tree fragment formed the shorter arc between two
endpoints. The shorter arc between points A and B is called the **minor arc** and
is written $\overset{\frown}{AB}$. The larger arc is called a **major arc** and is usually written
using three points, such as $\overset{\frown}{ACB}$. What do you know about $\overline{AB}$ if the minor
and major arcs are the same length? Explain how you know.

10-3.   In problem 10-1, folding the arc several times resulted in a point that seemed to be
the center of the circle. But how can we prove that the line bisecting an arc (or
chord) will pass through the center? To consider this, first assume that the
perpendicular bisector does *not* pass through the center. (This is an example of a
proof by contradiction.)

a.   According to our assumption, if the perpendicular
bisector does not pass through the center, then the
center, C, will be off the line in the circle, as shown at
right. Copy this diagram onto your paper.

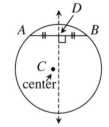

b.   Now consider $\triangle ACD$ and $\triangle BCD$. Are these two
triangles congruent? Why or why not?

c.   Explain why your result from part (b) contradicts the original assumption.
That is, explain why the center must lie on the perpendicular bisector of $\overline{AB}$.

10-4.   What if you know the lengths of two chords in a circle? How can you use the chords
        to find the center of the circle?

        a.   On the Lesson 10.1.1 Resource Page, locate the chords provided for ⊙P and
             ⊙Q. Work with your team to determine how to find the center of each circle.
             Then use a compass to draw the circles that contain the given chords. Tracing
             paper may be helpful.

        b.   Describe how to find the center of a circle without tracing paper. That is, how
             would you find the center of ⊙P with only a compass and a straightedge? Be
             prepared to share your description with the rest of the class.

10-5.   **Examine** the chord $\overline{WX}$ in ⊙Z at right. If $WX = 8$ and the radius
        of ⊙Z is 5, how far from the center is the chord? Draw the
        diagram on your paper and show all work.

 # ⓂETHODS AND MEANINGS

Circle Vocabulary

MATH NOTES

An **arc** is a part of a circle. Remember that a circle does not contain its
interior. A bicycle tire is an example of a circle. The piece of tire
between any two spokes of the bicycle wheel is an example of an arc.

Any two points on a circle create two arcs.
When these arcs are not the same length, the
larger arc is referred to as the **major arc**, while
the smaller arc is referred to as the **minor arc.**

To name an arc, an arc symbol is drawn over
the endpoints, such as $\overset{\frown}{AB}$. To refer to a major
arc, a third point on the arc should be used to
identify the arc clearly, such as $\overset{\frown}{ACB}$.

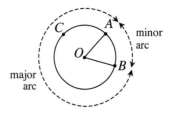

A **chord** is a line segment that has both
endpoints on a circle. $\overline{AB}$ in the diagram at
right is an example of a chord. When a chord
passes through the center of the circle, it is
called a **diameter.**

10-6.   A rectangular prism has a cylindrical hole removed, as shown at right. If the radius of the cylindrical hole is 2 inches, find the volume and total surface area of the solid.

10-7.   In the diagram below, $\overline{AD}$ is a diameter of $\odot B$.

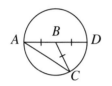

a.   If $m\angle A = 35°$, what is $m\angle CBD$?

b.   If $m\angle CBD = 100°$, what is $m\angle A$?

c.   If $m\angle A = x$, what is $m\angle CBD$?

10-8.   Lavinia started a construction at right. Explain what she is constructing. Then copy her diagram and finish her construction.

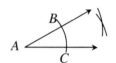

10-9.   A sector is attached to the side of a parallelogram, as shown in the diagram at right. Find the area and perimeter of the figure.

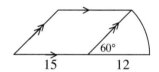

10-10.   On the same set of axes, graph both equations listed below. Then name all points of intersection in the form (x, y). How many times do the graphs intersect?

$$y = 4x - 7$$
$$y = x^2 - 2x + 2$$

10-11.   **Multiple Choice:** Dillon starts to randomly select cards out of a normal deck of 52 playing cards. After selecting a card, he does not return it to the deck. So far, he has selected a 3 of clubs, an ace of spades, a 4 of clubs, and a 10 of diamonds. Find the probability that his fifth card is an ace.

a. $\frac{1}{16}$ b. $\frac{3}{52}$ c. $\frac{1}{13}$ d. $\frac{1}{52}$

10.1.2 What's the relationship?

Angles and Arcs

In order to learn more about circles, we need to **investigate** the different types of angles and chords that are found in circles. In Lesson 10.1.1, you studied an application with a tree to learn about the chords of a circle. Today you will study a different application that will demonstrate the importance of knowing how to measure the angles and arcs within a circle.

10-12.

ERATOSTHENES' REMARKABLE DISCOVERY

Eratosthenes (who lived in the 3rd century B.C.) was able to determine the circumference of the Earth at a time when most people thought the world was flat! Since he knew that the Earth was round, he discovered that he could use a shadow to help calculate the Earth's radius.

Eratosthenes knew that Alexandria was located about 500 miles north of a town near the equator, called Syene. When the sun was directly overhead at Syene, a meter stick had no shadow. However, at the same time in Alexandria, a meter stick had a shadow due to the curvature of the Earth. Since the sun is so far away from the Earth, Eratosthenes assumed that the sun's rays were essentially parallel once they entered the Earth's atmosphere and realized that he could therefore use the stick's shadow to help calculate the Earth's radius.

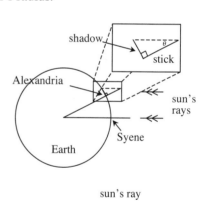

a.   Unfortunately, the precise data used by Eratosthenes was lost long ago. However, if Eratosthenes used a meter stick for his experiment today, then the stick's shadow in Alexandria would be 127 mm long. Determine the angle θ that the sunrays made with the meter stick. Remember that a meter stick is 1000 millimeters long.

b.   Assuming that the sun's rays are essentially parallel, determine the central angle of the circle if the angle passes through Alexandria and Syene. How did you find your answer?

Problem continues on next page →

10-12.   *Problem continued from previous page.*

c.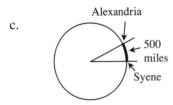
Since the distance along the Earth's surface from Alexandria to Syene is about 500 miles, that is the length of the arc between Alexandria and Syene. Use this information to approximate the circumference of the Earth.

d.   Use your result from part (c) to approximate the radius of the Earth.

10-13.   PARTS OF A CIRCLE, Part Two

In order to find the circumference of the Earth, Eratosthenes used an angle that had its vertex at the center of the circle. Like the angles in polygons that you studied in Chapter 8, this angle is called a **central angle**.

a.   An **arc** is a part of a circle. Every central angle has a corresponding arc. For example, in $\odot T$ at right, $\angle STU$ is a central angle and corresponds to $\overset{\frown}{SU}$. Since the measure of an angle helps us know what part of 360° the angle is, an arc can also be measured in degrees, representing what fraction of an entire circle it is. Thus, an **arc's measure** is equal to the measure of its corresponding central angle.

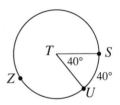

Examine the circle above. What is the measure of $\overset{\frown}{SU}$ (written $m\overset{\frown}{SU}$)? What is $m\overset{\frown}{SZU}$? Show how you got your answer.

b.   When Eratosthenes measured the distance from Syene to Alexandria, he measured the length of an arc. This distance is called **arc length** and is measured with units like centimeters or feet. One way to find arc length is to wrap a string about a part of a circle and then to straighten it out and measure its length. Calculate the arc length of $\overset{\frown}{SU}$ above if the radius of $\odot T$ is 12 inches.

10-14.   INSCRIBED ANGLES

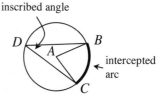
inscribed angle

In the diagram at right, ∡*BDC* is an example of an **inscribed angle**, because it lies in ⊙*A* and its vertex lies on the circle. It corresponds to central angle ∡*BAC* because they both intercept the same arc, $\overparen{BC}$. (An **intercepted arc** is an arc with endpoints on each side of the angle.)

intercepted arc

Investigate the measure of inscribed angles as you answer the questions below.

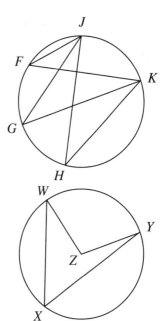

a.   In the circle at right, ∡*F*, ∡*G*, and ∡*H* are examples of inscribed angles. Notice that all three angles intercept the same arc ($\overparen{JK}$). Use tracing paper to compare their measures. What do you notice?

b.   Now compare the measurements of the central angle (such as ∡*WZY* in ⊙*Z* at right) and an inscribed angle (such as ∡*WXY*). What is the relationship of an inscribed angle and its corresponding central angle? Use tracing paper to test your idea.

10-15.   In problem 10-14, you found that the measure of an inscribed angle is always half of the measure of its corresponding central angle. Since the measure of the central angle always equals the measure of its intercepted arc, then the measure of the inscribed angle must be half of the measure of its intercepted arc.

Examine the diagrams below. Find the measures of the indicated angles. If a point is labeled *C*, assume it is the center of the circle.

a. 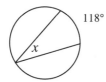   118°

b. *x* 41°

c. 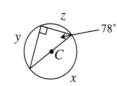 *z* *y* 78° *x*

d. *x* 56° *C* *y*

e. 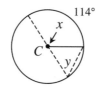 *x* 114° *C* *y*

f. *x* 28° *y*

10-16.   Reflect on what you have learned during this lesson. Write a Learning
          Log entry describing the relationships between inscribed angles and
          their intercepted arcs. Be sure to include an example. Title this entry
          "Inscribed Angles" and include today's date.

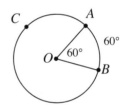

Review & Preview

10-17.   In $\odot A$ at right, $\overline{CF}$ is a diameter and $m\angle C = 64°$. Find:

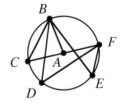

a. $m\angle D$ b. $m\overset{\frown}{BF}$ c. $m\angle E$

d. $m\overset{\frown}{CBF}$ e. $m\angle BAF$ f. $m\angle BAC$

10-18.   Find the area of a regular polygon with 100 sides and with a perimeter of 100 units.

10-19.   For each of the geometric relationships represented below, write and solve an equation for the given variable. For parts (a) and (b), assume that C is the center of the circle. Show all work.

a.

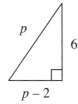

5m + 1

C

3m + 9

b.

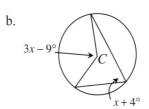

3x − 9°

C

x + 4°

c.

p

6

p − 2

d.

2t + 9°

3t + 1°

5t

8t − 10°

10-20.   On graph paper, plot △ABC if $A(-1, -1)$, $B(1, 9)$ and $C(7, 5)$.

a.   Find the midpoint of $\overline{AB}$ and label it D. Also find the midpoint of $\overline{BC}$ and label it E.

b.   Find the length of the midsegment, $\overline{DE}$. Use it to predict the length of $\overline{AC}$.

c.   Now find the length of $\overline{AC}$ and compare it to your prediction from (b).

10-21. $ABCDE$ is a regular pentagon inscribed in $\odot O$.

a.   Draw a diagram of $ABCDE$ and $\odot O$ on your paper.

b.   Find $m\angle EDC$. How did you find your answer?

c.   Find $m\angle BOC$. What relationship did you use?

d.   Find $m\overset{\frown}{EBC}$. Is there more than one way to do this?

10-22.   **Multiple Choice:** Jill's car tires are spinning at a rate of 120 revolutions per minute. If her car tires' radii are each 14 inches, how far does she travel in 5 minutes?

a.   140π   b.   8400π in   c.   3360π in   d.   16800π in

10.1.3 What more can I learn about circles?

Chords and Angles

As you **investigate** more about the parts of a circle, look for connections you can make to other shapes and relationships you have studied so far.

10-23.   WHAT IF IT'S A SEMICIRCLE?

What is the measure of an angle when it is inscribed in a **semicircle** (an arc with measure 180°)? Consider this as you answer the questions below.

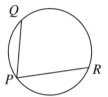

a.   Assume that the diagram at right is not drawn to scale. If $m\overarc{QR} = 180°$, then what is $m\angle P$? Why?

b.   Since you have several tools to use with right triangles, the special relationship you found in part (a) can be useful. For example, $\overline{UV}$ is a diameter of the circle at right. If $TU = 6$ and $TV = 8$, what is the radius of the circle? What is its area?

10-24.   In Lesson 10.1.1, you learned that a chord is a line segment that has its endpoints on a circle. What geometric tools do you have that can help find the length of a chord?

a.   **Examine** the diagram of chord $\overline{LM}$ in $\odot P$ at right. If the radius of $\odot P$ is 6 units and if $m\overarc{LM} = 150°$, find LM. Be ready to share your method with the class.

b.   What if you know the length of a chord? How can you use it to reverse the process? Draw a diagram of a circle with radius 5 units and chord $\overline{AB}$ with length 6 units. Find $m\overarc{AB}$.

10-25. Timothy asks, "What if two chords intersect inside a circle? Can triangles help me learn something about these chords?" Copy his diagram at right in which chords $\overline{AB}$ and $\overline{CD}$ intersect at point E.

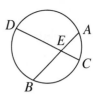

a. Timothy decided to create two triangles ($\triangle BED$ and $\triangle ACE$). Add line segments $\overline{BD}$ and $\overline{AC}$ to your diagram.

b. Compare $\angle B$ and $\angle C$. Which is bigger? How can you tell? Likewise, compare $\angle D$ and $\angle A$. Write down your observations.

c. How are $\triangle BED$ and $\triangle ACE$ related? **Justify** your answer.

d. If $DE = 8$, $AE = 4$, and $EB = 6$, then what is EC? Show your work.

10-26. Use the relationships in the diagrams below to solve for the variable. **Justify** your solution.

a.
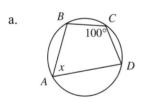

b. $\overline{KL}$ and $\overline{MP}$ intersect at Q and $KL = 8$ units

c. $\overline{RT}$ is a diameter

d.
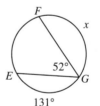

10-27. Look over your work from today. Consider all the geometric tools you applied to learn more about angles and chords of circles. In a Learning Log entry, describe which connections you made today. Title this entry "Connections with Circles" and include today's date.

Ⓜ ETHODS AND MEANINGS

Inscribed Angle Theorem

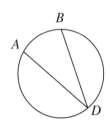

The measure of any inscribed angle is half of the measure of its intercepted arc. Likewise, any intercepted arc is twice the measure of any inscribed angles whose sides pass through the endpoints of the arc.

For example, in the diagram at right:

$$m\angle ADB = \tfrac{1}{2}m\widehat{AB} \text{ and } m\widehat{AB} = 2m\angle ADB.$$

Proof:

To prove this relationship, consider the relationship between an inscribed angle and its corresponding central angle. In problem 10-7, you used the isosceles triangle $\triangle ABC$ to demonstrate that if one of the sides of the inscribed angle is a diameter of the circle, then the inscribed angle must be half of the measure of the corresponding central angle. Therefore, in the diagram at right, $m\angle DAC = \tfrac{1}{2}m\widehat{DC}$.

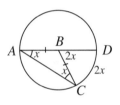

But what if the center of the circle instead lies in the interior of an inscribed angle, such as $\angle EAC$ shown at right? By constructing the diameter $\overline{AD}$, the work above shows that if $m\angle EAD = k$ then $m\widehat{ED} = 2k$ and if $m\angle DAC = p$, then $m\widehat{DC} = 2p$. Since $m\angle EAC = k + p$, then $m\widehat{EC} = 2k + 2p = 2(k + p) = 2m\angle EAC$.

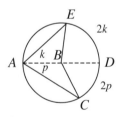

The last possible case to consider is when the center lies outside of the inscribed angle, as shown at right. Again, constructing a diameter $\overline{AD}$ helps show that if $m\angle CAD = k$ then $m\widehat{CD} = 2k$ and if $m\angle EAD = p$, then $m\widehat{ED} = 2p$. Since $m\angle EAC = p - k$, then $m\widehat{EC} = 2p - 2k = 2(p - k) = 2m\angle EAC$.

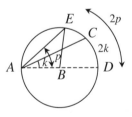

Therefore, an arc is always twice the measure of any inscribed angle that intercepts it.

10-28.   Assume point *B* is the center of the circle below. Match each item in the left column with the best description for it in the right column.

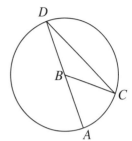

a.	$\overline{AB}$	1.	inscribed angle
b.	$\overline{CD}$	2.	semicircle
c.	$\overarc{AD}$	3.	radius
d.	$\angle CDA$	4.	minor arc
e.	$\overarc{AC}$	5.	central angle
f.	$\angle ABC$	6.	chord

10-29.   The figure at right shows two concentric circles.

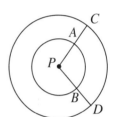

a.   Which arc has greater **measure**: $\overarc{AB}$ or $\overarc{CD}$? Explain.

b.   Which arc has greater **length**? Explain how you know.

c.   If $m\angle P = 60°$ and $PD = 14$, find the length of $\overarc{CD}$. Show all work.

10-30.   In $\odot Y$ at right, assume that $m\overarc{PO} = m\overarc{EK}$. Prove that $\overline{PO} \cong \overline{EK}$. Use the format of your choice.

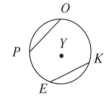

10-31.   While working on the quadrilateral hotline, Jo Beth got this call: "I need help identifying the shape of the quadrilateral flowerbed in front of my apartment. Because a shrub covers one side, I can only see three sides of the flowerbed. However, of the three sides I can see, two are parallel and all three are congruent. What are the possible shapes of my flowerbed?" Help Jo Beth answer the caller's question.

10-32.   For each pair of triangles below, decide if the triangles are similar or not and explain how you know. If the triangles are similar, complete the similarity statement $\triangle ABC \sim \triangle$_____ .

a.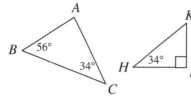

b.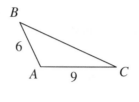

c.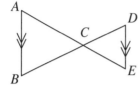

10-33.   **Multiple Choice:** Which equation below is perpendicular to $y = \frac{-2}{5}x - 7$ and passes through the point $(4, -1)$?

a. $2x - 5y = 13$ b. $2x + 5y = 3$ c. $5x - 2y = 22$

d. $5x + 2y = 18$ e.   None of these

10.1.4 What's the relationship?

Tangents and Chords

So far, you have studied about the relationships that exist between angles and chords (line segments) in a circle. Today you will extend these ideas to include the study of lines and circles.

10-34.   Consider all the ways a circle and a line can intersect. Can you **visualize** a line and a circle that intersect at exactly one point? What about a line that intersects a circle twice? On your paper, draw a diagram for each of the situations below, if possible. If it is not possible, explain why.

a.   Draw a line and a circle that do not intersect.

b.   Draw a line and a circle that intersect at exactly one point. When this happens, the line is called a **tangent**.

c.   Draw a line and a circle that intersect at exactly two points. A line that intersects a circle twice is called a **secant**.

d.   Draw a line and a circle that intersect three times.

Geometry Connections

10-35. A line that intersects a circle exactly once is called a **tangent**. What is the relationship of a tangent to a circle?

To **investigate**, carefully copy the diagram showing line *l* tangent to ⊙*A* at right onto tracing paper. Fold the tracing paper so that the crease is perpendicular to line *l* through point *P*. Your crease should pass through point *A*. What does this tell you about the tangent line?

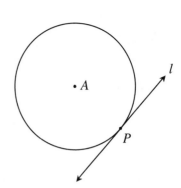

10-36. In the figure at right, $\overleftrightarrow{PA}$ is tangent to ⊙*R* at *E* and *PE* = *EA*. Is △*PER* ≅ △*AER* ? If so, prove it. If not, show why not.

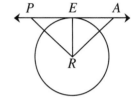

10-37. Use the relationships in the diagrams below to answer the following questions. Be sure to name what relationship(s) you used.

a. $\overrightarrow{PQ}$ is tangent to ⊙*C* at *P*. If *PQ* = 5 and *CQ* = 6, find *CP* and m∡*C*.

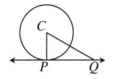

b. In ⊙*H*, $m\overarc{DR} = 40°$ and $m\overarc{GOR} = 210°$. Find $m\overarc{GD}$, $m\overarc{OR}$, and m∡*RGO*.

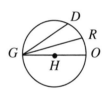

c. $\overline{AC}$ is a diameter of ⊙*E* and $\overline{BC}$ // $\overline{ED}$. Find the measure of $\overarc{CD}$.

d. $\overline{HJ}$ and $\overline{IK}$ intersect at *G*. If *HG* = 9, *GJ* = 8, and *GK* = 6, find *IG*.

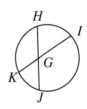

e. $\overline{AC}$ is a diameter of ⊙*E*, the area of the circle is 289π un², and *AB* = 16 units. Find *BC* and $m\overarc{BC}$.

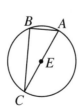

f. △*ABC* is inscribed in the circle at right. Using the measurements provided in the diagram, find $m\overarc{AB}$.

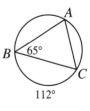

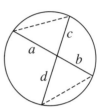
Review & Preview

10-38.   If $\overline{QS}$ is a diameter and $\overline{PO}$ is a chord of the circle at right, find the measure of the geometric parts listed below.

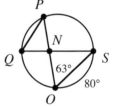

   a. $m\angle QSO$ b. $m\angle QPO$ c. $m\angle ONS$

   d. $m\widehat{PS}$ e. $m\widehat{PQ}$ f. $m\angle PQN$

10-39.   For each triangle below, solve for the given variables.

   a. 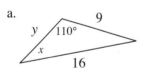   b.   c.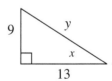

10-40.   The spinner at right is designed so that if you randomly spin the spinner and land in the shaded sector, you win $1,000,000. Unfortunately, if you land in the unshaded sector, you win nothing. Assume point C is the center of the spinner.

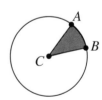

   a.   If $m\angle ACB = 90°$, how many times would you have to spin to reasonably expect to land in the shaded sector at least once? How did you get your answer?

   b.   What if $m\angle ACB = 1°$? How many times would you have to spin to reasonably expect to land in the shaded sector at least once?

   c.   Suppose $P(\text{winning } \$1,000,000) = \frac{1}{5}$ for each spin. What must $m\angle ACB$ equal? Show how you got your answer.

10-41.   Calculate the total surface area and volume of the prism at
          right. Assume that the base is a regular pentagon.

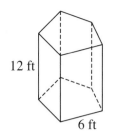

12 ft

6 ft

10-42.   Quadrilateral *ABCD* is graphed so that *A*(3, 2), *B*(1, 6),
          C(5, 8), and *D*(7, 4).

    a.   Graph *ABCD* on graph paper. What shape is *ABCD*?
          Justify your answer.

    b.   *ABCD* is rotated 180° about the origin to create *A′B′C′D′* . Then *A′B′C′D′* is
          reflected across the *x*-axis to form *A″B″C″D″* . Name the coordinates of *C′*
          and *D″* .

10-43.   **Multiple Choice:** Which graph below represents $y > -\frac{1}{2}x + 1$?

    a.   b.   c.   d.

10.1.5 How can I solve it?

Problem Solving with Circles

Your work today is focused on consolidating your understanding of the relationships between
angles, arcs, chords, and tangents in circles. As you work today, ask yourself the following
focus questions:

Is there another way?

What's the relationship?

10-44.   On a map, the coordinates of towns *A, B,* and *C*
          are $A(-3, 3)$, $B(5, 7)$, and $C(6, 0)$. City
          planners have decided to connect the towns with
          a circular freeway.

   a.   Graph the map of the towns on graph
         paper. Once the freeway is built, $\overline{AB}$,
         $\overline{BC}$, and $\overline{AC}$ will be chords of the circle.
         Use this information to find the center of
         the circle (called the **circumcenter** of the
         triangle because it is the center of the
         circle that circumscribes the triangle).

   b.   Use a compass to draw the circle connecting all three towns on your graph
         paper. Then find the radius of the circular freeway.

   c.   The city planners also intend to locate a new restaurant at the point that is an
         equal distance from all three towns. Where on the map should that restaurant
         be located? **Justify** your conclusion.

10-45.   An 8-inch dinner knife is sitting on a circular plate so that its ends
                                   are on the edge of the plate. If the minor arc that is intercepted by
                                   the knife measures 120°, find the diameter of the plate. Show all
                                   work.

10-46.   A cylindrical block of cheese has a 6-inch diameter and is 2
          inches thick. After a party, only a sector remains that has a
          central angle of 45°. Find the volume of the cheese that remains.
          Show all work.

10-47.   Dennis plans to place a circular hot tub in the corner
          of his backyard so that it is tangent to a fence on two
          sides, as shown in the diagram at right.

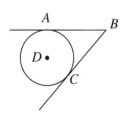

   a.   Prove that $\overline{AB} \cong \overline{CB}$.

   b.   The switch to turn on the air jets is located at point *B*. If the diameter of the
         hot tub is 6 feet and $AB = 4$ feet, how long does his arm need to be for him to
         reach the switch from the edge of the tub? (Assume that Dennis will be in the
         tub when he turns the air jets on and that the switch is level with the edge of
         the hot tub.)

MATH NOTES

METHODS AND MEANINGS

Points of Concurrency

In Chapter 9, you learned that the **centroid** of a triangle is the intersection of the three medians of the triangle, as shown at right. When three lines intersect at a single point, that point is called a **point of concurrency**.

Another point of concurrency, located where the perpendicular bisectors of each side of a triangle meet, is called the **circumcenter**. This point is the center of the circle that circumscribes the triangle. See the example at right. Note that the point that represents the location of the restaurant in problem 10-44 is a circumcenter.

The point where the three angle bisectors of a triangle meet is called the **incenter**. It is the center of the circle that is inscribed in a triangle. See the example at right.

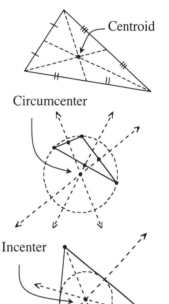

Centroid

Circumcenter

Incenter

Review & Preview

10-48.   In the diagram at right, $\odot M$ has radius 14 feet and $\odot A$ has radius 8 feet. $\overleftrightarrow{ER}$ is tangent to both $\odot M$ and $\odot A$. If $NC = 17$ feet, find ER.

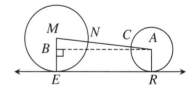

10-49.   Phinneus is going to spin both spinners at right once each. If he lands on the same color twice, he will go to tonight's dance. Otherwise, he will stay home. What is the probability that Phinneus will attend the dance?

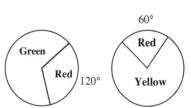

10-50.   In the figure at left, find the interior height (*h*)
          of the obtuse triangle. Show all work.

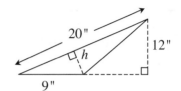

10-51.   A cylinder with volume 500π cm³ is similar to a
          smaller cylinder. If the scale factor is $\frac{1}{5}$, what is
          the volume of the smaller cylinder? Explain your
          reasoning.

10-52.   In the figure at right, $\overrightarrow{EX}$ is tangent to ⊙O at point *X*.
          OE = 20 cm and *XE* = 15 cm.

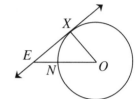

          a.   What is the area of the circle?

          b.   What is the area of the sector bounded by $\overline{OX}$
                and $\overline{ON}$?

          c.   Find the area of the region bounded by $\overline{XE}$, $\overline{NE}$, and $\overset{\frown}{NX}$.

10-53.   **Multiple Choice:** In the circle at right, $\overline{CD}$ is a
          diameter. If $AE = 10$, $CE = 4$, and $AB = 16$, what is
          the radius of the circle?

          a.   15   b.   16   c.   18

          d.   19   e.   None of these

10.2.1 What's the probability?

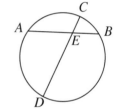

Designing Spinners

In Chapter 4, you played several games and learned how to determine if a game is fair. You
also used several models, such as tree diagrams and area models, to represent the probabilities
of the various outcomes.

Today you will look at probability from a new perspective. What if you want to design a game
that has particular, predictable outcomes? How can you design a spinner so that the result of
many spins matches a desired outcome?

10-54.   To review your understanding of probability, play the
          game as described below 50 times to determine a winner.
          You will need a paperclip and a Lesson 10.2.1 Resource
          Page. Then answer the questions that follow.

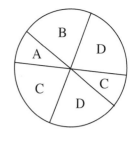

- Choose a different member of your
  team to be responsible for the
  following tasks:

  o Keeping track of time

  o Spinning a paperclip

  o Recording the result of each spin

  o Tallying the number of spins

- Assign each team member a region
  (or pair of regions) alphabetically by
  first name. That is, the team member
  whose name is first alphabetically
  will be assigned region A, the next
  person region B, and so on. You
  may want to color the spinner so that
  the four regions are different colors.

- Place the resource page on a flat, level surface. Then place a paperclip at
  the center of the spinner, hold it in place with the point of a pen or pencil,
  and spin it 50 times. Each time the spinner lands on a player's letter, that
  player gets a point. The person with the most points wins.

a.   Which player (A, B, C, or D) won the game? Is the result what you expected?
      Why or why not?

b.   Use a protractor to measure the central angles for each region. What is the
      probability that the spinner will land in each region?

c.   Calculate the percentage of the points scored for each region based on your
      results from playing the game and compare them to the probabilities you
      calculated in part (b). How closely did the results from spinning match the
      actual probabilities? Explain any large differences.

10-55. Do you have to collect data to predict the outcomes? For each spinner below, predict how many times a spinner would land in each region if you spun it 60 times randomly. Show all work.

a.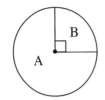

b.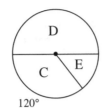

120°

c.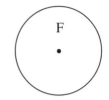

10-56. Your teacher will now spin a hidden spinner. Your team's task is to use the results to predict what the spinner looks like. Then, using the blank spinner on the resource page, use a protractor to design the spinner. As your teacher gives you the result of each spin, take careful notes! Your accuracy depends on it.

10-57. For a school fair, Donny is going to design a spinner with red, white, and blue regions. Since he has a certain proportion of three types of prizes, he wants the P(red) = 40% and P(white) = 10%.

a. If the spinner only has red, white, and blue regions, then what is P(blue)? Explain how you know.

b. Find the central angles of this spinner if it has only three sections. Then draw a sketch of the spinner. Be sure to label the regions accurately.

c. Is there a different spinner that has the same probabilities? If so, sketch another spinner that has the same probabilities. If not, explain why there is no other spinner with the same probabilities.

 ETHODS AND MEANINGS

Probability

MATH NOTES

While the information below was provided in Chapter 1, it is reprinted here for your reference during this section.

Probability is a measure of the likelihood that an event will occur at random. It is expressed using numbers with values that range from 0 to 1, or from 0% to 100%. For example, an event that has no chance of happening is said to have a probability of 0 or 0%. An event that is certain to happen is said to have a probability of 1 or 100%. Events that "might happen" have values somewhere between 0 and 1 or between 0% and 100%.

The probability of an event happening is written as the ratio of the number of ways that the desired outcome can occur to the total number of possible outcomes (assuming that each possible outcome is equally likely).

$$P(\text{event}) = \frac{\text{Number of Desired Outcomes}}{\text{Total Possible Outcomes}}$$

For example, on a standard die, P(5) means the probability of rolling a 5. To calculate the probability, first determine how many possible outcomes exist. Since a die has six different numbered sides, the number of possible outcomes is 6. Of the six sides, only one of the sides has a 5 on it. Since the die has an equal chance of landing on any of its six sides, the probability is written:

$$P(5) = \frac{1 \text{ side with the number five}}{6 \text{ total sides}} = \frac{1}{6} \text{ or } 0.1\overline{6} \text{ or approximately } 16.7\%$$

Review & Preview

10-58.   When the net at right is folded, it creates a die with values as shown.

	3		
1	5	2	1
	1		

a.   If the die is rolled randomly, what is P(even)? P(1)?

b.   If the die is rolled randomly 60 times, how many times would you expect an odd number to surface? Explain how you know.

c.   Now create your own net so that the resulting die has P(even) = $\frac{1}{3}$, P(3) = 0, and P(a number less than 5) = 1.

10-59.   In the diagram at right, $\overline{AB}$ is a diameter of $\odot L$. If $BC = 5$ and $AC = 12$, use the relationships shown in the diagram to solve for the quantities listed below.

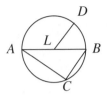

a. AB b.   radius of $\odot L$

c. $m\angle ABC$ d. $m\overset{\frown}{AC}$

10-60.   When Erica and Ken explored a cave, they each found a gold nugget. Erica's nugget is similar to Ken's nugget. They measured the length of two matching parts of the nuggets and found that Erica's nugget is five times longer than Ken's. When they took their nuggets to the metallurgist to be analyzed, they learned that it would cost $30 to have the surface area and weight of the smaller nugget calculated, and $150 to have the same analysis done on the larger nugget.

"I won't have that kind of money until I sell my nugget, and then I won't need it analyzed!" Erica says.

"Wait, Erica. Don't worry. I'm pretty sure we can get all the information we need for only $30."

a.   Explain how they can get all the information they need for $30.

b.   If Ken's nugget has a surface area of 110 cm², what is the surface area of Erica's nugget?

c.   If Ken's nugget weighs 56 g (about 1.8 oz), what is the weight of Erica's nugget?

10-61.   Find x if the angles of a quadrilateral are $2x$, $3x$, $4x$, and $5x$.

10-62.   A graph of an inequality is shown at right. Decide if each of the points (x, y) listed below would make the inequality true or not. For each point, explain how you know.

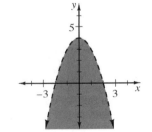

a. $(1, 1)$ b. $(-3, 2)$

c. $(-2, 0)$ d. $(0, -2)$

10-63.   **Multiple Choice:** Which expression below represents the length of the hypotenuse of the triangle at right?

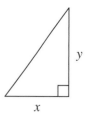

a. $\frac{y}{x}$ b. $\sqrt{x^2 + y^2}$ c. $x + y$

d. $\sqrt{y^2 - x^2}$ e.   None of these

10.2.2 What should I expect?

Expected Value

Around the world, different cultures have developed creative forms of games of chance. For example, native Hawaiians play a game called Konane, which uses markers and a board and is similar to checkers. Native Americans play a game called To-pe-di, in which tossed sticks determine how many points a player receives.

When designing a game of chance, much attention must be given to make sure the game is fair. If the game is not fair, or if there is not a reasonable chance that someone can win, no one will play the game. In addition, if the game has prizes involved, care needs to be taken so that prizes will be distributed based on availability. In other words, if you only want to give away one grand prize, you want to make sure the game is not set up so that 10 people win the grand prize!

Today your team will analyze different games to learn about **expected value**, which helps to predict the result of a game of chance.

10-64.   TAKE A SPIN

Consider the following game: After you spin the wheel at right, you win the amount spun.

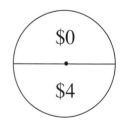

a.   If you play the game 10 times, how much money would you expect to win? What if you played the game 30 times? 100 times? Explain your process.

b.   What if you played the game *n* times? Write a rule that governs how much money one can expect to win after playing the game *n* times.

c.   If you were to play only once, what should you expect to earn according to your rule in part (b)? It is actually possible to win that amount? Explain why or why not.

10-65.   What if the spinner looks like the one at right instead?

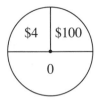

   a.   If you win the amount that comes up on each spin,
         how much would you expect to win after 4 spins?
         What about after 100 spins?

   b.   Find this spinner's **expected value**. That is, what is the expected amount you
         will win for each spin? Be ready to **justify** your answer.

   c.   Gustavo describes his thinking this way:
         "Half the time, I'll earn nothing. One-
         fourth the time, I'll earn $4 and the other
         one-fourth of the time I'll earn $100.
         So, for one spin, I can calculate
         $\frac{1}{2}(0)+\frac{1}{4}(\$4)+\frac{1}{4}(\$100)$. Calculate
         Gustavo's expression. Does his result
         match your result from part (b)?

10-66.   Jesse has created the spinner at right. This time, if you land on a
          positive number, you win that amount of money. However, if
          you land on a negative number, you lose that amount of money!
          Want to try it?

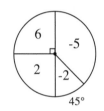

   a.   Before analyzing the spinner, predict whether a person
         would win money or lose money after many spins.

   b.   Now calculate the actual expected value. How does the result compare to your
         estimate from part (a)?

10-67.   Finding an expected value is similar to finding a **weighted average** because it takes
          into account the different probabilities for each possible outcome. For example, in
          problem 10-66, –5 is expected to result three times as often as –2. Therefore, in
          averaging these values, –5 must be weighted three times for every –2. However, the
          2 and the 6 have equal probabilities, so they must be averaged using the same
          weighting.

          To understand the effect of weighted
          averaging, consider the two spinners at right.
          Each has two sections, labeled $100 and $0.
          Which spinner has the greater expected
          value? How can you tell?

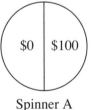

Spinner A   Spinner B

10-68.   In your Learning Log, explain what "expected value" means.
          What does it find? When is it useful? Be sure to include an
          example. Title this entry "Expected Value" and include today's
          date.

Geometry Connections

10-69. The spinner at right has three regions: A, B, and C. If it is spun 80 times, how many times would you expect each region to result? Show your work.

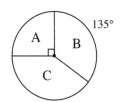

10-70. Review what you know about the angles and arc of circles below.

a. A circle is divided into nine congruent sectors. What is the measure of each central angle?

b. In the diagram at right, find $m\overarc{AD}$ and $m\angle C$ if $m\angle B = 97°$.

c. In $\odot C$ at right, $m\angle ACB = 125°$ and $r = 8$ inches. Find $m\overarc{AB}$ and the length of $\overarc{AB}$. Then find the area of the smaller sector.

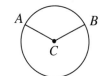

10-71. **Examine** the diagram at right. Use the given geometric relationships to solve for x, y, and z. Be sure to **justify** your work by stating the geometric relationship and applicable theorem.

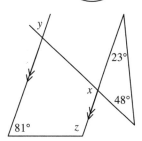

10-72. Solve each equation below for x. Check your work.

a. $\frac{x}{2} = 17$ b. $\frac{x}{4} = \frac{1}{3}$ c. $\frac{x+6}{2} + 2 = \frac{5}{2}$ d. $\frac{4}{x} = \frac{5}{8}$

10-73. Mrs. Cassidy solved the problem $(w-3)(w+5) = 9$ and got $w = 3$ and $w = -5$. Is she correct? If so, show how you know. If not, show how you know and find the correct solution.

10-74. **Multiple Choice:** Which expression represents the area of the trapezoid at right?

a. $\frac{c(a+b)}{4}$ b. $\frac{c(a+b)}{2}$

c. $\frac{bc}{2}$ d. $\frac{a+b+c}{2}$

e. None of these

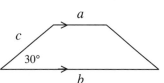

10.2.3 What can I expect?

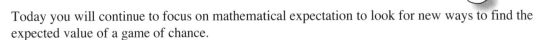

More Expected Value

Today you will continue to focus on mathematical expectation to look for new ways to find the expected value of a game of chance.

10-75.   Janine's teacher has presented her with this opportunity to raise her grade: She can roll a die and possibly gain points. If a positive number is rolled, Janine gains the number of points indicated on the die. However, if a negative roll occurs, then Janine loses that many points.

Janine does not know what to do! The die, formed when the net at right is folded, offers four sides that will increase her number of points and only two sides that will decrease her grade. She needs your help to determine if this die is **fair**.

		2	
2	−4	2
		1	
		−4	

a.   What are the qualities of a fair game? How can you tell if a game is fair? Discuss this with your team and be ready to share your ideas with the class.

b.   What is the expected value of one roll of this die? Show how you got your answer. Is this die fair?

c.   Change only one side of the die in order to make the expected value 0.

d.   What does it mean if a die or spinner has an expected value of 0? Elaborate.

10-76.   **Examine** the spinner at right. If the central angle of Region A is 7°, find the expected value of one spin **two different ways**. Be ready to share your methods with the class.

Region A

Region B

10-77.   Now reverse the process. For each spinner below, find x so that the expected value of the spinner is 3. Be prepared to explain your method to the class.

a.

b.

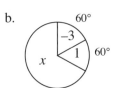

c.

Geometry Connections

10-78.   Revisit your work from part (c) of problem 10-77.

a.   To solve for x, Julia wrote the equation:

$$\frac{140}{360}(9) + \frac{40}{360}(18) + \frac{90}{360}(-3) + \frac{90}{360}x = 3$$

Explain how her equation works.

b.   She's not sure how to solve her equation. She'd like to rewrite the equation so that it does not have any fractions. What could she do to both sides of the equation to eliminate the fractions? Rewrite her expression and solve for x.

c.   Review the equation-rewriting techniques you learned in algebra by solving the equations below. You may benefit by reading the Math Notes box for this lesson.

(1) $\frac{4}{3} + \frac{x}{7} = 5$ (2) $\frac{1}{2}(5x - 3) + \frac{7}{4} = \frac{x}{2}$

10-79.   If you have not done so already, write an equation and solve for x for parts (a) and (b) of problem 10-75. Did your answers match those you found in problem 10-75?

10-80.   During this lesson, you examined two ways to find the expected value of a game of chance. What do these methods have in common? How are they different? Explain any connections you can find. Title this entry "Method for Finding Expected Value" and include today's date.

$\bigcirc$ ETHODS AND MEANINGS

Solving Equations by Rewriting (Fraction Busters)

Two equations are **equivalent** if they have the same solution(s). There are many ways to change one equation into a different, equivalent equation. If an equation contains a fraction, it may be easier to solve if it is first rewritten so that it has no fractions. This process is sometimes referred to as **fraction busters**.

Example: Solve for x: $\frac{x}{3} + \frac{x}{5} = 2$

The complicating issue in this problem is dealing with the fractions. We could add them by first writing them in terms of a common denominator, but there is an easier way.

There is no need to use the time-consuming process of adding the fractions if we can "eliminate" the denominators. To do this, we will need to find a common denominator of all fractions and multiply both sides of the equation by that common denominator. In this case, the lowest common denominator is 15, so we multiply both sides of the equation by 15. Be sure to multiply every term on each side of the equation!

$$\frac{x}{3} + \frac{x}{5} = 2$$

The lowest common denominator of $\frac{x}{3}$ and $\frac{x}{5}$ is 15.

$$15 \cdot \left(\frac{x}{3} + \frac{x}{5}\right) = 15 \cdot 2$$

$$15 \cdot \frac{x}{3} + 15 \cdot \frac{x}{5} = 15 \cdot 2$$

$$5x + 3x = 30$$

$$8x = 30$$

$$x = \frac{30}{8} = \frac{15}{4} = 3.75$$

The result is an equivalent equation without fractions! Now the equation looks like many you have seen before, and it can be solved using standard methods, as shown above. Note: If you cannot determine the common denominator, then multiply both sides by the product of the denominators.

Review & Preview

10-81.   For each spinner below, find the expected value of one spin.

a.

b.

c.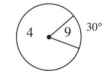

Geometry Connections

10-82.　For each equation below, write an equivalent equation that contains no fractions. Then solve your equation for x and check your answer.

　　a.　$\frac{2}{3}x - \frac{1}{4} = \frac{x}{2}$

　　b.　$\frac{7x}{1000} + \frac{2}{500} = \frac{11}{100}$

10-83.　The mat plan for a three-dimensional solid is shown at right.

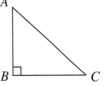

　　a.　On graph paper, draw <u>all</u> of views of this solid. (There are six views.) Compare the views. Are any the same?

　　b.　Find the volume and surface area of the solid. Explain your method.

　　c.　Do the views you drew in part (a) help calculate volume or surface area? Explain.

10-84.　For the triangle at right, find each trigonometric ratio below. The first one is done for you.

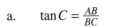

　　a.　$\tan C = \frac{AB}{BC}$　　b.　$\sin C$　　c.　$\tan A$

　　d.　$\cos C$　　e.　$\cos A$　　f.　$\sin A$

10-85.　Review circle relationships as you answer the questions below.

　　a.　On your paper, draw a diagram of $\odot B$ with arc $\overset{\frown}{AC}$. If $m\overset{\frown}{AC} = 80°$ and the radius of $\odot B$ is 10, find the length of chord $\overline{AC}$.

　　b.　Now draw a diagram of a circle with two chords, $\overline{EF}$ and $\overline{GH}$, that intersect at point K. If $EF = 15$, $EK = 6$, and $HK = 3$, what is GK?

10-86.　**Multiple Choice: Examine** $\odot L$ at right. Which of the mathematical statements below is not necessarily true?

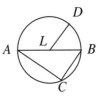

　　a.　$LD = AL$　　b.　$m\angle DLB = m\overset{\frown}{DB}$

　　c.　$\overline{LD} \parallel \overline{CB}$　　d.　$m\overset{\frown}{BC} = 2m\angle BAC$

　　e.　$2AL = AB$

10.3.1 What's the equation?

The Equation of a Circle

During Chapters 7 through 10, you studied circles *geometrically*, that is, based on the geometric shape of a circle. For example, the relationship of circles and polygons helped you develop a method to find the area and circumference of a circle, while geometric relationships of intersecting circles helped you develop constructions of shapes such as a rhombus and a kite.

However, how can circles be represented *algebraically* or *graphically*? And how can you use these representations to learn more about circles? Today your team will develop the equation of a circle.

10-87.   EQUATION OF A CIRCLE

We have equations for lines and parabolas, but what type of equation could represent a circle? On a piece of graph paper, draw a set of $x \rightarrow y$ axes. Then use a compass to construct a circle with radius 10 units centered at the origin (0, 0).

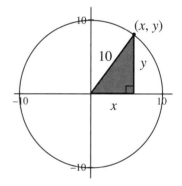

a.   What do all of the points on this circle have in common? That is, what is true about each point on the circle?

b.   Find all of the points on the circle where $x = 6$. For each point, what is the y-value? Use a right triangle (like the one shown at right) to **justify** your answer.

c.   What if $x = 3$? For each point on the circle where $x = 3$, find the corresponding y-value. Use a right triangle to **justify** your answer.

d.   Mia picked a random point on the circle and labeled it (x, y). Unfortunately, she does not know the value of x or y! Help her write an equation that relates x, y, and 10 based on her diagram above.

e.   Does your equation from part (d) work for the points (10, 0) and (0, 10)? What about (−8, −6)? Explain.

10-88.   In problem 10-87, you wrote an equation of a circle with radius 10 and center at $(0, 0)$.

   a.   What if the radius were instead 4 units long? Discuss this with your team and write an equation of a circle with center $(0, 0)$ and radius 4.

   b.   Write the equation of a circle centered at $(0, 0)$ with radius r.

   c.   On graph paper, sketch the graph of $x^2 + y^2 = 36$. Can you graph it without a table? Explain your method.

   d.   Describe the graph of the circle $x^2 + y^2 = 0$.

10-89.   What if the center of the circle is not at $(0, 0)$? On graph paper, construct a circle with a center $A(3, 1)$ and radius 5 units.

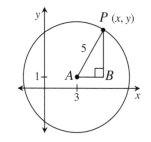

   a.   On the diagram at right, point P represents a point on the circle with no special characteristics. Add a point P to your diagram and then draw a right triangle like $\triangle ABC$ in the circle at right.

   b.   What is the length of $\overline{PB}$? Write an expression to represent this length. Likewise, what is the length of $\overline{AB}$?

   c.   Use your expressions for AB and BP, along with the fact that the radius of the circle is 5, to write an equation for this circle. (Note: You do not need to worry about multiplying any binomials.)

   d.   Find the equation of each circle represented below.

       (1)   The circle with center $(2, 7)$ and radius 1.

       (2)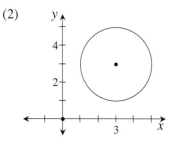

       (3)   The circle for which $(6, 0)$ and $(-6, 0)$ are the endpoints of a diameter.

10-90.   On graph paper, graph and shade the solutions for the inequalities below. Then find the area of each shaded region.

   a. $x^2 + y^2 \le 49$ b. $(x - 3)^2 + (y - 2)^2 \le 4$

10-91.   In a Learning Log entry, describe what you learned in this lesson
          about the equation of a circle. What connections did you make to
          other areas of algebra or geometry? Be sure to include an example
          of how to find the equation of a circle given its center and radius.
          Title this entry "Equation of a Circle" and include today's date.

10-92.   Jamika designed a game that allows some people to win money
          and others to lose money, but overall Jamika will neither win
          nor lose money. Each player will spin the spinner at right and
          will win the amount of money shown in the result. How much
          should each player pay to spin the spinner? Explain your
          reasoning.

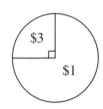

10-93.   For each diagram below, write an equation to represent the relationship between x and y.

a.

b.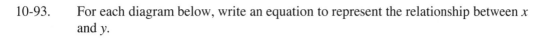

c.

d.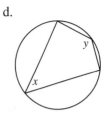

10-94.   For each triangle below, use the information in the diagram to decide if it is a right triangle. **Justify** each conclusion. Assume the diagrams are not drawn to scale.

a.

$64°$ $26°$

b.

8 17

15

10-95.   A cube (a rectangular prism in which the length, width, and depth are equal) has an edge length of 16 units. Draw a diagram of the cube and find its volume and surface area.

10-96.   For each pair of triangles below, decide if the pair is similar, congruent or neither. **Justify** your conclusion (such as with a similarity or congruence property like AA ~ or SAS ~ or the reasons why the triangles cannot be similar or congruent). Assume that the diagrams are not drawn to scale.

a.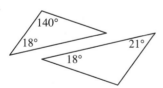

$140°$

$18°$ $21°$

$18°$

b.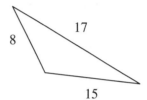

8 6

c.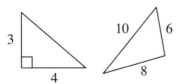

3 10 6

4 8

d.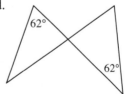

$62°$

$62°$

10-97.   **Multiple Choice:** $\triangle ABC$ is a right triangle and is graphed on coordinate axes. If $m\angle B = 90°$ and if the slope of $\overline{AB}$ is $-\frac{4}{5}$, what is the slope of $\overline{BC}$?

a. $\frac{4}{5}$ b. $\frac{5}{4}$ c. $-\frac{5}{4}$ d. $-\frac{4}{5}$

e.   Cannot be determined

Chapter 10 Closure   What have I learned?

The activities below offer you a chance to reflect on what you have learned during this chapter. As you work, look for concepts that you feel very comfortable with, ideas that you would like to learn more about, and topics you need more help with. Look for **connections** between ideas as well as **connections** with material you learned previously.

①   TEAM BRAINSTORM

With your team, brainstorm a list for each of the following three topics. Be as detailed as you can. How long can you make your list? Challenge yourselves. Be prepared to share your team's ideas with the class.

Topics:	What have you studied in this chapter? What ideas and words were important in what you learned? Remember to be as detailed as you can.
Problem Solving:	What did you do to solve problems? What different **strategies** did you use?
Connections:	How are the topics, ideas, and words that you learned in previous courses are **connected** to the new ideas in this chapter? Again, make your list as long as you can.

② MAKING CONNECTIONS

The following is a list of the vocabulary used in this chapter. The words that appear in bold are new to this chapter. Make sure that you are familiar with all of these words and know what they mean. Refer to the glossary or index for any words that you do not yet understand.

arc length	**arc measure**	center
central angle	**chord**	circle
circumference	**circumscribed**	diameter
expected value	**fair**	inscribed
major arc	measure	**minor arc**
perpendicular	probability	radius
secant	**semicircle**	similar
tangent	$x^2 + y^2 = r^2$	**weighted average**

Make a concept map showing all of the **connections** you can find among the key words and ideas listed above. To show a **connection** between two words, draw a line between them and explain the **connection**, as shown in the example below. A word can be **connected** to any other word as long as there is a **justified connection**. For each key word or idea, provide a sketch of an example.

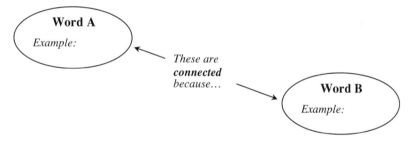

Your teacher may provide you with vocabulary cards to help you get started. If you use the cards to plan your concept map, be sure either to re-draw your concept map on your paper or to glue the vocabulary cards to a poster with all of the **connections** explained for others to see and understand.

While you are making your map, your team may think of related words or ideas that are not listed above. Be sure to include these ideas on your concept map.

③ SUMMARIZING MY UNDERSTANDING

This section gives you an opportunity to show what you know about certain math topics or ideas. Your teacher will give you directions for exactly how to do this.

④   WHAT HAVE I LEARNED?

This section will help you evaluate which types of problems you have seen with which you feel comfortable and those with which you need more help. This section will appear at the end of every chapter to help you check your understanding. Even if your teacher does not assign this section, it is a good idea to try these problems and find out for yourself what you know and what you need to work on.

Solve each problem as completely as you can. The table at the end of this closure section has answers to these problems. It also tells you where you can find additional help and practice on problems like these.

CL 10-98.   Copy the diagram at right onto your paper. Assume $\overrightarrow{AD}$ is tangent to $\odot C$ at D.

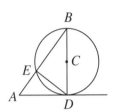

    a.   If $AD = 9$ and $AB = 15$, what is the area of $\odot C$?

    b.   If the radius of $\odot C$ is 10 and the $m\overset{\frown}{ED} = 30°$, what is $m\overset{\frown}{EB}$? AD?

    c.   If $m\overset{\frown}{EB} = 86°$ and if $BC = 7$, find EB.

CL 10-99.   A game is set up so that a person randomly selects a shape from the shape bucket shown at right. If the person selects a triangle, he or she wins $5. If the person selects a circle, he or she loses $3. If any other shape is selected, the person does not win or lose money.

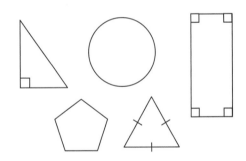

    a.   If a person plays 100 times, how much money should the person expect to win or lose?

    b.   What is the expected value of this game?

CL 10-100. Consider the solid represented by the mat plan at right.

3	1	0
0	1	1
0	2	3

Front   (Right)

    a.   Draw the front, right, and top view of this solid on graph paper.

    b.   Find the volume and surface area of this solid.

    c.   If this solid is enlarged by a linear scale factor of 3.5, what will be its new volume and surface area?

CL 10-101. Consider the descriptions of the different shapes below. Which shapes <u>must</u> be a parallelogram? If a shape does not have to be a parallelogram, what other shapes could it be?

    a.   A quadrilateral with two pairs of parallel sides.

    b.   A quadrilateral with two pairs of congruent sides.

    c.   A quadrilateral with one pair of sides that is both congruent and parallel.

    d.   A quadrilateral with two diagonals that are perpendicular.

    e.   A quadrilateral with four congruent sides.

CL 10-102. In $\odot C$ at right, $\overline{AB} \cong \overline{DE}$. Prove that $\angle ACB \cong \angle DCE$.

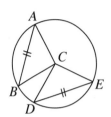

CL 10-103. Find the measure of x in each diagram below. Assume each polygon is regular.

    a.

    b.

    c.

CL 10-104. The circle at right is inscribed in a regular pentagon. Find the area of the shaded region.

CL 10-105. On graph paper, graph the equation $x^2 + y^2 = 100$.

    a.   What are the values of x when $y = 8$? Show how you know.

    b.   What are the values of y when $x = 11$? Show how you know.

CL 10-106. Check your answers using the table at the end of the closure section. Which problems do you feel confident about? Which problems were hard? Use the table to make a list of topics you need help on and a list of topics you need to practice more.

This course focuses on five different **Ways of Thinking**: investigating, examining, choosing a strategy/tool, visualizing, and reasoning and justifying. These are some of the ways in which you think while trying to make sense of a concept or to solve a problem (even outside of math class). During this chapter, you have probably used each Way of Thinking multiple times without even realizing it!

Choose three of these Ways of Thinking that you remember using while working in this chapter. For each Way of Thinking that you choose, show and explain where you used it and how you used it. Describe why thinking in this way helped you solve a particular problem or understand something new. Be sure to include examples to demonstrate your thinking.

Answers and Support for Closure Activity #4
What Have I Learned?

Problem	Solution	Need Help?	More Practice
CL 10-98.	a. 36π b. $m\widehat{EB} = 150°$, $AD = 20\tan15° \approx 5.36$ c. ≈ 9.55	Lessons 2.3.3, 5.1.2, 8.3.2, 10.1.1, 10.1.2, 10.1.3 Math Notes boxes	Problems 10-7, 10-15, 10-17, 10-23, 10-24, 10-26, 10-37, 10-38, 10-52, 10-59, 10-70, 10-86, 10-93
CL 10-99.	a. $140 should be won after 100 games b. $1.40 should be won per game	Lesson 10.3.1 Math Notes box	Problems 10-55, 10-64, 10-66, 10-67, 10-69, 10-75, 10-76, 10-77, 10-81, 10-92
CL 10-100.	a. Front Right Top b. $V = 11$ un^3, $SA = 42$ un^2 c. $V = 11(3.5)^3 = 471.625$ un^3, $42(3.5)^2 = 514.5$ un^2	Lessons 9.1.3 an 9.1.5 Math Notes boxes, problems 9-1, 9-2, and 9-14	Problems 9-3, 9-4, 9-5, 9-7, 9-13, 9-15, 9-16, 9-20, 9-25, 9-35, 9-37, 9-39, 9-40, 9-45, 9-46, 9-47, 9-48, 9-63, 9-73, 9-79

Problem	Solution	Need Help?	More Practice
CL 10-101.	Must be a parallelogram: (a), (c), and (e) (b) could be a kite or an isosceles trapezoid (d) could be a kite	Lessons 7.2.3, 8.1.2, and 9.2.2 Math Notes boxes, problem 7-47	Problems 7-101, 7-106, 7-116, 7-117, 7-121, 8-11, 8-56, 8-101, 9-50, 10-31
CL 10-102.	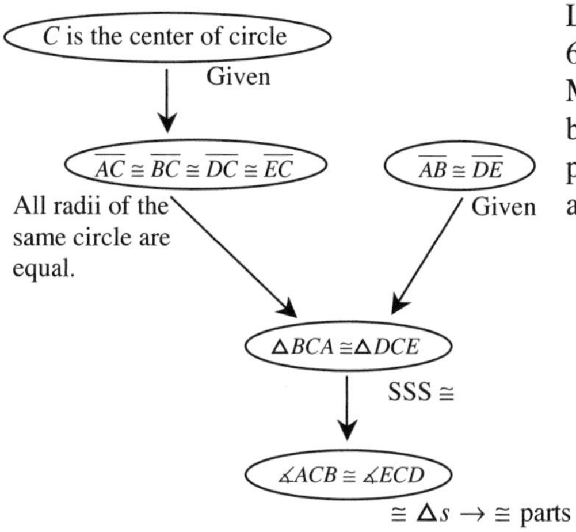 C is the center of circle — Given $\overline{AC} \cong \overline{BC} \cong \overline{DC} \cong \overline{EC}$ (All radii of the same circle are equal.) $\overline{AB} \cong \overline{DE}$ — Given $\triangle BCA \cong \triangle DCE$ — SSS $\cong$ $\angle ACB \cong \angle ECD$ — $\cong \triangle s \rightarrow \cong$ parts	Lessons 3.2.4, 6.1.3, and 7.1.2 Math Notes boxes, problems 7-56 and 7-79	Problems 7-61, 7-78, 7-85, 7-87, 7-96, 7-104, 7-105, 8-20, 8-28, 8-58, 8-79, 8-88, 9-12, 9-31, 9-71, 9-80, 10-30, 10-36, 10-47
CL 10-103.	a. 60° b. 135° c. 36°	Lessons 7.1.4, 8.1.1, and 8.1.4 Math Notes boxes	Problems 8-25, 8-29, 8-33, 8-34, 8-35, 8-40, 8-49, 8-55, 8-56, 8-87, 8-99, 8-109, 9-21, 9-50, 9-72, 9-83, 9-91, 10-61
CL 10-104.	Area of shaded region ≈ 8.37 un^2	Lessons 8.1.4, 8.3.1, and 8.3.2 Math Notes boxes	Problems 8-45, 8-47, 8-48, 8-67, 8-85, 8-103, 9-10
CL 10-105.	See graph at right. 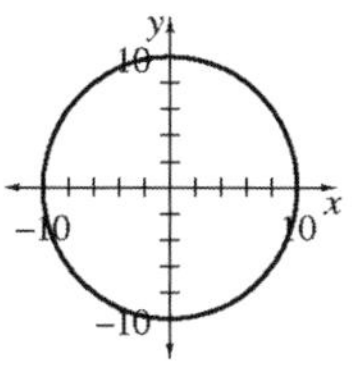 a. $x = 6$ or -6 because $6^2 + 8^2 = 100$ b. y does not exist when $x = 10$ because it is off the graph.	Problem 10-87	Problems 10-88, 10-89, 10-90

SOLIDS AND CIRCLES

11

CHAPTER 11

Solids and Circles

In Chapter 9, you learned how to find the volume and surface area of three-dimensional solids formed with blocks. Then you extended these concepts to include prisms and cylinders. In this chapter, you will complete your study of three-dimensional solids to include pyramids, cones, and spheres. You will learn how to identify the cross-sections of a solid and will **investigate** a special group of solids known as Platonic Solids.

As the word *geometry* literally means the "measurement of the Earth," it is only fitting that Section 11.2 focuses on developing the geometric tools that are used to learn more about the Earth. For example, by studying the height at which satellites orbit the Earth, you will get a chance to develop tools to work with the angle and arc measures that occur when two lines that are tangent to the same circle intersect each other.

Guiding Questions

Think about these questions throughout this chapter:

What's the relationship?

How can I measure it?

What information do I need?

Is there another way?

In this chapter, you will learn:

➢ How to find the volume and surface area of a pyramid, a cone, and a sphere.

➢ About the properties of special polyhedra, called Platonic Solids.

➢ How to find the cross-section of a solid.

➢ How to find the measures of angles and arcs that are formed by tangents and secants.

➢ About the relationships between the lengths of segments created when tangents or secants intersect outside a circle.

Chapter Outline

Section 11.1 In this section, you will learn how regular polygons can be used to form three-dimensional solids called "polyhedra." You will extend your knowledge of finding volume and surface area to include other solids, such as pyramids, cones, and spheres.

Section 11.2 By studying the coordinate system of latitude and longitude lines that help use refer to locations on the Earth, you will learn about great circles and how to find the distance between two points on a sphere. You will also **investigate** the geometric relationships created when tangents and secants intersect a circle.

11.1.1 How can I build it?

....

Platonic Solids

In Chapter 9, you explored three-dimensional solids such as prisms and cylinders. You developed methods to measure their sizes using volume and surface area and learned to represent three-dimensional solids using mat plans and two-dimensional (front, right, and top) views.

But what other types of three-dimensional solids can we learn about? During Section 11.1, you and your team will **examine** new types of solids in order to expand your understanding of three-dimensional shapes.

11-1.   EXAMINING A CUBE

In Chapter 9, you studied the volume and surface area of three-dimensional solids, such as prisms and cylinders. A **cube** is a special type of rectangular prism because each face is a square.

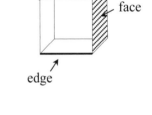

a.   What are some examples of cubes you remember seeing?

b.   Find the volume and surface area of a cube with an edge length of 10 units.

c.   A "flat side" of a prism is called a **face**, as shown in the diagram above. Notice that the line segment where two faces meet is called an **edge**, while the point where the edges meet is called a **vertex**. How many faces does a cube have? How many edges? How many vertices? ("Vertices" is plural for "vertex.")

d.   Confirm with your team that a cube has three square faces that meet at each vertex. Is it possible to have a solid where only two square faces meet at a vertex? Could a solid have four or more square faces at a vertex? Explain.

OTHER REGULAR POLYHEDRA

A three-dimensional solid made up of flat, polygonal faces is called a **polyhedron** (*poly* is the Greek root for "many," while *hedron* is the Greek root for "faces"). A cube, like the one you studied in problem 11-1, is an example of a **regular polyhedron** because all of the faces are congruent, regular polygons and the same number of faces meet at each vertex. In fact, you found that a cube is the *only* regular polyhedron with square faces.

But what if the faces are equilateral triangles? Or what if the faces are other regular polygons such as pentagons or hexagons?

Your Task: With your team, determine what other regular polyhedra are possible. First, obtain building materials from your teacher. Then work together to build regular polyhedra by testing how the different types of regular polygons can meet at a vertex. For example, what type of solid is formed when three equilateral triangle faces meet at each vertex? Four? Five? Six? Do similar tests for regular pentagons and hexagons. For each regular polyhedron, describe its shape and count its faces. Be ready to discuss your results with the class.

Discussion Points

- What is this task asking you to do?

- How should you start?

- How can your team organize the task among the members to complete the task efficiently?

Further Guidance

11-3.   For help in testing the various ways that congruent, regular polygons can build regular polyhedra, follow the directions below.

a.   Start by focusing on equilateral triangles. Attach three equilateral triangles so that they are adjacent and share a common vertex, as shown at right. Then fold and attach the three triangles so that they completely surround the common vertex. Complete the solid with as many equilateral triangles as needed so that each vertex is the intersection of three triangles. How would you describe this shape?

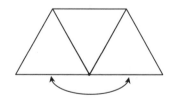

Fold so that these edges meet.

Problem continues on next page →

Geometry Connections

11-3.   *Problem continued from previous page.*

    b.   Now repeat your test to determine what solids are possible when 4, 5, or more equilateral triangles meet at each vertex. If a regular polyhedron is possible, describe it and state the number of faces it has. If a regular polyhedron is not possible, explain why not.

    c.   What if the faces are regular pentagons? Try building a regular polyhedron so that each vertex is the intersection of three regular pentagons. What if four regular pentagons meet at a vertex? Explain what happens in each case.

    d.   End your **investigation** by considering regular hexagons. Place three or more regular hexagons at a common vertex and explain what solids are formed. If no solid is possible, explain why.

11-4.   POLYHEDRA VOCABULARY

The regular polyhedra you discovered in problem 11-2 (along with the cube from problem 11-1) are sometimes referred to as **Plato's Solids** (or **Platonic Solids**) because the knowledge about them spread about 2300 years ago during the time of Plato, a Greek philosopher and mathematician.

In addition, polyhedra are classified by the number of faces they have. For example, a cube is a solid with six faces, so it can be called a regular hexahedron (because *hexa* is the Greek root meaning "six" and *hedron* is the Greek root for "face").

Plato (490 – 430 B.C.)

Examine the table of names below. Then return to your results from problem 11-2 and determine the name for each regular polyhedron you discovered.

4 faces	Tetrahedron	9 faces	Nonahedron
5 faces	Pentahedron	10 faces	Decahedron
6 faces	Hexahedron	11 faces	Undecahedron
7 faces	Heptahedron	12 faces	Dodecahedron
8 faces	Octahedron	20 faces	Icosahedron

11-5.   Find the surface area of each of Plato's Solids you built in problem 11-2 (the regular tetrahedron, octahedron, dodecahedron, and icosahedron) if the length of each edge is 2 inches. Show all work and be prepared to share your method with the class.

11-6.   DUAL POLYHEDRA

Ivan wonders, "What happens when the centers of adjacent faces of a regular polyhedron are connected?" These connections form the edges of a solid, which can be called a **dual polyhedron**.

To **investigate** dual polyhedra, first predict the results for each regular polyhedron with your team using spatial **visualization**. Then use a dynamic geometry tool to test your prediction of what solid is formed when the centers of adjacent faces of a Platonic Solid are connected. Be sure to test all five Platonic Solids (tetrahedron, cube, octahedron, dodecahedron, and icosahedron) and record the results.

11-7.   Reflect what you learned about Plato's Solids during this lesson. What connections did you make to previous material? Write an entry in your Learning Log explaining what is special about this group of solids. Name and describe each Platonic Solid. Title this entry "Plato's Solids" and include today's date.

11-8.   Draw a hexagon on your paper.

a.   Do all hexagons have an interior angle sum of 720°?

b.   Does every hexagon have an interior angle measuring 120°? Explain your **reasoning**.

c.   Does every hexagon have 6 sides? Explain your **reasoning**.

11-9.   The **lateral surface** of a cylinder is the surface connecting the bases. For example, the label from a soup can could represent the lateral surface of a cylindrical can. If the radius of a cylinder is 4 cm and the height is 15 cm, find the lateral surface area of the cylinder. Note: It may help you to think of "unrolling" a soup can label and finding the area of the label.

11-10.   For each of the relationships represented in the diagrams below, write and solve an equation for *x* and/or *y*. **Justify** your method. In part (a), assume that *C* is the center of the circle.

a.
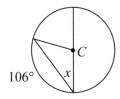

b.
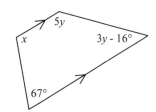

c.
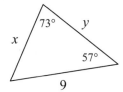

d.
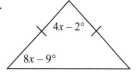

11-11.   Garland is having trouble with the copy machine. He's trying to copy a triangle with an area of 36 square units and a perimeter of 42 units.

a.   After he pressed the button to copy, Garland noticed the copier's zoom factor (the linear scale factor) was set to 200%. What is the area and perimeter of the resulting triangle?

b.   Now Garland takes the result from part (a) and accidentally shrinks it by a linear scale factor of $\frac{1}{3}$! What is the area and perimeter of the resulting triangle?

11-12.   Three flags are shown below on flagpoles. For each flag, determine what shape appears if the flag is spun very quickly about its pole. If you do not know the name of the shape, describe it.

a.

← pole

b.

pole

c.

pole

11-13.   **Multiple Choice:** $\triangle ABC$ has a right angle at *B*. If $m\angle A = 42°$ and $BC = 7$ mm, what is the approximate value of *AC*?

a.   9.4   b.   10.5   c.   7.8   d.   4.7

11-14.   Draw a tetrahedron on your paper.

      a.   How many faces does the tetrahedron have?

      b.   How many edges does it have?

      c.   How many vertices does it have?

11-15.   Mia found the volume of a rectangular prism to be 840 mm^3. As she was telling her father about it, she remembered that the base had a length of 10 mm and a width of 12 mm, but she could not remember the height. "Maybe there's a way you can find it by going backwards," her father suggested. Can you help Mia find the height of her prism? Explain your solution.

11-16.   On graph paper, graph a circle with center (4, 2) and radius 3 units. Then write its equation.

11-17.   **Examine** the diagram of the triangle at right.

      a.   Write an equation representing the relationship between x, y, and r.

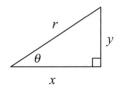

      b.   Write an expression for $\sin\theta$. What is $\sin\theta$ if $r = 1$?

      c.   Write an expression for $\cos\theta$. What is $\cos\theta$ if $r = 1$?

11-18.   In a circle, chord $\overline{AB}$ has length 10 units, while $m\overarc{AB} = 60°$. What is the area of the circle? Draw a diagram and show all work.

11-19.   **Multiple Choice:** Assume that the coordinates of $\triangle ABC$ are $A(5, 1)$, $B(3, 7)$, and $C(2, 2)$. If $\triangle ABC$ is rotated 90° clockwise (↻) about the origin, the coordinates of the image of B would be:

      a.   (–3, 7)   b.   (–7, 3)   c.   (7, –3)   d.   (7, 3)

11.1.2 How can I measure it?

. . . .

Pyramids

In Lesson 11.1.1, you explored Plato's five special solids: the tetrahedron, the octahedron, the cube (also known as the hexahedron), the dodecahedron, and the icosahedron. You discovered why these are the only regular polyhedra and developed a method to find their surface area.

Today you will **examine** the tetrahedron from a new perspective: as a member of the **pyramid** family. As you work today with your team, you will discover ways to classify pyramids by their shape and will develop new tools of measurement.

11-20.   A **pyramid** is a polyhedron with a polygonal base formed by connecting each point of the base to a single given point (the **apex**) that is above or below the flat surface containing the base. Each triangular lateral face of a pyramid is formed by the segments from the apex to the endpoints of a side of the base and the side itself. A tetrahedron is a special pyramid because any face can act as its base.

Obtain the four Lesson 11.1.2 Resource Pages, a pair of scissors, and either tape or glue from your teacher. Have each member of your team build one of the solids. When assembling each solid, be sure to have the printed side of the net on the exterior of the pyramid for reference later. Then answer the questions below.

a.   Sketch each pyramid onto your paper. What is the same about each pyramid? What is different? With your team, list as many qualities as you can.

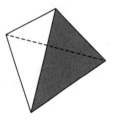

b.   A tetrahedron can also be called a **triangular-based pyramid**, because its base is always a triangle. Choose similar, appropriate names for the other pyramids that your team constructed.

c.   Find the surface area of pyramids **B** and **D**. Use a ruler to find the dimensions of the edges in centimeters.

d.   Compare pyramids **B** and **C**. Which do you think has more volume? **Justify** your **reasoning**.

11-21.   THE TRANSAMERICA BUILDING

The TransAmerica building in San Francisco is built of
concrete and is shaped like a square-based pyramid.
The building is periodically power-washed using one
gallon of cleaning solution for every 250 square meters
of surface. As the new building manager, you need to
order the cleaning supplies for this large task. The
problem is that you do not know the height of each
triangular face of the building; you only know the
vertical height of the building from the base to the top
vertex.

Your Task: Determine the amount of cleaning solution needed to wash the
TransAmerica building if an edge of the square base is 96 meters and the height of
the building is 220 meters. Include a sketch in your solution.

11-22.   Read the Math Notes box for this lesson, which introduces new vocabulary terms
such as "slant height" and "lateral surface area." Explain the difference between the
slant height and the height of a pyramid. How can you use one to find the other?

Ⓜ ETHODS AND MEANINGS

Pyramid Vocabulary

MATH NOTES

If a face of a pyramid (defined in problem 11-20) or prism is not a
base, it is called a **lateral face**.

The **lateral surface area** of a pyramid or prism is the sum of the areas of
all faces of the pyramid or prism, not including the base(s). The area of the
exterior of the TransAmerica building that needs cleaning (from problem
11-21) is an example of lateral surface area, since the exterior of the base of
the pyramid cannot be cleaned.

The **total surface area** of a pyramid or prism is the sum of the areas of all
faces, including the bases.

Sometimes saying the word "height"
for a pyramid can be confusing, since it
could refer to the height of one of the
triangular faces or it could refer to the
overall height of the pyramid.
Therefore, we call the height of each
lateral face a **slant height** to distinguish
it from the **height** of the pyramid itself.
See the diagram at right.

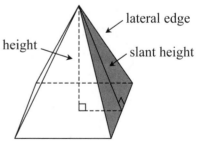

11-23.   Marina needs to win 10 tickets to get a giant stuffed panda
          bear. To win tickets, she throws a dart at the dartboard at right
          and wins the number of tickets listed in the region where her
          dart lands. Unfortunately, she only has enough money to play
          the game three times. If she throws the dart randomly, do you
          expect that she'll be able to win enough tickets? Assume that
          each dart will land on the dartboard.

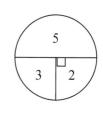

11-24.   **Examine** △*ABC* , △*ABD* , and △*ABE* in
          the diagram at right. If $\overline{CE}$ // $\overline{AB}$, explain
          what you know about the areas of the three
          triangles. **Justify** your statements.

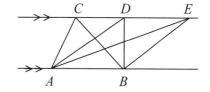

11-25.   Prove that when two lines that are tangent to
          the same circle intersect, the lengths between
          the point of intersection and the points of
          tangency are equal. That is, in the diagram
          at right, if $\overrightarrow{AB}$ is tangent to ⊙*P* at *B*, and $\overrightarrow{AC}$
          is tangent to ⊙*P* at *C*, prove that *AB* = *AC*. Use
          either a flowchart or a two-column proof.

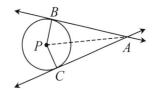

11-26.   Solve each equation below, if possible. Show all work.

          a. $\frac{3}{5} = \frac{2x}{3} - 8$ b. $\frac{9x}{5000} + \frac{2}{1000} = \frac{28}{5000}$

          c. $\frac{2x}{3} + \frac{x}{2} = \frac{2x}{3}$ d. $\frac{3}{2}(2x - 5) = \frac{1}{6}$

11-27.   On graph paper, plot the points *A*(4, 1) and *B*(10, 9).

          a.   Find the distance between points *A* and *B*. That is, find *AB*.

          b.   If point *C* is at (10, 1), find $m\angle CAB$. Show all work.

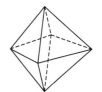

11-28.   **Multiple Choice:** Which net below **cannot** create a regular
          octahedron when folded, like the one at right?

          a.   b.   c.   d.

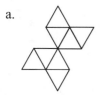

          e.   None of these

11.1.3 What's the volume?

. . . .

Volume of a Pyramid

Today, as you continue your focus on pyramids, look for and utilize connections to other geometry concepts. The models of pyramids that you constructed in Lesson 11.1.2 will be useful as you develop a method for finding the volume of a pyramid.

11-29.   GOING CAMPING

As Soraya shopped for a tent, she came across two models that she liked best, shown at right. However, she does not know which one to pick! They are both made by the same company and appear to have the same quality. She has come to you for help in making her decision.

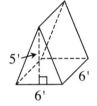

Tent A   *Tent B*

While she says that her drawings are not to scale, below are her notes about the tents:

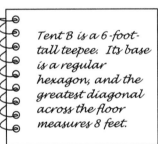

Tent A is a pup tent with a rectangular base. It has a height of 5 feet, a length of 6 feet, and a width of 6 feet.

Tent B is a 6-foot-tall teepee. Its base is a regular hexagon, and the greatest diagonal across the floor measures 8 feet.

With your team, discuss the following questions in any order. Be prepared to share your discussion with the class.

- What are the shapes of the two tents?

- Without doing any calculations, which tent do you think Soraya should buy and why?

- What types of measurement might be useful to determine which tent is better?

- What do you still need to know to answer her question?

11-30.   COMPARING SOLIDS

To analyze Tent B from problem 11-29, you need to know how to find the volume of a pyramid. But how can you find that volume?

To start, consider a simpler pyramid with a square base, such as pyramid **B** that your team built in Lesson 11.1.2. To develop a method to find the volume of a pyramid, first consider what shape(s) we can compare it to. For example, when finding the area of a triangle, you compared it to the area of a rectangle and figured out that the area of a triangle is always half the area of a rectangle with the same base and height. What shape(s) can you compare the volume of pyramid **B** to? Discuss this with your team and be prepared to share your thinking with the class.

11-31.   VOLUME OF A PYRAMID

Soraya thinks that pyramid **B** could be compared to a cube, like the one shown at right, since the base edges and heights of both are 6 cm.

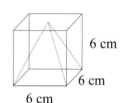
6 cm
6 cm
6 cm

a.   What fraction of the cube with edge length 6 cm is pyramid **B**? Discuss this with your team and make an estimate.

b.   Soraya remembers comparing pyramids **B** and **C** in Lesson 11.1.2. She decided to compare the volumes by using a model. She constructed a pyramid using foam layers as shown at right. What is the shape of each foam layer? Note: The name for the shape of a layer of a three-dimensional solid is called a **cross-section**.

c.   Soraya then slid all of the layers of the pyramid so that the top vertex was directly above one of the corners of the base, like Pyramid C from problem 11-20. Since she did not add or take away any foam layers, how does the volume of this pyramid compare with the pyramid in part (b) above?

d.   Test your estimate from part (a) by using as many pyramid **C**s as you need to assemble a cube. Was your estimate accurate? Now explain how to find the volume of a pyramid.

e.   Do you think your method for part (d) works with all pyramids? Why or why not?

11-32. When the top vertex of a pyramid is directly above (or below) the center of the base, the pyramid is called a **right pyramid**, while all other pyramids are referred to as **oblique pyramids**. **Examine** your models (**A**, **B**, **C**, and **D**) from problem 11-20 and decide which are right pyramids and which are oblique pyramids.

11-33. Now return to problem 11-29 and help Soraya decide which tent to buy for her backpacking trip. To make this decision, compare the volumes, base areas, and surface areas of both tents. Be ready to share your decision with the class.

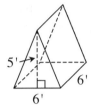
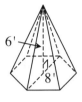

Tent A   *Tent B*

11-34. Write an entry in your Learning Log and explain how to find the volume of a pyramid. Be sure to include an example. Title this entry "Volume of a Pyramid" and include today's date.

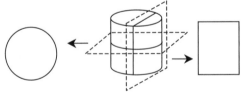

⊕ETHODS AND MEANINGS

Cross-Sections of Three-Dimensional Solids

MATH NOTES

The intersection of a three-dimensional solid and a plane is called a **cross-section** of the solid. The result is a two-dimensional diagram that represents the flat surface of a slice of the solid.

One way to **visualize** a cross-section is to imagine the solid sliced into thin slices like a ream of paper. Since a solid can be sliced in any direction and at any angle, you need to know the direction of the slice to find the correct cross-section. For example, the cylinder at right has several

A horizontal   A cylinder   A vertical
cross-section   cross-section
is a circle.   is a rectangle.

different cross-sections depending on the direction of the slice. When this cylinder is sliced vertically, the resulting cross-section is a rectangle, while the cross-section is a circle when the cylinder is sliced horizontally.

Geometry Connections

11-35.   Review the information about cross-sections in the Math Notes box for this lesson. Then answer the questions below.

   a.   Draw a cube on your paper. Is it possible to slice a cube and get a cross-section that is not a quadrilateral? Explain how.

   b.   Barbara has a solid on her desk. If she slices it horizontally at any level, the cross-section is a triangle. If she slices it vertically in any direction, the cross-section is a triangle. What could her shape be? Draw a possible shape.

11-36.   Find the volume and surface area of a square-based pyramid if the base edge has length 6 units and the height of the pyramid is 4 units. Assume the diagram at right is not to scale.

11-37.   The solid at right is a regular octahedron.

   a.   Trace the shape on your paper. How many faces does it have? How many edges? Vertices?

   b.   If an octahedron is sliced horizontally, what shape is the resulting cross-section?

11-38.   Assume that the prisms at right are similar.

   a.   Solve for x and y.

   b.   What is the ratio of the corresponding sides of Solid B to Solid A?

   c.   If the base area of Solid A is 27 square units, find the base area of Solid B.

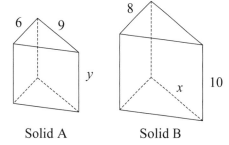

Solid A   Solid B

11-39.   In the diagram at right, assume that $m\angle ECB = m\angle EAD$ and point E is the midpoint of $\overline{AC}$. Prove that $\overline{AD} \cong \overline{CB}$.

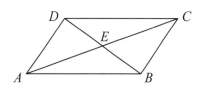

11-40.   **Multiple Choice:** The graph of $x^2 + y^2 = 4$ is:

      a.   a parabola with y-intercept $(0, 4)$

      b.   a circle with radius 4 and center $(0, 0)$

      c.   a parabola with x-intercepts $(-2, 0)$ and $(2, 0)$

      d.   a circle with radius 2 and center $(0, 0)$

      e.   None of these

11-41.   On graph paper, graph the equation $x^2 + (y-3)^2 = 25$. Name the x- and y-intercepts.

11-42.   While volunteering for a food sale, Aimee studied a cylindrical can of soup. She noticed that it had a diameter of 3 inches and a height of 4.5 inches.

      a.   Find the volume of the soup can.

      b.   If Aimee needs to fill a cylindrical pot that has a diameter of 14 inches and a height of 10 inches, how many cans of soup will she need?

      c.   What is the area of the soup can label?

11-43.   Find the area and circumference of $\odot C$ at right. Show all work.

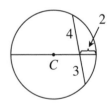

11-44.   STEP RIGHT UP!

At a fair, Cyrus was given the following opportunity. He could roll the die formed by the net at right one time. If the die landed so that a shaded die faced up, then Cyrus would win $10. Otherwise, he would lose $5. Is this game fair? Explain how you know.

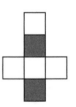

11-45.   Write and solve an equation from the geometric relationships provided in the diagrams below.

a.
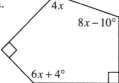

b.
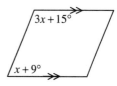

c.
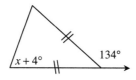

d.
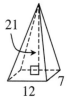

11-46.   **Multiple Choice:** Calculate the volume of the rectangle-based pyramid at right.

a.   84 un^3 b.   648 un^3 c.   882 un^3

d.   1764 un^3 e.   None of these

11.1.4 What if it's a cone?

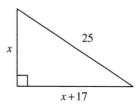

Surface Area and Volume of a Cone

Today you will continue to use what you know about the volume and surface area of prisms and pyramids and will extend your understanding to include a new three-dimensional shape: a cone. As you work with your team, look for connections to previous course material.

11-47.   Review what you learned in Lesson 11.1.3 by finding the volume of each pyramid below. Assume that the pyramid in part (a) corresponds to a rectangular-based prism and that the base of the pyramid and prism in part (b) is a regular hexagon.

a.
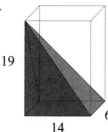

b.
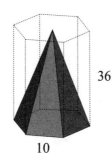

c.
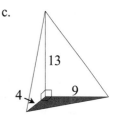

11-48. While finding the volumes of the pyramids in problem 11-47, Jamal asked, "But what if it is a cone? How would you find its volume?" Note that a **cone** is a three-dimensional figure that consists of a circular face, called the **base**, a point called the **apex**, that is not in the flat surface (plane) of the base, and the lateral surface that connects the apex to each point on the circular boundary of the base.

a. Discuss Jamal's question with your team. Then write a response explaining how to find the volume of a cone.

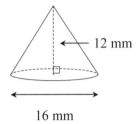
12 mm

16 mm

b. Find the volume of the cone at right. Show all work.

11-49. HAPPY BIRTHDAY!

Your class has decided to throw your principal a surprise birthday party tomorrow. The whole class is working together to create party decorations, and your team has been assigned the job of producing party hats. Each party hat will be created out of special decorative paper and will be in the shape of a cone.

Your Task: Use the sample party hat provided by your teacher to determine the size and shape of the paper that forms the hat. Then determine the amount of paper (in square inches) needed to produce one party hat and figure out the total amount of paper you will need for each person in your class to have a party hat.

11-50. The Math Club has decided to sell giant waffle ice-cream cones at the Spring Fair. Lekili bought a cone, but then she got distracted. When she returned to the cone, the ice cream had melted, filling the cone to the very top!

If the diameter of the base of the cone is 4 inches and the slant height is 6 inches, find the volume of the ice cream and the area of the waffle that made the cone.

11-51. Reflect on what you learned during this lesson and write an entry in your Learning Log on the surface area and volume of a cone. What connections did you make to previous material? Be sure to include an example. Title this entry "Surface Area and Volume of a Cone" and include today's date.

Geometry Connections

METHODS AND MEANINGS

Volume of a Pyramid

In general, the volume of a pyramid is one-third of the volume of the prism with the same base area and height. Thus:

$$V = \tfrac{1}{3}(\text{base area})(\text{height})$$

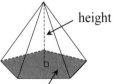

MATH NOTES

Review & Preview

11-52.   Find the volume and total surface area of each solid below. Show all work.

a.

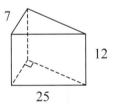

b.

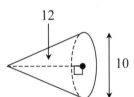

11-53.   **Examine** the diagram of the cone at right.

a.   How could you slice the cone so that the cross-section is a triangle?

b.   What cross-section do you get if you slice the cone horizontally?

c.   Lois is thinking of a shape. She says that no matter how you slice it, the cross-section will always be a circle. What shape is she thinking of? Draw and describe this shape on your paper.

11-54.   For each triangle below, decide if it is similar to the
          triangle at right. If it is similar, **justify** your
          conclusion and complete the similarity statement
          $\triangle ABC \sim \triangle$_____ . If the triangle is not similar,
          explain how you know. Assume that the diagrams
          are not drawn to scale.

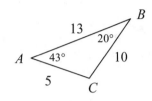

a.   b.   c.

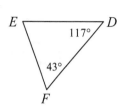   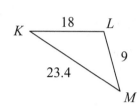   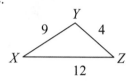

11-55.   On graph paper, graph the system of equations at right. $x^2 + y^2 = 25$
          Then list all points of intersection in the form (x, y). $y = x + 1$

11-56.   **Examine** the diagram at right. State the
          relationship between each pair of angles listed
          below (such as "vertical angles") and state
          whether the angles are congruent,
          supplementary, or neither.

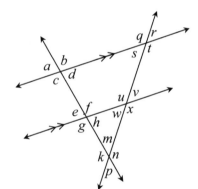

          a. $\angle e$ and $\angle a$

          b. $\angle t$ and $\angle u$

          c. $\angle v$ and $\angle x$

          d. $\angle g$ and $\angle v$

11-57.   **Multiple Choice:** In the diagram at right, the value of y is:

          a. $\sin\theta$ b. $\cos\theta$ c. $\tan\theta$

          d. x e.   None of these

11.1.5 What's the relationship?

Surface Area and Volume of a Sphere

This lesson will complete your three-dimensional shape toolkit. You will learn about a new shape that you encounter often in your daily life: a **sphere**. You will also make connections between a cylinder, cone, and sphere of the same radius and height.

As you work with your team, keep the following focus questions in mind:

What's the relationship?

What other tools or information do I need?

11-58.   Alonzo was blowing bubbles to amuse his little sister. He wondered, "Why are bubbles always perfectly round?"

   a.   Discuss Alonzo's question with the class. Why are free-floating bubbles always shaped like a perfectly round ball?

   b.   The shape of a bubble is called a **sphere**. What other objects can you remember seeing that are shaped like a sphere?

   c.   What shapes are related to spheres? How are they related?

11-59.   GEOGRAPHY LESSON, Part One

Alonzo learned in his geography class that about 70% of the Earth's surface is covered in water. "That's amazing!" he thought. This information only made him think of new questions, such as "What is the area of land covered in water?", "What percent of the Earth's surface is the United States?", and "What is the volume of the entire Earth?"

Discuss Alonzo's questions with your team. Decide:

   •   What facts about the Earth would be helpful to know?

   •   What do you still need to learn to answer Alonzo's questions?

11-60.   In order to answer his questions, Alonzo decided to get out his set of plastic geometry models. He has a sphere, cone, and cylinder that each has the same radius and height.

a.   Draw a diagram of each shape.

b.   If the radius of the sphere is r, what is the height of the cylinder? How do you know?

c.   Alonzo's models are hollow and are designed to hold water. Alonzo was pouring water between the shapes, comparing their volumes. He discovered that when he poured the water in the cone and the sphere into the cylinder, the water filled up the cylinder without going over! Determine what the volume of the sphere must be if the radius of the sphere is r units. Show all work.

11-61.   Now that Alonzo knows that spheres, cylinders, and cones with the same height and radius are related, he decides to **examine** the surface area of each one. As he paints the exterior of each shape, he notices that the lateral surface area of the cylinder and the surface area of the sphere take exactly the same amount of paint! If the radius of the sphere and cylinder is r, what is the surface area of the sphere?

11-62.   GEOGRAPHY LESSON, Part Two

Now that you have **strategies** for finding the volume and surface area of a sphere, return to problem 11-59 and help Alonzo answer his questions. That is, determine:

•   the area of the Earth's surface that is covered in water.

•   the percent of the Earth's surface that lies in the United States.

•   the volume of the entire Earth.

Don't forget that in Chapter 10, you determined that the radius of the Earth is about 4,000 miles! Alonzo did some research and discovered that the land area of the United States is approximately 3,537,438 square miles.

11-63.   Write an entry in your Learning Log describing the relationships between the volumes of a cube, cylinder, and sphere with the same radius and height. Also be sure to explain how to find the surface area and volume of a sphere and include an example of each. Title this entry "Surface Area and Volume of a Sphere" and include today's date.

METHODS AND MEANINGS

Volume and Lateral Surface of a Cone

MATH NOTES

Finding the volume of a cone (defined in problem 11-48) is very similar to finding the volume of a pyramid. The volume of a cone is one-third of the volume of the cylinder with the same radius and height. Therefore, the volume of a cone can be found using the formula shown below, where r is the radius of the base and h is the height of the cone.

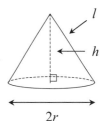

$$V = \tfrac{1}{3}(\text{Base Area})(\text{Height}) = \tfrac{1}{3}\pi r^2 h$$

To find the lateral surface area of a cone, imagine unrolling the lateral surface of the cone to create a sector. The radius of the sector would be the slant height, l, of the cone, and the arc length would be the circumference of the base of the cone, $2\pi r$.

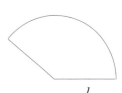

Therefore, the area of the sector (the lateral surface area of the cone) is:

$$LA = \tfrac{2\pi r}{2\pi l}\pi l^2 = \pi r l$$

Review & Preview

11-64.   As Shannon peeled her orange for lunch, she realized that it was very close to being a sphere. If her orange has a diameter of 8 centimeters, what is its approximate surface area (the area of the orange peel)? What is the approximate volume of the orange? Show all work.

11-65.   Review what you know about polyhedra as you answer the questions below. Refer to the table in problem 11-4 if you need help.

   a.   Find the total surface area of a regular icosahedron if the area of each face is 45 mm^2. Explain your method.

   b.   The total surface area of a regular dodecahedron is 108 cm^2. What is the area of each face?

   c.   A regular tetrahedron has an edge length of 6 inches. What is its total surface area? Show all work.

11-66.   Hokiri's ladder has two legs that are each 8 feet
          long. When the ladder is opened safely and
          locked for use, the legs are 4 feet apart on the
          ground. What is the angle that is formed at the
          top of the ladder where the legs meet?

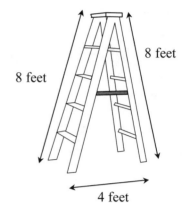

8 feet

8 feet

4 feet

11-67.   Find the area of the region that represents the
          solution of the inequality $x^2 + y^2 \le 72$.

11-68.   Solve each system of equations below. Write the
          solution in the form (x, y). Show all work.

          a. $y + 3x = 14$ b. $y = 6 - 3x$
                $y - 3x = 6$ $2x + y = 7$

11-69.   **Multiple Choice:** The probability of winning $3 on the spinner at
          right is equal to the chance of winning $5. Find the expected value
          for one spin.

$3
$6 $5

          a. $3.00   b. $4.50   c. $4.67

          d. $6.00   e.   None of these

11.2.1 Where's this location?

. . . .

Coordinates on a Sphere

As you learned in Chapter 1, the word *geometry* literally means "measurement of the Earth." In
fact, so far in this course, you have used your geometric tools to learn more about Earth. For
example, in Lesson 11.1.5, you learned that the United States only makes up 1.8% of the
Earth's surface. Also, in Lesson 10.1.1, you learned how Eristothenes used shadows to estimate
the Earth's radius.

Today, you will **examine** other earthly questions that can be answered using geometry. Since
we can approximate the shape of the earth as a sphere, you will be able to use many of the tools
you have used previously. As you work with your team, consider the following focus
questions:

                    What **strategy** or tool can I use?

                         Is there another way?

                    Does this **strategy** always work?

11-70.   YOU ARE HERE

In order to help a person describe a location on
the Earth, scientists have developed a reference
grid on the planet's surface, referred to as
longitude and **latitude lines**. While these
reference markings are referred to as "lines,"
they are technically circles that wrap around the
Earth.

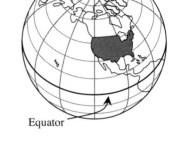

Lines of longitude extend north and south, while
lines of latitude extend east and west, as shown in
the diagrams below. These lines help to mark off

Equator

arc measures on the planet's surface. In the
diagrams below, the lines of latitude are marked every 15° while the lines of
longitude are marked every 30°. The most famous line of latitude is the **equator**,
which separates the Earth into two **hemispheres** (half a sphere).

a.   The equator is an example of
a **great circle**, which means
that it is a circle that lies on
the sphere and has the same
diameter as the sphere.
Compare the equator with the
other lines of latitude. What
do you notice?

Lines of Latitude   **Lines of Longitude**

b.   Is it possible for two great circles
on the same sphere to intersect?
If so, draw an example on your
paper. If not, explain why not.

North Pole

c.   On the sphere provided by your teacher,
carefully draw circles to represent the lines of
latitude (every 30°) and longitude (every 30°)
on the Earth. Highlight the equator by making
it darker or a different color than the other
lines of latitude. Also choose one line of
longitude to represent 0° (called the **prime
meridian**, which passes through Greenwich,
England, on the eastern edge of London) and
highlight it as well.

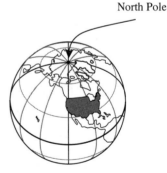

d.   Norman is exactly 1 mile north of Sula. If they both travel at the same rate due
west, will their paths cross? Why or why not? Assume that people can travel
over water and all types of terrain.

e.   Erin is exactly 1 mile east of Wilber. If they both travel due south at the same
rate, will their paths cross? Why or why not?

11-71.   DEAR PEN-PAL

Brianna, who lives in New Orleans, has
been writing to her pen-pal in
Jacksonville, Florida. "Gosh," she
wonders, "How far away is my friend?"

a.   On your "globe" from problem 11-70,
     locate Brianna's home. (New Orleans,
     LA, is approximately 90° west of the
     prime meridian and 30° north of the
     equator.) Mark it with a pushpin.

b.   Now, with a second pushpin, mark the location of Brianna's friend, if
     Jacksonville is 82° west of the prime meridian and 30° north of the equator.
     Use a rubber band to locate the circle with the smallest radius that passes
     through these two locations.

c.   What is the measure of the arc connecting these two cities? Show how you
     know.

d.   Brianna thinks that if she knew the
     circumference of the circle marked with the
     rubber band, then she could use the arc measure
     to approximate the distance between the two
     cities. The shaded circle in the diagram at right
     represents the cross-section of the earth 30°
     above the equator. If the radius of the earth is
     approximately 4000 miles, find the
     circumference of the shaded circle.

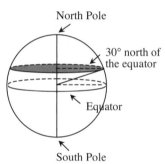

e.   Find the distance between New Orleans and Jacksonville.

11-72.   Obtain a Lesson 11.2.1 Resource Page from your teacher.
           On it, mark and label the following locations.

a.   London, England, which is on the prime meridian
     and is approximately 51° **north** of the equator.

b.   Narsarssuaq, Greenland, which is approximately
     45° **west** of the prime meridian and 61° **north** of
     the equator.

c.   Quito, Equador, which is on the equator and is
     approximately 79° **west** of the prime meridian.

d.   Cairo, Egypt, which is approximately 31° **east** of the prime meridian and 30°
     north of the equator.

e.   Buenos Aires, Argentina, which is approximately 58° **west** of the prime
     meridian and 35° **south** of the equator.

11-73.   EXTENSION

a.   If a polar bear travels 1 mile south from the North Pole, travels one mile east, and then travels one mile north, where does it end up? Explain what happens and why.

b.   Is there another location the polar bear could have started from so that it still ends up where it started after following the same directions? Explain.

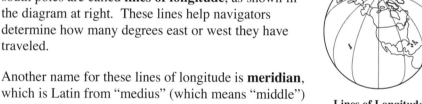

OOKING DEEPER

MATH NOTES

Meridian and Time Zones

The reference lines connecting the north and south poles are called **lines of longitude**, as shown in the diagram at right. These lines help navigators determine how many degrees east or west they have traveled.

Another name for these lines of longitude is **meridian**, which is Latin from "medius" (which means "middle") and "diem" (which means "day"). Meridian also used to refer to noon, since it was the time the sun was directly overhead. In the morning, it was "ante meridian" or before noon. This is where the abbreviation **a.m.** comes from. Likewise, **p.m.** is short for "post meridian," which means "after noon."

Lines of Longitude

11-74.   The moon is an average distance of 238,900 miles away from the Earth. While that seems very far, how far is it?

   a.   Compare that distance with the circumference of the Earth's equator. Assume that the Earth's radius is 4000 miles. How many times greater than the Earth's circumference is the distance to the moon?

   b.   One way to estimate the distance between the Earth and the sun is to consider the triangle formed by the sun, Earth, and moon when the moon appears to be half-full. (See the diagram at right.) When the moon appears from earth to be half-full, it can be assumed that the moon forms a 90° angle with the sun and the Earth.

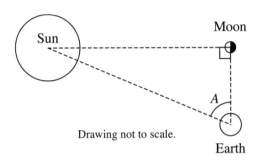

Drawing not to scale.

   Using special equipment, Ray found the measure of angle A to be 89.95°. If the moon is 238,900 miles away from the Earth, then how far is the sun from the Earth?

11-75.   The length of chord $\overline{AB}$ in $\odot D$ is 9 mm. If the $m\overarc{AB} = 32°$, find the length of $\overarc{AB}$. Draw a diagram.

11-76.   On your paper, draw a diagram of a square-based pyramid if the side length of the base is 9 cm and the height of the pyramid is 12 cm.

   a.   Find the volume of the pyramid.

   b.   If a smaller pyramid is similar to the pyramid in part (a), but has a linear scale factor of $\frac{1}{3}$, find its volume.

11-77.   **Examine** the spinner at right. Assume that the probability of spinning a –8 is equal to that of spinning a 0.

   a.   Find the spinner's expected value if the value of region A is 8.

   b.   Find the spinner's expected value if the value of region A is –4.

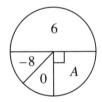

   c.   What does the value of region A need to be so that the expected value of the spinner is 0?

11-78.   In the picture of a globe at right, the lines of
          latitude are concentric circles. Where else might
          you encounter concentric circles?

11-79.   **Multiple Choice:** The volume of a solid is V. If the
          solid is enlarged proportionally so that the perimeter
          increases by a factor of 9, what is the volume of the
          enlarged solid?

          a. $9V$ b. $\frac{81}{4}V$ c. $81V$ d. $729V$

11.2.2 What's the relationship?

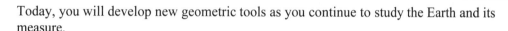

Tangents and Arcs

Today, you will develop new geometric tools as you continue to study the Earth and its
measure.

11-80.   EYE IN THE SKY

          Did you know that as of 1997, over 8000 operating
          satellites orbited the Earth performing various
          functions such as taking photographs of our planet?
          One way scientists learn more about the Earth is to
          carefully **examine** photographs that are taken by an
          orbiting satellite.

          However, how much of the Earth can a satellite see?
          What does this depend on? In other words, what
          information would you need to know in order to figure
          out how much of the planet is in view of a satellite in
          space? Discuss this with your team and be ready to
          share your ideas with the rest of the class.

          Satellite

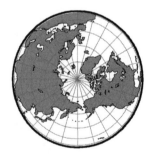

11-81.   On the Lesson 11.2.2 Resource Page obtained from your teacher, locate Satellites A,
          B, and C.

          a.   On the resource page, draw an angle from Satellite A that shows the portion of
                the Earth's equator that is visible from the satellite. What is the relationship of
                the sides of the angle and the circle that represents the equator of the Earth?

Problem continues on next page →

11-81.   *Problem continued from previous page.*

    b.   Draw a quadrilateral *ADEF* that connects Satellite A, the points of tangency, and the center of the Earth (point *E*). If the measure of the angle at Satellite A is 90°, what is the measure of the equator's arc that is in view? Explain how you know.

    c.   What is the relationship of *AD* and *AF*? Prove the relationship using congruent triangles.

    d.   If $m\angle A = 90°$ and the radius of the Earth is 4000 miles, how far above the surface of the planet is Satellite A?

11-82.   What if the satellite is placed higher in orbit? Consider this as you answer the questions below.

    a.   Using a different colored pen or pencil, draw the viewing angle from Satellite B on the Lesson 11.2.2 Resource Page. Label the points of tangency *G* and *H*. Will Satellite B see more or less of the Earth's equator than Satellite A?

    b.   If $m\angle B = 60°$, find the length of the equator in view of Satellite B. Assume that the radius of the Earth is 4000 miles.

    c.   Use a third color to draw the viewing angle from Satellite C on the resource page. Label the points of tangency *J* and *K*. If $m\angle C = 45°$, find the $m\widehat{JK}$ and $m\widehat{JZK}$.

    d.   Is it possible for a satellite to see 50% of the Earth's equator? Why or why not?

11-83.   HOW ARE THEY RELATED?

    In problems 11-81 and 11-82, you found the measures of angles and arcs formed by two tangents to a circle that intersect each other.

    a.   Copy the diagram at right onto your paper. Using intuition, describe how the measure of the angle formed by the tangents ($m\angle A$) seems to be related to the measures of the major and minor arcs formed by the points of tangency ($\widehat{BD}$ and $\widehat{BZD}$).

    b.   If $m\angle A = a$, find $m\widehat{BD}$ and $m\widehat{BZD}$ in terms of *a*. Compare the measure of the angle with the measure of the major and minor arcs. What do you notice?

    c.   Write an entry in your Learning Log describing the relationship between the angles and arcs formed by two intersecting tangents to a circle. Also record what you found out about the lengths of the tangents from the point of tangency to their point of intersection. Title this entry "Tangents and Arcs" and include today's date.

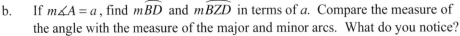

Geometry Connections

ETHODS AND MEANINGS

Volume and Lateral Surface of a Sphere

A **sphere** is a three-dimensional solid formed by points that are equidistant from its center.

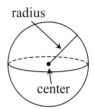

radius

center

The **volume of a sphere** is twice the volume of a cone with the same radius and height. Since the volume of a cone with radius r and height $2r$ is $V = \frac{1}{3}\pi r^2(2r) = \frac{2}{3}\pi r^3$, the volume of a sphere with radius r is:

$$V = \frac{4}{3}\pi r^3$$

The **surface area of a sphere** is four times the area of a circle with the same radius. Thus, the surface area of a sphere with radius r is:

$$SA = 4\pi r^2$$

Review & Preview

11-84.   While making his lunch, Alexander sliced off a portion of his grapefruit. If the area of the cross-section of the slice (shaded at right) was 3 in², and if the diameter of the grapefruit was 5 inches, find the distance between the center of the grapefruit and the slice. Assume the grapefruit is a sphere.

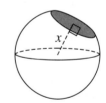

11-85.   The approximate surface areas of the seven Earth continents are shown in the table at right. If the radius of the Earth's moon is approximately 1080 miles, how would its surface area compare with the size of the continents?

Continent	Area (sq. miles)
Asia	17,212,048.1
Africa	11,608,161.4
North America	9,365,294.0
South America	6,879,954.4
Antarctica	5,100,023.4
Europe	3,837,083.3
Australia/Oceania	2,967,967.3

11-86.   Find the area of the regular decagon if the length of each side is 20 units.

11-87.   The solid at right is an example of a **truncated pyramid**. It is formed by slicing and removing the top of a pyramid so that the slice is parallel to the base of the pyramid. If the original height of the square-based pyramid at right was 12 cm, find the volume of this truncated pyramid. (Hint: you may find your results from problem 11-76 useful.)

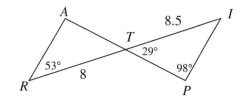

3 cm

9 cm

9 cm

11-88.   **Examine** the triangles at right. Are they similar? Are they congruent? Explain how you know. Then write an appropriate similarity or congruence statement.

A

8.5

I

T

29°

53°   8

98°

R   P

11-89.   **Multiple Choice:** Which shape below has the <u>least</u> area?

a.   A circle with radius 5 units

b.   A square with side length 9 units

c.   A trapezoid with bases of length 8 and 10 units and height of 9 units

d.   A rhombus with side length 9 units and height of 8 units.

11.2.3 What is the measure?

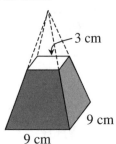

Secant and Tangent Relationships

In Lesson 11.2.2, you studied the angles and arcs formed by tangents when a satellite orbits the Earth, as shown in the diagram at right. Today, you will consider a related question: What if the sides of the angle intersect the circle more than once? What are the relationships between the angles and arcs formed when this happens? And what can you learn about the lengths of the segments created by the points of intersection?

Satellite

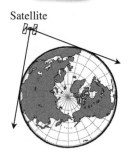

As you work with your team, carefully record your team's conjectures. And while you work, keep the following questions in mind:

What patterns do I see?

Is this relationship always true?

11-90.   Review what you learned in Lesson 11.2.2 by solving for the given variables in the diagrams below. Show all work.

a.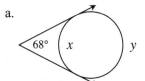

b.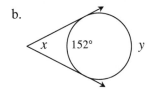

c.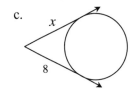

11-91.   While a tangent is a line that intersects a circle (like $\odot C$ in the diagram at right) at exactly one point, a **secant** is a line that intersects a circle twice. $\overleftrightarrow{PR}$ is an example of a secant, while $\overleftrightarrow{QS}$ is an example of a tangent.

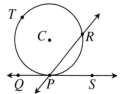

a.   What happens to the measure of the angles and arcs when a secant intersects the circle at the point of tangency? Namely, how are the angles located at P in the diagram above related to $m\overparen{PR}$ and $m\overparen{PTR}$? First make an educated guess. Then test your ideas out using a dynamic geometry tool. Write a conjecture and be ready to share it with the class.

b.   Uri wants to prove his conjecture from part (a) for a non-special secant (meaning that $\overleftrightarrow{PR}$ is not a diameter). He decided to extend a diameter from point P and to create an inscribed angle that intercepts $\overparen{PR}$. With your team, **examine** Uri's diagram carefully and consider all the relationships you can identify that could be useful.

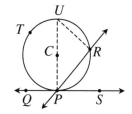

To get you started, use the list below.

i.   What is $m\angle URP$?

ii. How is $m\angle RUP$ related to $m\overparen{PR}$?

iii. What is the sum of the angles in $\triangle PRU$?

iv. How are $\angle UPR$ and $\angle RPS$ related?

c.   Using the relationships you explored in part (b), prove that if $m\angle RPS = x$ in the diagram from part (b), then $m\overparen{PR} = 2x$. Remember to **justify** each step.

11-92.   Uri now has this challenge for you:
What happens when secants and tangents intersect outside a circle? To consider this, you need to **examine** two separate cases: One is when a secant and tangent intersect outside a circle (case *i* below). The other is when two secants intersect outside a circle (case *ii* below). As with your earlier **investigations**,

- First make a prediction about the relationship between the measures of *x*, *a*, and *b* for each case.

- Then use your dynamic geometry tool to test your conjectures.

- For each case, write an algebraic statement (equation) that relates *x*, *a*, and *b*. Be ready to share each equation with the rest of the class.

i.   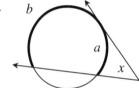   ii.

11-93.   Now prove your conjectures from problem 11-92. For each diagram, add a line segment that will help to create an inscribed angle. Then use angle relationships (such as the sum of the angles of a triangle must be 180°) to then find the measures of all the angles in terms of *x*, *a*, and *b*. Be sure to show that in each case, $x = \frac{b-a}{2}$. Remember to **justify** each statement.

i.   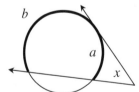   ii.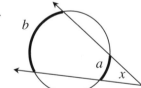

11-94.　Camille is interested in the lengths of segments that
are created by the points of intersection. She
remembers proving in Lesson 11.2.2 that the lengths
of the tangents between their intersection and the
points of tangency are equal, as shown in the
diagram at right. She figures that there must be
some relationships in the lengths created by the
intersections of secants, too.

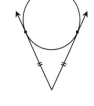

a.　The first case she wants to consider is when a
tangent and secant intersect outside a circle,
as shown in the figure at right. Copy this
diagram onto your paper.

b.　"To find a relationship, I think we need to add
some line segments to create some inscribed
angles and triangles," Camille tells her team.
She decides to add the line segments shown at
right. Show why the angles marked x and y
must be congruent.

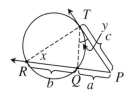

c.　"If some of these triangles are similar, I can use that to find a relationship
between these side lengths," Camille explains. Help her prove that
$\triangle PQT \sim \triangle PTR$.

d.　Use the fact that $\triangle PQT \sim \triangle PTR$ to write a proportion using a, b, and c.
Simplify this equation as much as possible to find an equation that helps you
understand the relationship between a, b, and c.

e.　Use the same process to find the
relationship between the lengths created
when two secants intersect outside a circle.
Two extra segments have been added to the
diagram to help create similar triangles. Be
ready to **justify** your relationship.

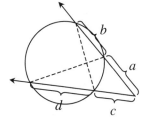

11-95.   Use all your circle relationships to solve for the variables in each of the diagrams below.

a.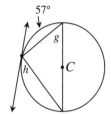

b.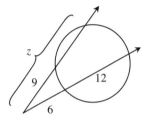

c.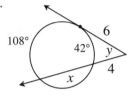

d.   C is the center

e.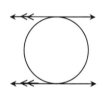

f.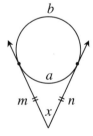

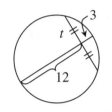

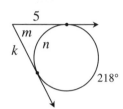

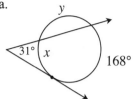

Ⓜ︎ETHODS AND MEANINGS

Intersecting Tangents

Two lines that are tangent to the same circle may not intersect. When this happens, the tangent lines are parallel, as shown in the diagram at right. The arcs formed by the points of tangency are both 180°.

However, when the lines of tangency intersect outside the circle, some interesting relationships are formed. For example, the lengths m and n from the point of intersection to the points of tangency are equal.

The angle and arcs are related to the angle outside the circle as well. If x is the measure of the angle formed by the intersection of the tangents, a represents the measure of the minor arc, and b represents the measure of the major arc, then:

$$a = 180° - x \quad \text{and} \quad b = 180° + x$$

11-96.   Solve for the variables in each of the diagrams below. Assume point *C* is the center of the circle in part (b).

a.   b.   c.

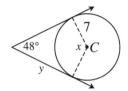

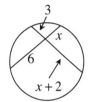

11-97.   In part (c) of problem 11-96, you used the relationship between the segment lengths formed by intersecting chords to find a missing length. But how are the arc measures of two random intersecting chords related? **Examine** the diagram at right.

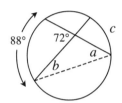

a.   Solve for *a*, *b*, and *c* using what you know about inscribed angles and the sum of the angles of a triangle.

b.   Compare the result for *c* with 88° and 72°. Is there a relationship?

11-98.   Perhaps you think the Earth is big? Consider the sun!

a.   Assume that the radius of the Earth 4000 miles. The sun is approximately 109 times as wide. Find the sun's radius.

b.   The distance between the Earth and the moon is 238,900 miles. Compare this distance with the radius of the sun you found in part (a).

c.   If the sun were hollow, how many Earths would fill the inside of it?

11-99.   Write the equation for the graph at right.

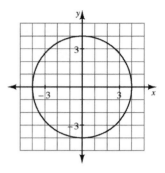

11-100.   On your paper, draw a diagram of a square-based
          pyramid. If the base has side length 6 units and the
          height of the pyramid is 10 units, find the total
          surface area. Show all your work.

11-101.   **Multiple Choice:** Which of the following cannot be the measure of an exterior angle
          of a regular polygon?

          a.   18°   b.   24°   c.   28°   d.   40°

11-102.   Solve for the variables in each of the diagrams below. Assume that point *C* is the
          center of the circle in part (b).

          a.   b.   c.

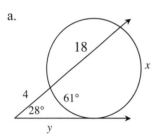
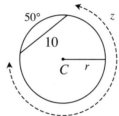
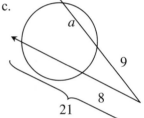

11-103.   Which has greater volume: a cylinder with radius 38 units and height 71 units or a
          rectangular prism with dimensions 34, 84, and 99 units? Show all work and support
          your **reasoning**.

11-104.   Copy the diagram at right onto your paper. Use the
          process from problem 11-97 to find the measure of *x*.
          Show all work.

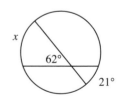

11-105.   Mr. Kyi has placed 3 red, 7 blue, and 2 yellow beads in a hat. If a person selects a
           red bead, he or she wins $3. If that person selects a blue bead, he or she loses $1. If
           the person selects a yellow bead, he or she wins $10. What is the expected value for
           one draw?

11-106.   A solar eclipse occurs
           when the moon passes
           between the Earth and
           the sun and is perfectly
           aligned so that it
           blocks the Earth's
           view of the sun.

Note: This diagram is not to scale.

           How do scientists figure out what areas of the Earth will see an eclipse? To find out,
           copy the diagram above onto your paper. Then use tangents representing the sun's
           rays to find the portion of the Earth's equator that will see the total eclipse.

11-107.   **Multiple Choice:** The Mona Lisa, by
           Leonardo da Vinci, is arguably the most
           famous painting in existence. The rectangular
           artwork, which hangs in the Musée du Louvre,
           measures 77 cm by 53 cm. When the museum
           created a billboard with an enlarged version of
           the portrait for advertisement, they used a linear
           scale factor of 20. What was the area of the
           billboard?

           a.   4081 cm^2 b.   32,638,000 cm^2

           c.   81,620 cm^2 d.   1,632,400 cm^2

           e.   None of these

Chapter 11 Closure   What have I learned?

Reflection and Synthesis

The activities below offer you a chance to reflect on what you have learned during this chapter. As you work, look for concepts that you feel very comfortable with, ideas that you would like to learn more about, and topics you need more help with. Look for **connections** between ideas as well as **connections** with material you learned previously.

①   TEAM BRAINSTORM

With your team, brainstorm a list for each of the following three topics. Be as detailed as you can. How long can you make your list? Challenge yourselves. Be prepared to share your team's ideas with the class.

Topics:	What have you studied in this chapter? What ideas and words were important in what you learned? Remember to be as detailed as you can.
Problem Solving:	What did you do to solve problems? What different **strategies** did you use?
Connections:	How are the topics, ideas, and words that you learned in previous courses are **connected** to the new ideas in this chapter? Again, make your list as long as you can.

MAKING CONNECTIONS

The following is a list of the vocabulary used in this chapter. The words that appear in bold are new to this chapter. Make sure that you are familiar with all of these words and know what they mean. Refer to the glossary or index for any words that you do not yet understand.

arc	base	circle
cone	**cross-section**	**cube**
cylinder	diameter	**edge**
equator	**face**	**great circle**
height	**hemisphere**	lateral face
latitude	**longitude**	oblique
octahedron	**platonic solid**	**polyhedron**
pyramid	radius	secant
slant height	**sphere**	surface area
tangent	**tetrahedron**	volume

Make a concept map showing all of the **connections** you can find among the key words and ideas listed above. To show a **connection** between two words, draw a line between them and explain the **connection**, as shown in the example below. A word can be **connected** to any other word as long as there is a **justified connection**. For each key word or idea, provide a sketch of an example.

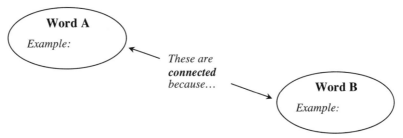

Your teacher may provide you with vocabulary cards to help you get started. If you use the cards to plan your concept map, be sure either to re-draw your concept map on your paper or to glue the vocabulary cards to a poster with all of the **connections** explained for others to see and understand.

While you are making your map, your team may think of related words or ideas that are not listed above. Be sure to include these ideas on your concept map.

③ SUMMARIZING MY UNDERSTANDING

This section gives you an opportunity to show what you know about certain math topics or ideas. Your teacher will give you directions for exactly how to do this.

④   WHAT HAVE I LEARNED?

This section will help you evaluate which types of problems you have seen with which you feel comfortable and those with which you need more help. This section will appear at the end of every chapter to help you check your understanding. Even if your teacher does not assign this section, it is a good idea to try these problems and find out for yourself what you know and what you need to work on.

Solve each problem as completely as you can. The table at the end of this closure section has answers to these problems. It also tells you where you can find additional help and practice on problems like these.

CL 11-108. Use all your circle relationships to solve for the variables in each of the diagrams below. Assume that C is the center of the circle for parts (b) and (c).

a.

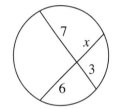

b.   The area of $\odot C$ is 25π un^2

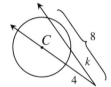

c.

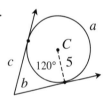

d.

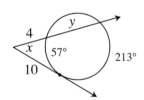

CL 11-109. The radius of the moon is approximately 1738 km. Draw a diagram of the moon on your paper.

    a.   If all the Earth's water were distributed on the surface of the moon, it would be about 33.6 km deep! How much water is on the Earth?

    b.   If all of this water were to be collected and reshaped into a gigantic spherical drop out in space, what would its radius be?

CL 11-110. On the same set of axes, graph the equations below. Name all points of intersection.

$$x^2 + y^2 = 100$$
$$y = \tfrac{1}{2}x + 5$$

CL 11-111. **Examine** the spinner at right.

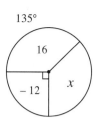

a. Find the expected value of the spinner if $x = 4$.

b. Find the expected value of the spinner if $x = -8$.

c. Find x so that the expected value of the spinner is 6.

CL 11-112. Find the volume of a pyramid if its base is a regular pentagon with perimeter 20 units and if its height is 7 units.

CL 11-113. **Examine** the triangles below. Based on the markings and measurements provided in the diagrams, which are similar to $\triangle ABC$ at right? Which are congruent? Are there any that you cannot determine? **Justify** your conclusion and, if appropriate, write a similarity or congruence statement. **Note:** The diagrams are not drawn to scale.

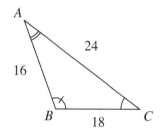

a.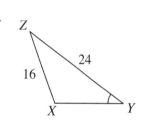

b.

c.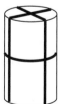

CL 11-114. Talila is planning on giving her geometry teacher a gift. She has two containers to choose from:

- A cylinder tube with diameter 6 inches and height 10 inches
- A rectangular box with dimensions 5 inches by 6 inches by 9 inches

a. Assuming that her gift can fit in either box, which will require the least amount of wrapping paper?

b. She plans to tie two loops of ribbon about the package as shown at right. Which package will require the least amount of ribbon? Ignore any ties or bows.

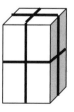

CL 11-115. A big warehouse carrying tents has a miniature model that is similar to the full-sized tent. The tent is a triangular-based prism and the miniature model has dimensions shown in the diagram at right.

3 ft

2 ft

3 ft

a. How much fabric does the small tent use? That is, what is its surface area?

b. What is the volume of the small model?

c. If the volume of the full-sized model is 72 ft^3, how tall is the full-sized tent?

d. How much fabric does the full-sized tent use?

CL 11-116. Check your answers using the table at the end of the closure section. Which problems do you feel confident about? Which problems were hard? Use the table to make a list of topics you need help on and a list of topics you need to practice more.

⑤   HOW AM I THINKING?

This course focuses on five different **Ways of Thinking**: investigating, examining, visualizing, choosing a strategy/tool, and reasoning and justifying. These are some of the ways in which you think while trying to make sense of a concept or to solve a problem (even outside of math class). During this chapter, you have probably used each Way of Thinking multiple times without even realizing it!

Choose three of these Ways of Thinking that you remember using while working in this chapter. For each Way of Thinking that you choose, show and explain where you used it and how you used it. Describe why thinking in this way helped you solve a particular problem or understand something new. Be sure to include examples to demonstrate your thinking.

Answers and Support for Closure Activity #4
What Have I Learned?

Problem	Solution	Need Help?	More Practice
CL 11-108.	a. $x = 3.5$ b. $k = 7$ c. $a = 240°$, $b = 60°$, $c = 5\sqrt{3}$ d. $x = 78°$, $y = 21$	Lessons 10.1.2, 10.1.3, 10.1.4, and 11.2.3 Math Notes box, problem 10-35	Problems 10-7, 10-15, 10-26, 10-37, 10-52, 10-59, 10-93, 11-43, 11-95, 11-96, 11-97, 11-102, 11-104
CL 11-109.	a. $1,300,222,453$ km b. 677.1 km	Lesson 11.2.2 Math Notes box	Problems 11-62, 11-64, 11-98
CL 11-110.	$(6, 8)$ and $(-10, 0)$	Lesson 10.3.1	Problems 10-88, 10-89, 10-90, 11-16, 11-40, 11-41, 11-55, 11-67, 11-99
CL 11-111.	a. 4.5 b. 0 c. $x = -8$	Lesson 10.3.1 Math Notes box	Problems 10-55, 10-64, 10-66, 10-67, 10-69, 10-75, 10-76, 10-77, 10-81, 10-92, 11-23, 11-69, 11-77, 11-105
CL 11-112.	$V \approx 64.23$ un^3	Lessons 8.1.4, 8.1.5, 8.3.1, 11.1.2, and 11.1.4 Math Notes boxes	Problems 8-45, 8-47, 8-48, 8-67, 8-85, 9-9, 10-41, 11-31, 11-36, 11-46, 11-76

Problem	Solution	Need Help?	More Practice
CL 11-113.	a. $\triangle ABC \sim \triangle RTS$ (AA $\sim$) b. $\triangle ABC \cong \triangle MPK$ (AAS $\cong$) c. Cannot be determined because there are two possible triangles when SSA is given.	Lessons 3.2.1, 3.2.2, and 3.2.5 Math Notes boxes	Problems 7-6, 7-14, 7-28, 7-53, 7-77, 7-87, 7-104, 8-32, 8-54, 8-80, 9-12, 9-61, 10-32, 10-96, 11-54
CL 11-114.	a. The cylinder needs less paper ($SA = 78\pi$ in^2) b. The prism requires less ribbon (80 inches)	Lessons 8.3.2, 9.1.2, and 9.1.3 Math Notes boxes	Problems 9-15, 9-17, 9-25, 9-26, 9-27, 9-28, 9-33, 9-40, 9-59, 9-89, 10-6, 10-41, 10-95, 11-9, 11-15, 11-42, 11-103
CL 11-115.	a. SA ≈ 31.0 ft^2 b. V ≈ 9 ft^3 c. linear scale factor = 2, height = 6 ft. d. $SA \approx 124$ ft^2	Lessons 9.1.2, 9.1.3, 9.1.5 Math Notes boxes, problems 9-35 and 9-36	Problems 9-37, 9-39, 9-40, 9-45, 9-46, 9-47, 9-48, 9-73, 10-51, 10-60, 11-11, 11-38, 11-76, 11-107

CHAPTER 12   Conics and Closure

As this course draws to a close, it is appropriate to reflect on what you have learned so far and to look for connections between topics in both algebra and geometry.

For example, in Section 12.1, you will extend your understanding of the cross-sections of a cone, called "conic sections." You will learn about the geometric properties of conic sections and will discover how to represent them algebraically.

Then, in Section 12.2, four activities offer a chance for you to apply your geometric **tools** in new ways. You will find new connections between familiar geometric ideas and learn even more special properties about familiar shapes.

Guiding Questions

Think about these questions throughout this chapter:

What's the cross-section?

How can I draw it?

What's the relationship?

What information do I need?

Is there another way?

In this chapter, you will learn:

➤ How to identify the cross-section of a solid.

➤ How to represent the cross-sections of a cone (the "conic sections") algebraically.

➤ The geometric definitions of a circle, a parabola, an ellipse, and a hyperbola.

Chapter Outline

Section 12.1 By studying the different cross-sections of a cone, you will discover how geometry and algebra can each define a shape. You will learn how to construct several conic sections using tools, such as string and tracing paper, and will develop geometric definitions for a circle, parabola, ellipse, and hyperbola.

Section 12.2 As you complete the course closure activities, you will apply the mathematics you have learned throughout this course. For example, you will discover a special ratio that often occurs in nature, find new relationships that exist in basic polyhedra, learn about a shape created when the midpoints of the sides of a quadrilateral are connected, and determine where a goat should be tethered to a barn so that it has the lowest probability of eating a poisoned weed.

12.1.1 What's the cross-section?

· ·

Introduction to Conic Sections

In Section 11.1, you explored the cross-sections of several types of solids. In fact, you learned that a sphere is special because it is the only solid that has a circular cross-section no matter which direction it is sliced. But what about the cross-sections of a cone? What different cross-sections can be found? And what can be learned about these cross-sections?

In Section 12.1, you and your team will explore the various cross-sections of a cone. As you explore, look for connections with other mathematical concepts you have studied previously.

12-1.   CONIC SECTIONS

Obtain the Lesson 12.1.1 Resource Page from your teacher and construct a cone using scissors and tape. Then, with your team, explore the different cross-sections of a cone (called **conic sections**). Imagine slicing a cone as many different ways as you can. Draw and describe each cross-section on your paper. Do you know the names for any of these shapes?

12-2.   **Examine** your conic sections from problem 12-1.

    a.   If you have not done so already, determine how you can slice the cone so that the cross-section is a ray.

    b.   Can the cross-section of a cone be a single point? Explain.

    c.   If you have not done so already, describe the cross-section of the cone shown at right when it is sliced vertically through the top vertex.

12-3. Interestingly, one of the types of conic sections is a curve you studied in algebra: the **parabola**. What more can be learned about the geometry of a parabola?

a. In Chapter 9, you constructed a parabola using tracing paper (see problem 9-78). With your team, reconstruct a parabola by carefully drawing a line (*l*) and a point not on the line (*P*) on tracing paper. Then fold the tracing paper so that line *l* passes through point *P*. Unfold the tracing paper and fold it again at a different point on line *l* so that line *l* still passes through point *P*. Continue this process until you have at least 20 creased lines forming a parabola.

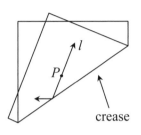

b. **Examine** where the parabola lies in relationship with the original point *P* and line *l*. Where does the point lie? Where does the line lie? Do these relationships seem to hold for the other parabolas constructed by your teammates?

12-4. FOCUS AND DIRECTRIX OF A PARABOLA

Since the point and the line help to determine the parabola, there are special names that are used to refer to them. The point is called the **focus** of the parabola, while the line is called the **directrix**.

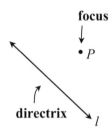

a. Together, the focus and the directrix determine the parabola. For example, can you **visualize** the parabola formed by the focus and directrix shown at right? Trace the point and line on your paper and sketch the parabola.

b. What is the relationship between the points on the parabola and its focus and directrix? Mark a point on the parabola and label it *A*. Fold the tracing paper so that *l* passes through *P* and the crease passes through *A*. Compare the distance between *P* and *A* and the distance between *A* and *l*. What do you notice? Does this work for all points on the parabola?

c. How does the distance between the focus (the point) and the directrix (the line) affect the shape of the parabola? Explore this using a dynamic geometry tool, if possible. (If a dynamic tool is not available, use tracing paper to test several different distances between the focus and directrix.) Explain the result.

12-5. Write an entry in your Learning Log describing what you learned during this lesson. Include information about the cross-sections of a cone and the geometric relationships in a parabola. What questions do you have about the other conic sections? Title this entry "Conic Sections" and include today's date.

Geometry Connections

12-6.   Cawker City, Kansas, claims to have the
world's largest ball of twine. Started in
1953 by Frank Stoeber, this ball has been
created by wrapping more than 1300 miles
of twine. In fact, this giant ball has a
circumference of 40 feet. Assuming the
ball of twine is a sphere, find the surface
area and volume of the ball of twine.

12-7.   The equations below are the types of equations that you will need to be able to solve
automatically in a later course. Try to solve these in 10 minutes or less. The
solutions are provided after problem 12-11 for you to check your answers.

a. $2x - 5 = 7$ b. $x^2 = 16$ c. $2(x - 1) = 6$

d. $\frac{x}{5} = 6$ e. $2x^2 + 5 = x^2 + 14$ f $(x - 3)(x + 5) = 0$

12-8.   **Examine** the pen or pencil that you are using right now. Imagine slicing it in
different directions. On your paper, draw at least three different cross-sections of the
pen or pencil.

12-9.   On graph paper, graph $x^2 + y^2 = 9$.

a.   Consider the inequality $x^2 + y^2 \leq 9$. Does the point (0, 0) make this inequality
true? What is the graph of $x^2 + y^2 \leq 9$? Explain.

b.   Now consider the inequality $x^2 + y^2 > 9$. Does the point (0, 0) make this
inequality true? What region is shaded? Describe the graph of this inequality.

12-10.   Remember that the **absolute value function** finds the distance on a number line
between a number and zero. For example, the absolute value of –6 (written $|-6|$)
equals 6, while $|2| = 2$.

On graph paper, copy and complete the table below and graph the function
$y = |x| + 2$.

x	−4	−3	−2	−1	0	1	2	3	4
y	6				2				

12-11.   **Multiple Choice:** In the diagram at right, the value of x is:

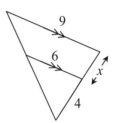

a.   1   b.   2   c.   3

d.   4   e.   None of these

Solutions to problem 12-7:   a: 6, b: 4 or –4, c: 4, d: 30, e: 3 or –3, f: 3 or –5

12.1.2 How can I graph it?

Graphing Parabolas Using The Focus and Directrix

In Lesson 12.1.1, you **investigated** the geometric properties of a parabola, one of the cross-sections of a cone. Today you and your team will explore other conic sections as you continue to find ways to connect geometry and algebra.

12-12.   GRAPHING WITH A FOCUS AND DIRECTRIX

In the past, you have graphed conics, such as circles and parabolas, using rectangular graph paper and an equation. However, another way to graph conic sections is to use **focus-directrix paper**, that is designed with lines and concentric circles like the example shown in Figure A at right.

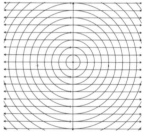

Figure A:
Focus and directrix paper

How can you graph parabolas using this paper? Obtain at least two sheets of focus-directrix paper from your teacher and follow the directions below.

a.   In problem 12-4, you discovered that each point on a parabola is an equal distance from the focus and directrix. To graph a parabola, highlight the center of the concentric circles on your focus-directrix grid with a colored marker. This will be the focus of the parabola. Then highlight a line that is two units away from the focus, as shown in the Figure B at right.

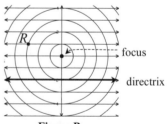

Figure B:
Point R on grid

b.   **Examine** point R on the focus-directrix grid in Figure B. Notice that the circles help you count the distance between R and the focus (the center of the circles). Explain how you know that the point R is 3 units from the focus and 3 units from the directrix.

Problem continues on next page →

12-12.   *Problem continued from previous page.*

      c.   Use the circles and lines to plot a point that is 1 unit away from the focus and the directrix. Is there another point that is also 1 unit away from both the focus and directrix?

      d.   Likewise, find two points that are 2 units away from both the focus and the directrix. Continue plotting points that are equidistant from the focus and the directrix until the parabola appears. Compare your parabola with those of your teammates to double-check for accuracy.

12-13.   What else can you learn about graphing parabolas using focus-directrix paper? The left column of the table below contains several **investigative** questions to explore using focus-directrix paper. For each question, first **visualize** the resulting parabola and discuss your prediction with your team. Then **test** the situation on a fresh focus-directrix grid. The right-hand column contains a suggested way to test your idea, although your team may design its own way to **investigate** the question.

Investigative Questions	Try it out!
a. What would happen to the parabola from problem 12-12 if the directrix were moved so that it is above the focus?	On a new focus-directrix grid, place the directrix on the line that is 2 units above the focus. Then plot the points of the resulting parabola, making sure each point is the same distance from the focus and directrix. Describe what happens.
b. What would happen to the parabola from problem 12-12 if the directrix were moved so that it is <u>farther</u> from the focus?	On a new focus-directrix grid, place the directrix so that it is 6 units away from the focus. Then plot the points of the resulting parabola, making sure each point is the same distance from the focus and directrix. Describe what happens.
c. What would happen to the parabola from problem 12-12 if the directrix were moved so that it is <u>closer</u> to the focus?	On a new focus-directrix grid, place the directrix so that it is 1 unit away from the focus. Then plot the points of the resulting parabola, making sure each point is the same distance from the focus and directrix. Describe what happens.

12-14.   With your team, brainstorm your own **investigative** question you would like to explore. You may want to start the questions with "What if…" or "What happens when…" to help you get started. Share your questions with the class.

12-15.   Celia asks this question: "What if the points are *closer* to the focus than the directrix?" For example, what if the distance between each point and the focus is <u>half</u> the distance between that point and the directrix? Consider this as you answer the questions below.

   a.   On a new sheet of focus-directrix paper, highlight the center (focus) and a line that is 6 units away from the center.

   b.   For every unit a point is away from the focus, it needs to be 2 units away from the directrix. Find the first point that is 2 units away from the focus and 4 units away from the directrix. Then find another point that is 1 *more* unit away from the focus and 2 *more* units away from the directrix. (This point should be a total of 3 units away from the focus and 6 units away from the directrix.) Continue this pattern until the graph is complete. What shape do you see?

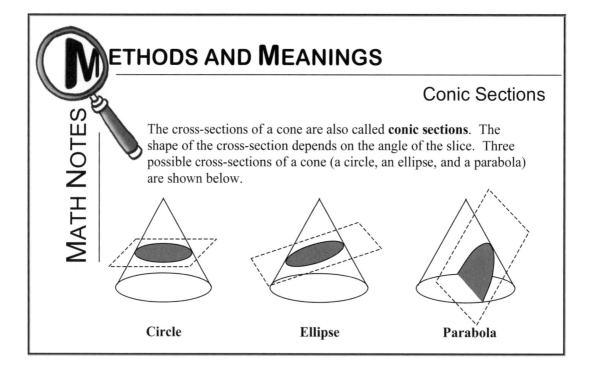

12-16.   A solid with volume 820 cm^3 is reduced proportionally with a linear scale factor of $\frac{1}{2}$. What is the volume of the result?

12-17.   On graph paper, graph the equations below.

  a. $x^2 + y^2 = 4.5^2$ b. $x^2 + y^2 = 75$

12-18.   Use the relationships in each diagram below to solve for the given variables.

  a.   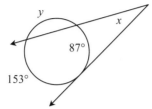   b.   The area of $\odot K$ is 36π un^2.

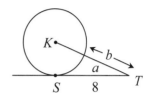

  c.   The diameter of $\odot C$ is 13 units.   d.
        w is the length of $\overline{AB}$.

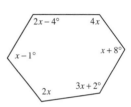

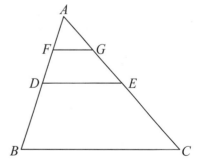

12-19.   In the triangle at right, $\overline{DE}$ is a midsegment of
          $\triangle ABC$ and $\overline{FG}$ is a midsegment of $\triangle ADE$. If
          $DE = 7$ cm, find BC and FG.

12-20.   Find the volume of a cone if the
          circumference of the base is 28π inches and
          the height is 18 inches.

12-21.   **Multiple Choice:** As Carol shopped for a spring picnic, she spent $2.00 for each liter
          of soda and $3.50 for each bag of chips. In all, she bought 18 items for a total of
          $43.50. Assuming she only bought chips and soda, how many bags of chips did she
          buy?

  a.   9   b.   5   c.   15   d.   3

12.1.3 What shape does it make?

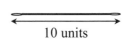

Circles and Ellipses

In Lesson 12.1.1, you **investigated** the geometric properties of a parabola, one of the cross-sections of a cone. Today you and your team will explore other conic sections as you look for ways to connect geometry and algebra.

12-22.   What shapes can you make using string? To find out, obtain a piece of cardboard, some thumbtacks, a ruler, and a piece of string from your teacher. Then follow the directions below.

$\longleftrightarrow$
10 units

a.   Attach a piece of graph paper to the cardboard. Form two loops on the string so that, when pulled apart, the ends of the loops are 10 units apart. Attach one loop to the center of the cardboard using a thumbtack. If you place your pencil in the other loop and keep the string taut (tight), what shape will you create? Explain how you know and then test your idea by drawing the shape on the graph paper.

b.   What if the string is attached to the cardboard at the ends of both loops? Attach both loops of the string from part (a) to the cardboard using two thumbtacks spaced 8 units apart. Predict what shape a pencil will create as it pulls the string tight in all directions.

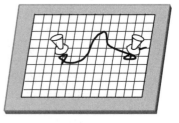

Then test your prediction by drawing the shape on your graph paper. Describe the result. Where have you seen this shape before?

c.   What happens to the shape if the thumbtacks are moved so that they are farther apart? What happens when the thumbtacks are closer together? What happens when the thumbtacks are at the same point? Explore these questions with your team and be prepared to share your conclusions with the class.

Geometry Connections

12-23. ELLIPSE

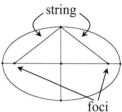
string

foci

The shape that you created in problem 12-22 is called an **ellipse**. Each thumbtack represents a **focus** of the ellipse, so while a parabola only has one focus, an ellipse has two **foci** (plural for "focus").

Examine the ellipse that you created in problem 12-22. How wide is it (in graph paper units)? How tall? How are these lengths related to the length of the string (between the loops)?

12-24. On focus-directrix paper, graph each set of points that are described below. Use a new focus-directrix grid for each part.

 a. Graph the set of points that are 6 units away from the focus.

 b. The directrix is 8 units away from the focus. Graph the set of set of points that are equidistant (i.e. the same distance) from the focus and the directrix.

 c. The directrix is 12 units away from the focus. Graph the result if the distance between each point of the graph and the focus is half the distance between that same point on the graph and the directrix.

Review & Preview

12-25. Find the volume of each shape below. Assume that all corners in part (b) are right angles.

 a. cone

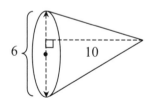
6   10

 b.

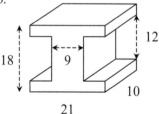
18   9   12
          10
     21

12-26. **Examine** the graph of the circle at right.

 a. Find the equation of the circle.

 b. On graph paper, sketch the graph of the equation $x^2 + y^2 = 49$. What is the radius?

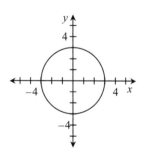

12-27.   **Examine** the diagram at right.

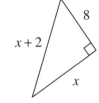

     a.   Write an equation using the geometric relationships in the diagram. Then solve your equation for x.

     b.   Find the measures of the acute angles of the triangle. What **tool** did you use?

12-28.   Jinning is going to flip a coin. If the result is "heads," he wins $4. If the result is "tails," he loses $7.

     a.   What is his expected value per flip?

     b.   If he flips the coin 8 times, how much should he win or lose?

12-29.   Use the diagram of $\odot C$ at right to answer the questions below.

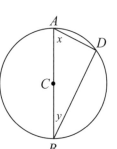

     a.   If $m\angle x = 28°$, what is $m\overarc{AD}$?

     b.   If $AD = 5$ and $BD = 5\sqrt{3}$, what is the area of $\odot C$?

     c.   If the radius of $\odot C$ is 8 and if $m\overarc{BD} = 100°$, what is BD?

12-30.   **Multiple Choice:** Based on the markings in the diagrams at right, which statement is true?

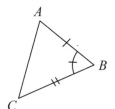
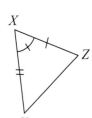

     a. $\triangle ABC \cong \triangle XYZ$

     b. $\triangle ABC \cong \triangle YXZ$

     c. $\triangle ABC \cong \triangle ZXY$

     d. $\triangle ABC \cong \triangle ZYX$

     e.   None of these

12.1.4 How can I construct it?

The Hyperbola

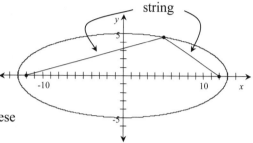

12-31.   Review what you have learned about ellipses as you answer the questions below. You may find the information in the Math Notes box useful.

a.   Maya used string and thumbtacks to draw the ellipse at right. How wide is her ellipse? How long is the string she used? How can you tell?

b.   Maya's thumbtacks are placed at (12, 0) and (–12, 0). What are these points called?

c.   How could Maya change her thumbtacks so that the ellipse is 26 units tall and 10 units wide?

12-32.   CONSTRUCTING AN ELLIPSE

Maya thinks that an ellipse is closely related to a circle. For one thing, they are both cross-sections of a cone. Also, when the two foci of an ellipse coincide (lie on top of each other), the ellipse becomes a circle. Therefore, Maya suspects that there must be a way to use a circle to construct an ellipse with tracing paper.

a.   Use a compass to construct a circle with a radius of approximately 2 inches on a piece of tracing paper. Label its center C.

b.   Find and label a point P inside the circle (other than the center). Then fold the tracing paper so that the circle passes through P. Unfold and then fold in a different location so that the circle still passes through P. Continue this process until an ellipse emerges.

c.   Where are the foci of the ellipse?

d.   Why does this construction work? Pick a point on the circle and label it A. Draw the radius $\overline{AC}$. Study what happens as you fold the circle so that point A lies on point P. Explain why the sum of the distances between each point on the ellipse and the foci must be constant.

12-33.   Maya asks, "What if the point is outside the circle?"

Your Task: Repeat the folding process that you used in problem 12-32 to explore Maya's question with your team. Each time you fold the tracing paper, be sure a different point on the circle passes through point P. Use fresh tracing paper for each construction. Write down what happens and be prepared to share your findings with the class.

12-34.   THE HYPERBOLA

The shape you constructed in problem 12-33
is called a **hyperbola**. A hyperbola is
sometimes described as having two curves
facing away from each other.

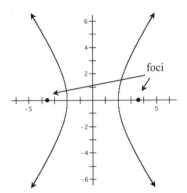

Example of a hyperbola

a.   A hyperbola has two foci. **Examine** the
hyperbola that you constructed in
problem 12-33. Where do you think the
foci might lie? Explain your thinking.

b.   In an ellipse, the sum of the distances
between each point on the ellipse and its
foci must be constant. One way to
remember this is to imagine constructing
the ellipse with a string that is attached
at both foci.

But what about a hyperbola? How are the distances between
each point on the hyperbola and the foci related? Explore this
idea with a dynamic geometry tool, if possible. (If a dynamic
tool is not available, ask your teacher for the Lesson 12.1.4
Resource Page or download a copy from www.cpm.org.)
Describe the relationship between the distances from each focus to each point
on the hyperbola.

12-35.   Maya is still wondering about her
construction. "What if the point lies
<u>on</u> the circle? What conic would be
created then?" Use the dynamic
geometry tool to **investigate** her
question. Explain what happens.

METHODS AND MEANINGS

MATH NOTES

Ellipses

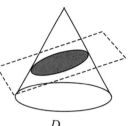

One of the cross-sections of a cone is an **ellipse**. An ellipse is often described as an oval or a circle that is "stretched."

An ellipse has two **foci** that lie on its longest line of symmetry. The sum of the distances between each point on the ellipse and the foci must be constant. For example, if A and B are the foci of the ellipse at right, then the sum of the lengths of $\overline{AD}$ and $\overline{BD}$ must be constant.

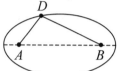

The sum of the distances between each point on the ellipse and the foci is also equal to the length of the widest distance across the ellipse.

Review & Preview

12-36.   Find the surface area of the solids below. Assume that the solid in part (a) is a prism with a regular octagonal base and the pyramid in part (b) is a square-based pyramid. Show all work.

a.

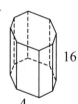

16

4

b.

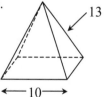

13

←——10——→

12-37.   Solve each equation below for the given variable. Check your solution by verifying that your solution makes the original equation true.

a. $\frac{2}{3}(15u - 6) = 14u$

b. $(5 - x)(3x + 8) = 0$

c. $2(k - 5)^2 = 32$

d. $2p^2 + 7p - 9 = 0$

12-38.   **Examine** the diagram at right.

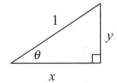

a.   Explain why $y = \sin\theta$ and $x = \cos\theta$.

b.   According to this diagram, what is $(\sin\theta)^2 + (\cos\theta)^2$? Explain how you know.

c.   Is this relationship true for all angles? Use your calculator to find $(\sin 23°)^2 + (\cos 23°)^2$ and $(\sin 81°)^2 + (\cos 81°)^2$. Write down your findings.

12-39.   For each geometric relationship below, determine whether a or b is larger, or if they are equal. Assume that the diagrams are not drawn to scale. If there is not enough information, explain what information is missing.

a.   b.   c.

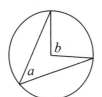

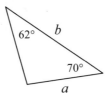

area of the   area of the
triangle is a.   square is b.

12-40.   On graph paper, graph the function $y = x^2 - 3x - 4$.

a.   What is y when $x = 3$? -2? $\frac{1}{2}$?   b.   What is x when $y = -4$? 0?

12-41.   **Multiple Choice:** What is the measure of each interior angle of a regular octagon?

a. $135°$ b. $120°$ c. $180°$ d. $1080°$

12.1.5 What will it look like?

Conic Equations and Graphs

To complete this study of conics, today you will examine the types of equations that represent the different conic sections. You will review your understanding of equations for circles and parabolas and will extend your understanding to include equations for ellipses and hyperbolas. Then you will have an algebraic and geometric understanding of each of the conic sections.

12-42. How can you tell what an equation will look like when it is graphed? In this problem, you and your team will review what you already know about the equations of conic sections.

a. **Examine** the equations below. Which one is the equation of a line? Of a parabola? Of a circle? How can you tell?

(1) $x^2 + y^2 = 25$ (2) $y = x^2 - 5$ (3) $y = x + 4$

b. On a piece of graph paper, graph the equations at right. Then name all points of intersection in the form (x, y).

$x^2 + y^2 = 25$

$y = x^2 - 5$

c. Is the circle you graphed in part (b) a function? What about the parabola? You may want to review the idea of "function" in the Math Notes box for this lesson.

12-43. GRAPH INVESTIGATION

Since the graph of the equation $x^2 + y^2 = 4$ will be a circle with radius 2 and center at (0, 0), what happens when the equation is changed to become $5x^2 + y^2 = 4$ or $x^2 - 2y^2 = 4$?

Your Task: Use a graphing tool such as a graphing calculator to investigate the graphs that can be found by altering the circle equation. Start with the equation $ax^2 + by^2 = 1$ and find out what happens as you change the values of a and/or b. Write down any ideas or conjectures you find during this **investigation** and be ready to share them with the class.

12-44.   GRAPHING HYPERBOLAS

In Lesson 12.1.2, you graphed a parabola by using the fact that each point on the graph was an equal distance to both the focus and directrix. Then, when you graphed a curve so that each point was closer to the focus than the directrix, you got an ellipse! What happens when each point is twice as far from the focus as it is from the directrix?

Obtain a sheet of focus-directrix paper from your teacher. Highlight the center of the circles to be the focus, and highlight a line that is three units away from the focus to be the directrix, as shown in the graph at right.

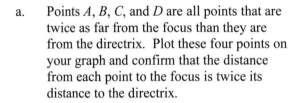

a.   Points A, B, C, and D are all points that are twice as far from the focus than they are from the directrix. Plot these four points on your graph and confirm that the distance from each point to the focus is twice its distance to the directrix.

b.   Continue plotting points that fit this pattern. Remember that each point you plot must be twice as far from the focus than the directrix. Also, notice in the case of point D above, the points can lie on both sides of the directrix. What curve appears?

12-45.   RETURN TO THE CONE

The hyperbola seems linked to the conic sections you found in problem 12-1. For example, a hyperbola can be graphed using a focus and directrix, just like an ellipse and parabola. A hyperbola can also be constructed using a circle and a point outside a circle, much like the construction of an ellipse (which uses a circle and a point inside the circle). Even the equation of a hyperbola (such as $x^2 - y^2 = 1$) looks a lot like the equation of a circle or an ellipse.

The reason why a hyperbola did not appear when you found the cross-sections of a cone is because a hyperbola needs to have two branches curving away from each other. A way to get this cross-section is by using a **double-cone**, a shape created with two cones placed in opposite directions with vertices together, as shown at right. With your team, explain how to slice this double-cone to create a cross-section that is a hyperbola.

Geometry Connections

METHODS AND **M**EANINGS

MATH NOTES

Functions and Relations

A **relation** establishes a correspondence between its inputs and outputs (in math language called "sets"). For equations, it establishes the relationship between two variables and determines one variable when given the other. Some examples of relations are:

$$y = x^2, \ y = \frac{x}{x+3}, \ y = -2x + 5$$

A **relation** is called a **function** if there exists <u>no more</u> <u>than</u> <u>one</u> output for each input. If a relation has two or more distinct outputs for a single input value, it is not a function. The relation graphed at right is not a function because, for example, there are two y-values for each x-value greater than -3.

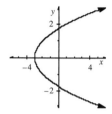

12-46. A silo (a structure designed to store grain) is designed as a cylinder with a cone on top, as shown in the diagram at right. Assume that the base of the cylinder is on the ground.

a. If a farmer wants to paint the silo, how much surface area must be painted?

b. What is the volume of the silo? That is, how many cubic meters of grain can the silo hold?

12-47. A regular pentagram is a 5-pointed star that has congruent angles at each of its outer vertices. Use the fact that all regular pentagrams can be inscribed in a circle to find the measure of angle a at right.

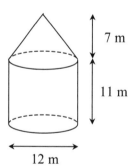

12-48.   In Chapter 7, you discovered that the midsegment of a triangle is not only parallel to
          the third side, but also half its length. But what about the midsegment of a
          trapezoid?

          The diagram at right shows a midsegment
          of a trapezoid. That is, $\overline{EF}$ is a
          midsegment because points E and F are
          both midpoints of the non-base sides of
          trapezoid $ABCD$.

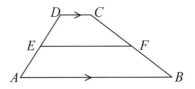

          a.   If $A(0,0)$, $B(9,0)$, $C(5,6)$, and $D(2,6)$, find the coordinates of E and F.
               Then compare the lengths of the bases ($\overline{AB}$ and $\overline{CD}$) with the length of the
               midsegment $\overline{EF}$. What seems to be the relationship?

          b.   See if the relationship you observed in part (a) holds if $A(-4,0)$, $B(9,0)$,
               $C(0,2)$, and $D(-2,2)$.

          c.   Write a conjecture about the midsegment of a trapezoid.

12-49.   For each diagram below, solve for x. Show all work.

          a.   b.   c.

                      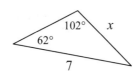

12-50.   For each relationship below, write and solve an equation for x. **Justify** your method.

          a.   b.   c.

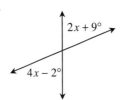   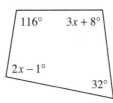   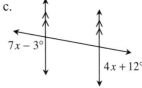

12-51.   **Multiple Choice:** The volume of the square-based pyramid
          with base edge 9 units and height 48 units is:

          a.   324 un^3 b.   1296 un^3

          c.   3888 un^3 d.   not enough information

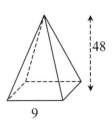

12.2.1 What's the shape?

· ·

Using Coordinate Geometry and Construction to Explore Shapes

In today's activity, you will learn more about quadrilaterals as you review what you know about coordinate geometry, construction, and proof.

12-52.   Review what you have learned about the midsegment of a triangle as you answer the questions below. Assume that $\overline{DE}$ is a midsegment of $\triangle ABC$.

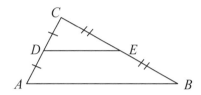

     a.   What is the relationship between $\overline{DE}$ and $\overline{AB}$?

     b.   What is the relationship between $\angle CDE$ and $\angle DAB$? How do you know?

     c.   What is the relationship between $\triangle ABC$ and $\triangle DEC$? **Justify** your conclusion.

     d.   If $DE = 4x + 7$ units and $AB = 34$ units, what is x?

12-53.   QUIRKY QUADRILATERALS

Quinn decided to experiment with the midpoints of the sides of a quadrilateral one afternoon. With a compass, he located the midpoint of each side of a quadrilateral. He then connected the four midpoints together to create a new quadrilateral inside his original quadrilateral.

     a.   Without knowing anything about Quinn's original quadrilateral and without trying the construction yourself, **visualize** the result. What can you predict about Quinn's resulting quadrilateral? Share your ideas with your team.

     b.   Use a compass and straightedge to repeat Quinn's experiment on an unlined piece of paper. Make sure each member of your team starts with a differently-shaped quadrilateral. Describe your results. Did the results of you and your teammates match your prediction from part (a)?

     c.   Does it matter if your starting quadrilateral is convex or not? Start with a non-convex quadrilateral, like the one shown at right, and repeat Quinn's experiment. On your paper, describe your results.

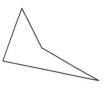

12-54.   Quinn decided to graph his quadrilateral on a set of coordinate axes and prove that his inner quadrilateral is, in fact, a parallelogram. His quadrilateral $ABCD$ uses the points $A(-3, -2)$, $B(-5, 4)$, $C(5, 6)$, and $D(1, -4)$.

   a.   On graph paper, graph the quadrilateral $ABCD$.

   b.   If the midpoint of $\overline{AB}$ is E, the midpoint of $\overline{BC}$ is F, the midpoint of $\overline{CD}$ is G, and the midpoint of $\overline{DA}$ is H, find and label points E, F, G, and H on $ABCD$.

   c.   Connect the midpoints of the sides you found in part (b). Then find the slope of each side of quadrilateral $EFGH$ and use these slopes to prove that Quinn's inner quadrilateral is a parallelogram.

   d.   Quinn wondered if his parallelogram is also a rhombus. Find EF and FG, and then decide if $EFGH$ is a rhombus. Show all work.

12-55.   PROVING THE RESULT FOR ALL QUADRILATERALS

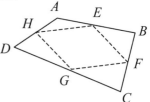

In problem 12-54, you proved that Quinn's inner quadrilateral was a parallelogram when $A(-3, -2)$, $B(-5, 4)$, $C(5, 6)$, and $D(1, -4)$. However, what about other, random quadrilaterals?

To prove this works for all quadrilaterals, start with a diagram of a generic quadrilateral, like the one above. Assume that the midpoint of $\overline{AB}$ is E, the midpoint of $\overline{BC}$ is F, the midpoint of $\overline{CD}$ is G, and the midpoint of $\overline{DA}$ is H. Prove that $EFGH$ is a parallelogram by proving that its opposite sides are parallel. It may help you to draw diagonal $\overline{AC}$ and consider what you know about $\triangle ABC$ and $\triangle ACD$. Use any format of proof.

12-56.   In $\triangle PQR$ at right, what is $m\angle Q$? Explain how you found your answer.

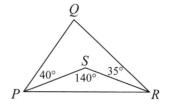

12-57.   The United State Department of Defense is located in a building called the Pentagon because it is in the shape of a regular pentagon. Known as "the largest office building in the world," it's exterior edges measure 921 feet. Find the area of land enclosed by the outer walls of the Pentagon building.

12-58.   **Examine** the triangles below. Decide if each one is a right triangle. If the triangle is a right triangle, **justify** your conclusion. Assume that the diagrams are not drawn to scale.

a.
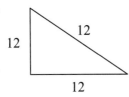

b.
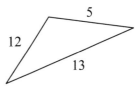

12-59.   **Examine** the Venn diagram at right. In which region should the figure below be placed? Show all work to **justify** your conclusion.

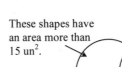

These shapes have an area more than 15 un².

These shapes have perimeter more than 20 units.

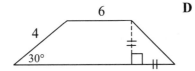

12-60.   For each equation below, decide if the equation has any real number for the given variable. For each problem, explain how you know.

a. $4(x-3)=11$

b. $x^2 = -10$

c. $3x^2 - 18 = 0$

d. $-7 = |x-6|$

12-61.   **Multiple Choice:** Which number below could be the length of the third side of a triangle with sides of length 29 and 51?

a.   10   b.   18   c.   23   d.   81

12.2.2 What's the pattern?

Euler's Formula for Polyhedra

Throughout this course, you have developed your skills of exploring a pattern, forming a conjecture, and then proving your conjecture. Today, with the assistance of materials such as toothpicks and gumdrops, you will review what you know about basic polyhedra (with no holes) as you look for a relationship between the number faces, edges, and vertices each basic polyhedron has. Once you have written a conjecture, your class will discuss how to prove that it must be true.

As you work today, consider the following questions:

Which types of basic polyhedra have we not tested yet?

What relationship can I find between the number of edges, faces, and vertices of a basic polyhedron?

Does this relationship always hold true?

12-62.   POLYHEDRA PATTERNS

Does a basic polyhedron (with no holes) usually have more faces, edges, or vertices? And if you know the number of faces and vertices of a basic polyhedron, how can you predict the number of edges? Today you will answer these questions and more as you **investigate** polyhedra.

Your Task: Obtain the necessary building materials from your teacher, such as toothpicks (for edges) and gumdrops (for vertices). Your team should build *at least* six distinctly different polyhedra and each person in your team is responsible for building *at least* one polyhedron. Be sure to build some regular polyhedra (such as a tetrahedron and an octahedron), basic prisms, pyramids, and unnamed polyhedra.

Create a table like the one at right to hold your data. Once you have recorded the number of vertices, edges, and faces for

Polyhedron	Faces (F)	Vertices (V)	Edges (E)

your team's polyhedra, look for a relationship between the numbers in each row of the table. Try adding, subtracting (or both) the numbers to find a pattern. Write a conjecture (equation) using the variables F, V, and E.

12-63.   EULER'S FORMULA FOR POLYHEDRA

The relationship you discovered in problem 12-62
between the number of faces, vertices, and edges of
a basic polyhedron is referred to as **Euler's
Formula for Polyhedra**, after Leonhard Euler
(pronounced "oiler"), a mathematician from
Switzerland.[1] It states that if V is the number of
vertices, F is the number of faces, and E is the
number of edges of a basic polyhedron, then
$V + F - E = 2$.

Use Euler's Formula to answer the following questions about basic polyhedra.

a.   If a polyhedron has 5 faces and 6 vertices, how many edges must it have?

b.   What if a polyhedron has 36 edges and 14 faces? How many vertices must it
     have?

c.   Could a polyhedron have 10 faces, 3 vertices, and 11 edges? Explain why or
     why not.

12-64.   If V represents the number of vertices, F represents the number of faces, and
         E represents the number of edges of a basic polyhedron, how can you prove that
         $V + F - E = 2$? First think about this independently. Then, as a class, prove Euler's
         Formula.

12-65.   For each situation below, decide if a is greater, b is greater, if they are the same
         value, or if not enough information is given.

a. a is the measure of a central angle of an equilateral triangle; b is the measure
     of an interior angle of a regular pentagon.

b.   c.
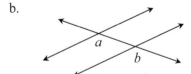   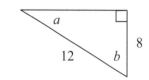

d. $a = b + 3$ e.

[1] While it is widely believed that Euler independently discovered this relationship, it has been recorded that René
Descartes (pronounced "Day-cart") found the relationship over 100 years earlier.

12-66.   In the diagram at right, $ABCD \sim DCFE$.
          Solve for x and y. Show all work.

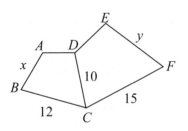

12-67.   On graph paper, make a table and graph the
          function $f(x) = -2(x-1)^2 + 8$.

    a.   Label the x- and y-intercepts and state their coordinates.

    b.   Name the vertex.

    c.   Find $f(100)$ and $f(-15)$.

12-68.   Find the area of the graph of the solution region of $x^2 + y^2 \le 49$.

12-69.   Find the volume and surface area of the
          box formed by the shaded net at right.

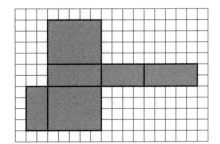

12-70.   **Multiple Choice:** The radius of the front
          wheel of Gavin's tricycle is 8 inches. If
          Gavin rode his tricycle for 1 mile in a
          parade, approximately how many rotations
          did his front wheel make?
          (Note: 1 mile = 5280 feet).

    a.   50   b.   1260   c.   660   d.   42,240

12.2.3 What's special about this ratio?

The Golden Ratio

In Chapter 8, you discovered an important irrational number: π. Pi (π) is the ratio of any circle's circumference to its diameter. However, there is another special ratio that appears not only in geometry, but also in nature. Today, you will discover this number and will examine several different contexts in which this number appears.

12-71. While doodling one day, Alex drew the diagram at right. He started with a large rectangle (*A*). He divided this rectangle into a square (*B*) and a smaller rectangle. Then he divided the smaller rectangle into a square (*C*) and a rectangle (*D*). He noticed that *D* had a height of 1 unit (as shown in the diagram).

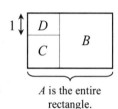

A is the entire rectangle.

   a. Draw Alex's diagram on a piece of graph paper.

   b. If the length of each side of square *C* is 3 units, what is the area of square *B*? What are the dimensions of rectangle *A*? **Justify** your answers.

   c. If the longest dimension of rectangle *A* is 9 units, what are the dimensions of squares *B* and *C*? Show how you know.

12-72. THE GOLDEN RATIO

   As Alex looked at his diagram from problem 12-71, he noticed that the dimensions of the large composite rectangle seemed to be subdivided proportionally. In other words, Figures *E* and *F*, shown at right, appeared to be similar.

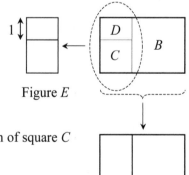

Figure *E*

Figure *F*

   a. If the height of *D* is 1 unit and the side length of square *C* is *x*, what is the side length of square *B*?

   b. If Figures *E* and *F* are similar, what is *x*? Be prepared to share your solution with the class.

   c. The value for *x* that you found in part (b) has a special name: the **Golden Ratio**. It is often represented by the greek letter phi (ϕ), pronounced "fee." Read the Math Notes box about the golden ratio before moving on to problem 12-73.

12-73.   GOLDEN SPIRALS

Each non-square rectangle in Alex's diagram
from problem 12-71 is an example of a **golden
rectangle** because the ratio of the longer length
to the shorter length is ϕ, the golden ratio. In
addition, Alex's process of subdividing each
golden rectangle into a square and smaller
golden rectangle can be **iterated** (repeated over
and over) creating an infinite series of nested
squares and rectangles.

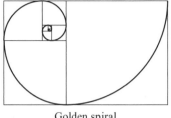

Golden spiral

When connected arcs are placed in
each of the squares, a spiral forms,
like the one shown above. One place
you can find this spiral is the human
ear, as shown at right.

Use a compass to draw a golden
spiral on the Activity 12.2.3
Resource Page provided by your
teacher. Where else (outside of
class) have you seen a spiral like
this?

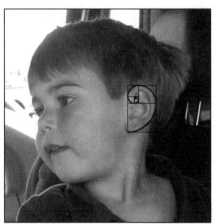

12-74.   Alex wonders where else the number phi (ϕ) shows up.
Look for phi (ϕ) as you analyze the following situations.

a.   **Examine** the regular decagon at right. If the side length is
1 unit, find the radius of the decagon. What do you notice?

b.   Each central triangle in the regular decagon from part (a) is a **golden triangle**
because the ratio of the congruent sides to the base of each triangle is phi (ϕ).
What are the angles of a golden triangle?

c.   In problem 12-73, you learned about nested golden rectangles (where each
golden rectangle is subdivided into a square and a smaller golden rectangle).
But what about nested expressions?

Consider the expression at right. The
"…" signifies that the pattern within the
expression continues infinitely. With
your team, find a way to approximate
the value of this expression. Try to find
the most accurate approximation you
can. What do you notice?

$$1+\sqrt{1+\sqrt{1+\sqrt{1+\sqrt{1+\sqrt{1+\ldots}}}}}$$

Geometry Connections

12-75. What if three golden rectangles intersect perpendicularly so that their centers coincide, as shown at right? If each vertex of the golden rectangles is connected with the five closest vertices, what three-dimensional shape appears? First **visualize** the result. Then, if you have a model available, test your idea with string.

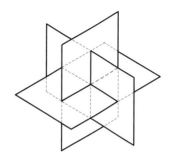

LOOKING **D**EEPER

The Golden Ratio

The number phi (ϕ), pronounced "fee," has a value of $\frac{1+\sqrt{5}}{2} \approx 1.618$ and is often referred to as the **golden ratio**. This special number is often found when comparing dimensions of geometric shapes and by comparing measurements of objects in nature.

For example, phi can be found multiple ways in a regular pentagon. The ratio of the length of any diagonal of a regular pentagon to the length of a side is phi (ϕ). This can be shown by assuming the side length is 1 unit and finding the length of the diagonal. Since each interior angle of the pentagon must be 108°, then the length of the diagonal must be:

$$d = \sqrt{1^2 + 1^2 - 2(1)(1)\cos 108°} \approx 1.618$$

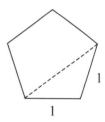

Regular pentagon

Phi also appears in nature. For example, the human body has many ratios that seem to relate to phi (ϕ). As shown in the picture at right, the ratio of the distance between the eyes and the corners of the mouth to the length of the nose is often phi (ϕ). That is, if the nose is 1 unit long, the vertical distance between the eyes and the corners of the mouth is often $\frac{1+\sqrt{5}}{2} \approx 1.618$ units.

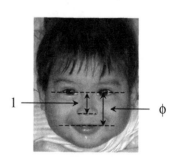

12-76.   If $\triangle ABC$ is equilateral, and if $A(3,2)$ and $B(7,2)$, find all possible coordinates of vertex C. **Justify** your answer.

12-77.   Find the area of the shaded region of the regular pentagon at right. Show all work.

4

12-78.   On graph paper, graph the following system of inequalities. Be sure that your shaded region represents all of the points that make both inequalities true.

$$y < \tfrac{2}{3}x - 2$$
$$y \ge -5x - 2$$

12-79.   Jamila solved the quadratic $x^2 + 3x - 10 = 8$ (see her work below). When she checked her solutions, they did not make the equation true. However, Jamila cannot find her mistake. Explain her error and then solve the quadratic correctly.

$$x^2 + 3x - 10 = 8$$
$$(x + 5)(x - 2) = 8$$
$$x + 5 = 8 \quad \text{or} \quad x - 2 = 8$$
$$x = 3 \quad \text{or} \quad x = 10$$

12-80.   If the sum of the interior angles of a regular polygon is $2160°$, how many sides must it have?

12-81.   **Multiple Choice:** Assume that $A(6,2)$, $B(3,4)$, and $C(4,-1)$. If $\triangle ABC$ is rotated $90°$ counterclockwise ($\circlearrowleft$) to form $\triangle A'B'C'$, and if $\triangle A'B'C'$ is reflected across the x-axis to form $\triangle A''B''C''$, then this is the coordinates of C''.

a.   (1, 4)   b.   (–4, 1)   c.   (1, –4)   d.   (4, 1)

12.2.4 What's the probability?

Using Geometry to Find Probability

In this final activity, you will **connect** and **apply** much of your knowledge from throughout the course to solve a challenging problem.

12-82.   ZOE AND THE POISON WEED

Dimitri is getting his prize sheep, Zoe, ready for the county fair. He keeps Zoe in the pasture beside the barn and shed. What he does not know is that there is a single locoweed in this pasture, which will make Zoe too sick to go to the fair if she eats it, and she can eat it in one bite. Zoe takes about one bite of grass or plant every three minutes for six hours a day.

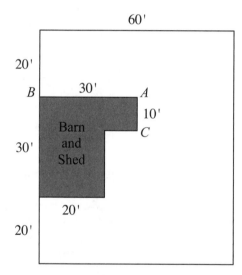

The layout of the field and building is provided at left and on the Activity 12.2.4 Resource Page provided by your teacher. Assume that the entire field (the unshaded region) has plants growing on it and that each square foot of field provides enough food for 40 bites. Also assume that each corner of the barn and field is a right angle.

Dimitri is worried that Zoe will get into trouble unless she is tethered with a rope to the building. He has decided to tether Zoe at point *A* with a 20-foot rope. Zoe is unable to enter the barn or shed while on her tether.

Your Task: If the locoweed lies in Zoe's grazing area, what is the probability that Zoe will get sick in one day?

Discussion Points

What is the problem asking you to find?

What does Zoe's grazing region look like?

What do you need to figure out in order to find the probability?

12-83.   To help find the probability that Zoe will eat the single locoweed, first consider the grazing region if she is tethered to point *A* with a 20-foot rope.

   a.   On your Activity 12.2.4 Resource Page, draw and label the region that Zoe can roam. Then find the area of that region.

   b.   Since each square foot of the field contains 40 bites of food, how many bites of food lie within Zoe's reach?

   c.   How many bites of food does Zoe eat each day? Show your calculations.

   d.   If the single locoweed is within this area, what is the probability that she eats the weed in one day? Be prepared to explain your answer to the class.

————————   *Further Guidance*   ————————
                  section ends here.

12-84.   FAMILY DISCUSSION

When Dimitri discussed his idea with his family, he received other ideas! Analyze each of the ideas given below and then report back to Dimitri about which of them has the least probability that Zoe will eat the locoweed. Your analysis should include:

   •   A diagram of each proposed region on the Activity 12.2.4 Resource Page (or use the figure in problem12-82).

   •   All calculations that help you determine the probability that Zoe will eat the poisoned weed for each proposed region.

Assume that a single locoweed lies somewhere in each region that is proposed.

   a.   **Dimitri's Father:** "Dimitri! Why do you need to waste rope? All you need is to tether your sheep with a 10-foot rope attached at point *A*. Take it from me: Less area to roam means there is less chance that the sheep will eat the terrible locoweed!"

   b.   **Dimitri's Sister:** "I don't agree. I think you should consider using a 30-foot rope attached to point *B*. The longer rope will give Zoe more freedom, but the building and fences will still limit her region. This is the best way to reduce the chance that Zoe gets sick before the fair."

Problem continues on next page →

12-84.   *Problem continued from previous page.*

c.   **Dimitri's Mother:** "Both of those
regions really restrict Zoe to the north-
eastern part of the field. That means she
won't be able to take advantage of the
grass grown in the southern section of
the field that is rich in nutrients because
of better sunlight. I recommend that you
use a 30-foot rope attached to point *C*.
You won't be disappointed!"

12-85.   Find the area of each quadrilateral below. Show all work.

a.   Kite   b.   Rhombus

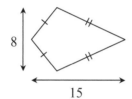

12-86.   A spinner is divided into two regions. One region, red, has a central angle of 60°.
The other region is blue.

a.   On your paper, sketch a picture of this spinner.

b.   If the spinner is spun twice, what is the probability that both spins land on
blue?

c.   If the radius of the spinner is 7 cm, what is the area of the blue region?

d.   A different spinner has three regions: purple, mauve, and green. If the
probability of landing on purple is $\frac{1}{4}$ and the probability of landing on mauve
is $\frac{2}{3}$, what is the central angle of the green region?

12-87. Perry threw a tennis ball up into the air from the edge of a cliff. The height of the ball was $y = -16x^2 + 64x + 80$, where y represents the height in feet of the ball above ground at the bottom of the cliff, and x represents the time in seconds after the ball is thrown.

  a. How high was the ball when it was thrown? How do you know?

  b. What was the height of the ball 3 seconds after it was thrown? What was its height $\frac{1}{2}$ a second after it was thrown? Show all work.

  c. When did the ball hit the ground? Write and solve an equation that represents this situation.

12-88. **Examine** the triangles below. Which, if any, are similar? Which are congruent? For each pair that must be similar, state how you know. Remember that the diagrams are not drawn to scale.

  a.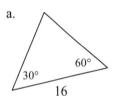
  b.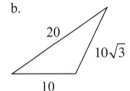
  c.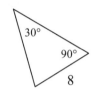
  d.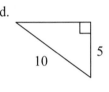

12-89. **Examine** the mat plan of a three-dimensional solid at right.

  a. On your paper, draw the front, right, and top views of this solid.

  b. Find the volume of the solid.

  c. If each edge of the solid is multiplied by 5, what will the new volume be? Show how you got your answer.

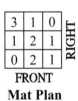

3	1	0
1	2	1
0	2	1

FRONT

Mat Plan

12-90. Prove that the base angles of an isosceles triangle must be congruent. That is, prove that if $\overline{BC} \cong \overline{AC}$ in the triangle at right, then $\angle A \cong \angle B$. (Hint: Is there a convenient auxiliary line that can be added to the diagram that divides the triangle into two congruent triangles?)

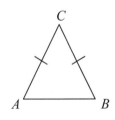

12-91. Find the area and perimeter of the shape at right. Assume that any non-straight portions of the shape are part of a circle. Show all work.

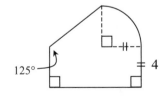

12-92. **Examine** the diagrams below. For each one, use the geometric relationships to solve for the given variable.

a. $\overrightarrow{PR}$ is tangent to $\odot C$ at P and $m\overparen{PMQ} = 314°$. Find QR.

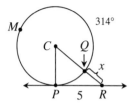

b. $\overline{AB}$ and $\overline{CD}$ intersect at E.

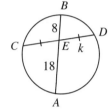

c. Radius = 7 cm

12-93. **Examine** the net at right.

a. Describe the solid that is formed by this net. What are its dimensions?

b. Find the surface area and volume of the solid formed by this net.

c. If all the dimensions of this solid are multiplied by 3, what is the SA of the resulting solid? What is the volume?

12-94. Polly has a pentagon with measures $3x - 26°$, $2x + 70°$, $5x - 10°$, $3x$, and $2x + 56°$. Find the probability that if one vertex is selected at random, then the measure of its angle is more than or equal to 90°.

12-95. Find at least three different shapes that can be cross-sections of a cube, like the one at right. For each one, draw the resulting cross-section and explain how you sliced the cube.

12-96. **Multiple Choice:** A square based pyramid has a slant height of 10 units and a base edge of 10 units. What is the height of the pyramid?

a. 5   b. $5\sqrt{3}$ c. 6   d. 8

Appendix A What's the intersection?

Points, Lines, and Planes

In this course, you will study many shapes that are "built" with other shapes. For example, the rectangle at right is formed with four sides that are called **line segments**. Understanding the parts of shapes will help you not only *understand* the shapes better, but will also help you *describe* the shapes accurately.

A-1.   PLANES

A **plane** is a two-dimensional flat surface that extends without end. Even though a plane has no thickness, you can visualize it as the top surface of your desk or tabletop (if all sides of your desk extend infinitely). A graph with x- and y-axes that continue forever is a representation of a plane (see diagram at right).

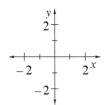

a.   A plane is often represented by drawing a parallelogram, such as the one shown at right. Although the diagram seems to show edges, it is important to remember that a plane extends without end. The plane at right is named Q.

Use two pieces of cardboard to represent two planes. Remembering that planes extend infinitely and do not have borders (unlike your cardboard, which has edges), try to place your two planes so that they will never intersect (cross). Is it possible? If so, explain how the planes can be placed so that they do not intersect (or draw a diagram showing this situation). If not, explain why it is not possible.

b.   What happens when two planes intersect? What does their intersection look like? How could you describe the intersection?

c.   What about three planes? Is it possible to have all three planes intersect each other? Is it possible to have two intersect, with a third plane somewhere else, never intersecting the other two? And is it possible for all three planes not to intersect anywhere? Consider these questions with your team members. For each case, explain how the planes are arranged and what their intersection(s) look(s) like. Be prepared to share your ideas with the class.

A-2.   LINES

In problem A-1, you discovered that when two planes intersect, their intersection is a line. While a plane is two-dimensional, a **line** is only one-dimensional. A line extends without end in two directions and has no thickness.

a.   Where have you used or seen a line before? In what contexts?

b.   If two lines intersect, how can you describe the intersection? Assume that the lines do not coincide (that is, assume that the lines do not lie directly on top of each other). Draw an example of two lines intersecting, and describe the intersection on your paper.

c.   When two lines intersect, their intersection is a **point**. A point has no dimension (and therefore, no thickness). Is it possible for two lines not to intersect? If so, is there more than one way for this to happen? To help you investigate this, use straight objects (such as two pencils) to represent lines. Be sure to keep in mind that lines extend infinitely in two directions. Explain what you find.

A-3.   Lines extend infinitely in two directions. However, what if a line extends infinitely in only one direction?

a.   On your paper, draw an example of a part of a line that extends infinitely in only one direction. This geometric figure is called a **ray**. A ray has one dimension, like a line. Where have you seen a ray before?

Ray

b.   A ray can be named using two points: the endpoint and one other point on the ray. For example, the ray above is called $\overrightarrow{AB}$. However, another way to represent a ray is on a number line, like the one shown below. This ray could be named $x \geq 2$.

On your paper, draw a number line. Then draw the ray $x \geq 3$. If this ray is named $\overrightarrow{EF}$, what is the coordinate of point E?

c.   Examine the ray at right. Name this ray at least three different ways.

A-4.   What about the points represented by the expression $-1 \le x \le 2$? What shape would these points form?

a.   On your paper, graph this shape on a number line. Then describe this shape.

b.   A portion of a line between two points is called a **line segment**. A line segment can be named using the two endpoints. For example, the line segment at right could be called $\overline{AB}$ or $\overline{BA}$.

Using this notation, name the sides of the triangle at right.

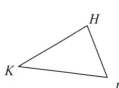

c.   Examine the number line below. Name the line segment in as many ways as you can. Be ready to share your names with the class.

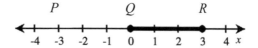

A-5.   In your Learning Log, write an entry describing what you learned today about points, lines, rays, line segments, and planes. Be sure to draw an example of each. Title this entry "Points, Lines, and Planes" and include today's date.

METHODS AND MEANINGS

Points, Lines, and Planes

MATH NOTES

A **point** is an undefined term in geometry. It has no dimension. It can be labeled with a capital letter (like point A at right) or with a coordinate (like 1).

A **line** has one dimension and extends without end in two directions. A number line is an example of a line. It is often named using two points on the line (like $\overleftrightarrow{BC}$ at right).

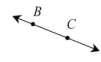

A **line segment** is a portion of a line between (and including) two points. It can be named using its endpoints (like $\overline{DE}$ at right) or with an interval on a number line (such as $-2 \leq x \leq 4$).

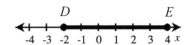

A **plane** is two-dimensional and extends without end. One way to think of a plane is as a flat surface, like the top of a table, that extends infinitely in all directions.

Review & Preview

A-6. The ray at right can be named $\overrightarrow{AB}$ or $x \leq -3$. For each shape represented on the number lines below, write the name at least two different ways.

a.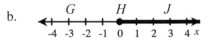

b.

c.

d.

610

Geometry Connections

A-7.   In problem A-2, you showed that when two lines intersect and do not coincide (lie directly on top of each other), their intersection is a point.

    a.   Now consider three lines. If none of the lines coincide, how many arrangements are possible? Draw each case.

    b.   How can a line and a plane be arranged? For each case, describe the arrangement. If they intersect, describe the intersection.

A-8.   Draw a number line for $-4 \leq x \leq 4$ on your paper.

    a.   Label point A if its coordinate is 3.

    b.   Draw $\overline{EF}$ if it can be represented as $0 \geq x \geq 2$ and the coordinate of E is 0.

    c.   Explain where ray $\overrightarrow{FA}$ would exist. Then write another name for this ray.

A-9.   Part of this course is learning about geometry on a coordinate plane. What is the difference between how a point is represented on a number line and how it is represented on a coordinate plane?

    a.   To think about this question, first consider point A on the number line at right. What is the coordinate of A?

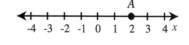

    b.   What happens when a y-axis is added to the diagram? Now how is point A represented with coordinates? Why is there a difference?

    c.   On graph paper, plot and label ray $\overrightarrow{AB}$ if A is (2, 3) and B is (4, 1).

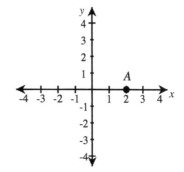

A-10.   On graph paper, find the intersection of line $\overleftrightarrow{AB}$ and ray $\overrightarrow{CD}$ if $A(4, 3)$, $B(-3, -4)$, $C(3, -4)$, and $D(1, -3)$.

Appendix B What can I assume?

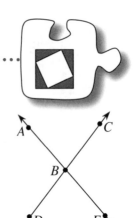

Euclidean Geometry

So far in this course, you have developed geometric ideas by assuming that certain geometric principals are true. For example, when you state that lines $\overline{AE}$ and $\overline{DC}$ intersect at point B, how can you be sure that they only intersect once? What are you assuming?

And when you became convinced that vertical angles are congruent (such as $\angle ABC$ and $\angle DBE$ at right), what assumptions did you make? What would be true if these assumptions were false?

To clarify issues like this, mathematicians have developed a geometric system of definitions, axioms, and theorems that includes assumptions as the basis for the mathematical ideas in the system. Perhaps the person most famous for this work is Euclid, a Greek mathematician who lived between 325 BC and 265 BC.

B-1.   THE ELEMENTS

Euclid is often referred to as the "Father of Geometry," in large part because of a famous set of books he wrote, collectively called *Elements*. In this set of books, he provided critical definitions for many basic geometric figures and relationships, such as parallel lines and obtuse angles. In the left column of part (b) below are four of the geometric shapes defined by Euclid in *Elements*.

a.   Complete this definition, which is a translation of Euclid's Definition 10: "*And when a straight-line stood upon another straight-line makes adjacent angles which are equal to one another, each of the equal angles is a right-angle, and the former straight line is called _____ to that upon which it stands.*" (Translation by Richard Fitzpatrick.)

b.   Match each geometric term in the left column with its definition (translated into English from Greek) in the right column. One definition in the right column will be left over. Decide what shape Euclid was defining in the unmatched definition.

a. Line	1. Greater than a right angle.
b. Obtuse angle	2. A quadrilateral figure that is right-angled and equilateral.
c. Boundary	3. That of which there is no part.
d. Point	4. Length without breadth.
	5. That which is the extremity of something.

B-2.   In addition to definitions, Euclid offered five **postulates** (also sometimes called **axioms**), which are statements that are assumed to be true. Euclid recognized that if these postulates were widely recognized as true, many other geometric properties and relationships could be proven true as well.

a.   One of Euclid's postulates is translated, "*A straight line segment can be drawn by joining any two points.*" Draw two points on your paper and label them *A* and *B*. How many straight lines can connect points *A* and *B*? Draw as many as you can.

b.   Another translated postulate states, "*All right angles are equal to one another.*" Do you agree with this assumption? If this postulate were not true, what would that mean?

c.   Perhaps Euclid's most famous and studied postulate was his fifth postulate, which has been translated as, "*If a straight line falling across two other straight lines makes internal angles on the same side that are less than two right angles, then these two lines, when extended to infinity, will intersect on the same side as the angles and do not meet on the other side.*"

Draw a diagram that could have accompanied this postulate. Is this assumption reasonable? What if this postulate were false?

B-3.   Another part of Book 1 of Euclid's *Elements* contains a series of Propositions (also called **theorems**), which are statements that are proven always to be true. By using definitions and postulates, Euclid set out to prove new geometric statements that, in turn, could be used to prove other new mathematical statements.

Proposition 15 may look very familiar. Euclid proved that, "*If two straight lines cut one another, the angles that are vertically opposite are equal to one another.*" He then proved this by using a combination of several propositions and postulates.

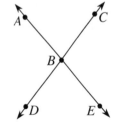

a.   In his argument, Euclid states that if $\overleftrightarrow{DC}$ intersects $\overleftrightarrow{AE}$ at *B*, then ∡*ABC* and ∡*CBE* together form two right angles. What do you think he meant by that statement?

b.   Euclid also used a common notion that "*things that are equal to the same thing are equal to each other.*" This is now referred to as the **Transitive Property**. Where have you used this property before?

OOKING DEEPER

MATH NOTES

Euclid's *Elements*

In his set of books, collectively called *Elements*, Euclid defined many geometric terms to help fellow mathematicians and to create a common language about geometry. In addition, he also shared what he saw as five "common notions," listed below. Consider how you have used each of these notions in the past.

- If two things are equal to the same thing, then they are equal to each other.
- When equal things are added to equal things, then the sums are equal.
- When equal things are subtracted from equal things, then the differences are equal.
- When things coincide with one another, they are also equal to one another.
- The whole is greater than the part.

B-4.   On graph paper, draw $\triangle ABC$ if $A(2,4)$, $B(9,5)$, and $C(4,10)$.

   a.   Verify that $D(3,7)$ is a midpoint of $\overline{AC}$.

   b.   Find the equation of the line through points D and B.

   c.   Is $\overline{BD}$ a height of $\triangle ABC$? Use slope to show that $\overline{BD}$ is perpendicular to $\overline{AC}$.

B-5.   Draw a number line on your paper. If point A is at -1 and B is at 2, draw $\overrightarrow{AB}$. Describe the result.

B-6.   Consider points *A* and *B* on the number line at right. For the coordinates provided below, find the length of $\overline{AB}$. Assume that the coordinate of *B* is greater than or equal to the coordinate of *A*.

   a.   *A* is 3 and *B* is 10   b.   *A* is –2 and *B* is 8

   c.   *A* is 6 and *B* is *x*   d.   *A* is *x* and *B* is *y*

B-7.   Examine △*ABC* on the graph at right.

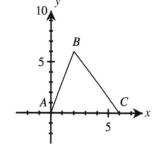

   a.   Find the midpoint of $\overline{AB}$ and call it *D*.

   b.   Find the midpoint of $\overline{BC}$ and call it *E*.

   c.   Find the equation of $\overleftrightarrow{DE}$.

   d.   Is $\overleftrightarrow{DE}$ parallel to $\overleftrightarrow{AC}$? Use slope to justify your conclusion.

B-8.   On a piece of graph paper, graph $y = -\frac{1}{2}x + 4$ and label its *x*-intercept *A* and its *y*-intercept *B*. Label the origin (0, 0) point *C*.

   a.   Find the area of △*ABC*.

   b.   Draw a height to $\overline{AB}$ through *C*. What is the slope of this height? Show how you found this slope.

   c.   If the height you drew in part (b) intersects side $\overline{AB}$ at point *D*, find the equation of the line through *C* and *D*.

Appendix C What if parallel lines intersect?

Non-Euclidean Geometry

In this course, you have learned many of the theorems set out in Euclid's book, *Elements*. Euclid, who lived between 325 BC and 265 BC, set up a system of definitions, postulates (also called axioms), and common notions that form much of the basis for plane geometry. This geometric system is now called **Euclidean geometry**, in reference to Euclid's work.

Today you will investigate a different system of geometry, called **spherical geometry**. Because it deviates from Euclidean geometry, it is a form of **non-Euclidean geometry**.

C-1.　　LINES ON A SPHERE

The paths chosen by pilots often surprise travelers. For example, someone traveling from Seattle, Washington to Athens, Greece may assume that because Greece is south of Seattle, the shortest path is one that travels south as shown with the dashed line on the map at right.

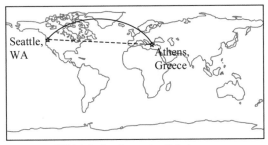

Which is the shortest path between Seattle, WA and Athens, Greece?

However, it turns out that the shortest path actually first takes a traveler *north*, over Greenland and very close to the North Pole, as shown by the solid curve on the diagram above. This is because the shape of the Earth is close to that of a sphere, and the shortest distance between two points on a sphere is along the great circle that passes through the two points. See this path at right.

Use your model of the Earth to help you think about each of the situations below. Assume the Earth's surface is smooth and spherical.

a.　　Since the path on a great circle is the shortest path on the sphere between the two points, it can also be referred to as a "line." Lines of longitude are examples of lines on the surface of a sphere.

On your spherical model, think about all possible lines on the surface. Are there any two great circles that do not intersect? With your team, try to find two lines (great circles) that do not intersect. Explain your conclusion(s).

Problem continues on next page →

C-1.   *Problem continued from previous page.*

   b.   In part (a), you determined that <u>any</u> two lines on the surface of a sphere must intersect. This contradicts plane geometry (also referred to as Euclidean geometry), in which it is possible for two lines not to intersect. Since the geometry on a sphere behaves differently than the geometric system set up by Euclid, it is referred to as **non-Euclidean geometry**. The study of geometry on a sphere is also referred to as **spherical geometry**.

   In a plane, a line has infinite length. How long is a line on a sphere? Explain.

   c.   In a plane, only one line can be drawn between two points. But what about on a sphere? Examine the poles of your spherical model. How many lines pass through two poles of the sphere? What are these lines called?

C-2.   COMPARING EUCLIDEAN AND NON-EUCLIDEAN GEOMETRY

   Look for differences between Euclidean and spherical geometry as you answer the questions below.

   a.   In a plane, two lines that do not coincide intersect at one point (at most). Two lines on a sphere that do not coincide intersect each other at how many points? Use your spherical model and rubber bands to help you visualize this situation.

   b.   In a plane, two lines can be perpendicular. Is it possible for two lines on a sphere to be perpendicular? If so, provide an example. If not, explain why not.

   c.   Is it possible to have a ray in spherical geometry? Why or why not?

   d.   The diagram of the Earth at right shows many lines of latitude. Are all of these lines of latitudes considered to be lines in spherical geometry? Why or why not?

Lines of Latitude

C-3.   TRIANGLES ON A SPHERE

   Consider the triangle drawn on the model of the Earth at right. Note that point P is the North Pole, while points A and B lie on the Equator.

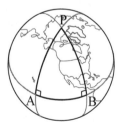

   a.   Without knowing $m\angle APB$, what can you state about the sum of the angles of $\triangle APB$? Compare this with what you know about the sum of the angles of a triangle in a plane.

   b.   Using rubber bands and your spherical model, think about the kinds of triangles you can create. Is it possible to form a triangle for which the sum of the angles is only 90°? Can the sum be 180°? What about 270°? 300°? 360°? Describe the possible sums.

C-4.   In your Learning Log, explain what you learned today about non-
        Euclidean geometry. What surprised you? How is spherical
        geometry different than Euclidean geometry? Title this entry "Non-
        Euclidean Geometry" and include today's date.

C-5.   When considering new plans for a covered baseball
        stadium, Smallville looked into a design that used a
        cylinder with a dome in the shape of a hemisphere.
        The radius of the proposed cylinder is 200 feet. See
        a diagram of this at right.

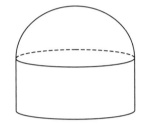

        a.   One of the concerns for the citizens of
             Smallville is the cost of heating the space inside
             the stadium for the fans. What is the volume of
             this stadium? Show all work.

        b.   The citizens of Smallville is also interested in having the outside of the new
             stadium painted in green. What is the surface area of the stadium? Do not
             include the base of the cylinder.

C-6.   An ice-cream cone is filled with ice-cream. It also has a semi-
        spherical scoop of ice-cream on top. It turns out that the volume
        of ice-cream inside the cone equals the volume of the scoop on
        top. If the height of the cone is 6 inches and the radius of the
        scoop of ice-cream is 1.5 inches, find the height of the extra
        scoop on top. Ignore the thickness of the cone.

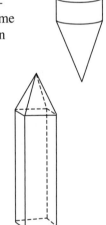

C-7.   The campanile at Iowa State University is a beautiful
        tower that is composed of a square-based pyramid atop
        a tall rectangular prism. The prism sits on a 16 by 16
        foot square base and extends 90 feet. The pyramid
        extends another 20 feet vertically. Find the surface area
        of the tower. Show all work.

C-8.   On graph paper, graph the inequality $y \geq x - 2$. Shade the region that makes the
        inequality true.

        a.   Is the point $(3, 0)$ a solution to this inequality? Why or why not?

        b.   Is the point $(3, 1)$ a solution to this inequality? Why or why not?

C-9.   Examine the diagram at right. Find the values of *a* and *b*. Show all work.

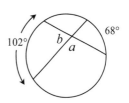

C-10.   **Multiple Choice:** Which point below will ray $\overrightarrow{AB}$ pass through?

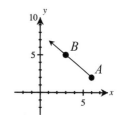

a. $(0, 8)$

b. $(3, 5)$

c. $(9, -1)$

d. $(5, 0)$

e.   None of these

Glossary

AA ~ (Triangle Similarity)   If two angles of one triangle are congruent to the two corresponding angles of another triangle, then the triangles are similar. For example, given $\triangle ABC$ and $\triangle A'B'C'$ with $\angle A \cong \angle A'$ and $\angle B \cong \angle B'$, then $\triangle ABC \sim \triangle A'B'C'$. You can also show that two triangles are similar by showing that *three* pairs of corresponding angles are congruent (which would be called AAA~), but two pairs are sufficient to demonstrate similarity. (pp. 151 and 171)

AAS ≅ (Triangle Congruence)   If two angles and a non-included side of one triangle are congruent to the corresponding two angles and non-included side of another triangle, the two triangles are congruent. Note that AAS ≅ is equivalent to ASA ≅. (p. 299)

absolute value   The absolute value of a number is the distance of the number from zero. Since the absolute value represents a distance, without regard to direction, it is always non-negative. Thus the absolute value of a negative number is its opposite, while the absolute value of a non-negative number is just the number itself. The absolute value of x is usually written $|x|$. For example, $|-5| = 5$ and $|22| = 22$. (pp. 453 and 575)

acute angle   An angle with measure greater than 0° and less than 90°. One example is shown at right. (p. 24)

adjacent angles   For two angles to be adjacent, they must satisfy these three conditions: (1) The two angles must have a common side; (2) They must have a common vertex; and (3) They can have no interior points in common. This means that the common side must be between the two angles; no overlap between the angles is permitted. In the example at right, $\angle ABC$ and $\angle CBD$ are adjacent angles.

adjacent leg   In a right triangle, the leg adjacent to an acute angle is the side of the angle that is not the hypotenuse. For example, in $\triangle ABC$ shown at right, $\overline{AB}$ is the leg adjacent to $\angle A$. (p. 241)

alpha (α)   A Greek letter that is often used to represent the measure of an angle. Other Greek letters used to represent the measure of an angle include theta (θ) and beta (β). (p. 190)

alternate interior angles   Angles between a pair of lines that switch sides of a third intersecting line (called a transversal). For example, in the diagram at right the shaded angles are alternate interior angles. If the lines intersected by the transversal are parallel, the alternate interior angles are congruent. Conversely, if the alternate interior angles are congruent, then the two lines intersected by the transversal are parallel. (p. 91)

altitude   See *height*.

ambiguous   Information is ambiguous when it has more than one interpretation or conclusion. (pp. 270)

angle   In general, an angle is formed by two rays joined at a common endpoint. Angles in geometric figures are usually formed by two segments, with a common endpoint (such as the angle shaded in the figure at right). (pp. 22 and 24)

angle measure   *See* measurement.

angle of depression   When an object (*B*) is below the horizontal line of sight of the observer (*A*), the angle of depression is the angle formed by the line of sight to the object and the horizontal line of sight.

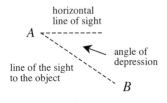

angle of elevation   When an object (*B*) is above the horizontal line of sight of the observer (*A*), the angle of elevation is the angle formed by the line of sight to the object and the horizontal line of sight.

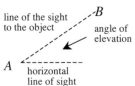

apex   In a cone or pyramid, the apex is the point that is the farthest away from the flat surface (plane) that contains the base.   In a pyramid, the apex is also the point at which the lateral faces meet. An apex is also sometimes called the "vertex" of a pyramid or cone. (p. 533)

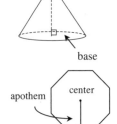

apothem   A segment that connects the center of a regular polygon to a point on one of its sides, and is perpendicular to that side. (p. 410)

approximate   Close, but not exact.   An approximate answer to a problem is often a rounded decimal.   For example, if the exact answer to a problem is $\sqrt{2}$, then 1.414 and 1.41 are approximate answers.   (p. 123)

arc   A connected part of a circle.   Because a circle does not contain its interior, an arc is more like a portion of a bicycle tire than a slice of pizza. (pp. 426 and 485)

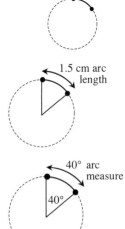

arc length   The length of an arc (in inches, centimeters, feet, etc.) is the distance from one of the arc's endpoints to the arc's other endpoint, measured around the circle.   Note that arc length is different from arc measure.   Arcs of two different circles may have the same arc measure (like 90°), but have different arc lengths if one circle has a larger radius than the other one.   For an example, see concentric circles.   (p. 490)

arc measure   The measure in degrees of an arc's central angle.   Note that arc measure is measured in degrees and is different from arc length. (p. 490)

area   On a flat surface (plane), the number of non-overlapping square units needed to cover the region.   *Also see* surface area.   (p. 490)

Geometry Connections

area model   A tool that uses the area of rectangles to represent the probabilities of possible outcomes when considering two independent events. For example, suppose you are going to randomly select a student from a classroom. If the probability of selecting a female student from the room is $\frac{2}{3}$ and the probability of selecting a 9th grader from the room is $\frac{1}{4}$, the area of the shaded rectangle at right represents the probability that a randomly selected student is a female 9th grader ($\frac{2}{3} \cdot \frac{1}{4} = \frac{1}{6}$). A generic area model is a tool that enables you to find the probabilities of outcomes without drawing a diagram to scale. (p. 219)

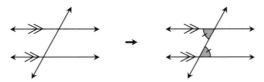

Area model

arrow diagram   A pictorial representation of a conditional statement. The arrow points toward the conclusion of the conditional (the "then" part of the "If… then" statement). For example, the conditional statement *"If two lines cut by a transversal are parallel, then alternate interior angles are congruent"* can be represented by the arrow diagrams below. (p. 84)

Lines cut by a transversal are parallel ➜ *alternate interior angles are congruent.*

arrowheads   To indicate a line on a diagram, we draw a segment with arrowheads on its ends. The arrowheads show that our diagram indicates a line, which extends indefinitely. Marks on pairs of lines or segments such as ">>" and ">>>" indicate that the lines or segments are parallel. Both types of marks are used in the diagram above. (p. 54)

ASA ≅ (Triangle Congruence)   If two angles and the included side of one triangle are congruent to the corresponding two angles and included side of another triangle, the triangles are congruent. Note that ASA ≅ is equivalent to AAS ≅. (p. 299)

auxiliary lines   Segments and lines added to existing figures. Auxiliary lines are usually added to a figure to allow us to prove something about the figure.

axioms   Statements accepted as true without proof. Also known as "postulates."

base   (a) Triangle: Any side of a triangle to which a height is drawn. There are three possible bases in each triangle (p. 112); (b) Trapezoid: the two parallel sides (p. 107); (c) Parallelogram (and rectangle, rhombus, and square): Any side to which a height is drawn. There are four possible bases (p. 106); (d) Solid: *See* cone, cylinder, prism, and pyramid.

binomial   An expression that is the sum or difference of exactly two terms, each of which is a monomial. For example, $-2x + 3y^2$ is a binomial. (p. 104)

bisect   To bisect a geometric object is to divide it into two congruent parts. (p. 317)

center of a circle   On a flat surface, the fixed point from which all points on the circle are equidistant. *See* circle. (p. 341)

central angle An angle with its vertex at the center of a circle. (p. 490)

centroid The point at which the three medians of a triangle intersect. The centroid is also the center of balance of a triangle. The other points of concurrency studied in this course are the circumcenter and the incenter. (pp. 471 and 501)

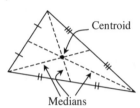
Centroid

Medians

chord A line segment with its endpoints on a circle. A chord that passes through the center of a circle is called a "diameter." *See* circle. (p. 485)

circle The set of all points on a flat surface that are the same distance from a fixed point. If the fixed point (center) is O, the symbol $\odot O$ represents a circle with center O. If r is the length of a circle's radius and d is the length of its diameter, the circumference of the circle is $C = 2\pi r$ or $C = \pi d$. The area of the circle is $A = \pi r^2$. The equation of a circle with radius length r and center $(0, 0)$ is $x^2 + y^2 = r^2$. (pp. 341, 485, and 490)

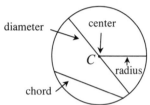
diameter   center

C

radius

chord

circular angle An angle with a measure of 360°. (pp. 22 and 24)

360°

circumcenter The center of the circle that passes through the vertices of a triangle. It can be found by locating the point of intersection of the perpendicular bisectors of the sides of the triangle. The other points of concurrency studied in this course are the centroid and the incenter. (p. 501)

circumference The perimeter of (distance around) a circle. (p. 426)

circumscribe A circle circumscribes a polygon when it passes through all of the vertices of the polygon. (p. 500)

$C\bullet$

$\odot C$ circumscribes the pentagon.

clinometer A device used to measure angles of elevation and depression. (p. 202)

clockwise Clockwise identifies the direction of rotation shown in the diagram at right. The word literally means "in the direction of the rotating hands of a clock." The opposite of clockwise is counter-clockwise. (p. 34)

compass (a) A tool used to draw circles; (b) A tool used to navigate the Earth. A compass uses the Earth's magnetic field to determination which direction is north.

complementary angles Two angles whose measures add up to 90°. Angles T and V are complementary because $m\angle T + m\angle V = 90°$. Complementary angles may also be adjacent, like $\angle ABC$ and $\angle CBD$ in the diagram at far right. (p. 74)

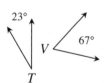
23°

V 67°

A C

T B D

concentric circles Circles that have the same center. For example, the circles shown in the diagram at right are concentric. (p. 460)

624

concurrency (point of)   The single point where two or more lines intersect on a plane. (p. 501)

conditional statement   A statement written in "If …, then …" form. For example, *"If a rectangle has four congruent sides, then it is a square"* is a conditional statement. (p. 108)

cone   A three-dimensional figure that consists of a circular face, called the "base," a point called the "apex," that is not in the flat surface (plane) of the base, and the lateral surface that connects the apex to each point on the circular boundary of the base. (pp. 542 and 547)

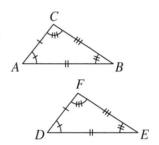

congruent   Two shapes are congruent if they have exactly the same shape and size. Congruent shapes are similar and have a scale factor of 1. The symbol for congruence is $\cong$. (pp. 159 and 291)

congruence statement   A statement indicating that two figures are congruent. The order of the letters in the names of the shapes indicates which sides and angles are congruent to each other. For example, if $\triangle ABC \cong \triangle DEF$, then $\angle A \cong \angle D$, $\angle B \cong \angle E$, $\angle C \cong \angle F$, $\overline{AB} \cong \overline{DE}$, $\overline{BC} \cong \overline{EF}$, and $\overline{AC} \cong \overline{DF}$. (p. 346)

conic section   A curve that is the intersection of a plane with a double cone. Conic sections include parabolas, circles, ellipses, and hyperbolas. Conic sections that degenerate (collapse) into a point or line are excluded. (p. 578)

conjecture   An educated guess. Conjectures often result from noticing a pattern during an investigation. Conjectures are also often written in conditional ("If…, then…") form. Once a conjecture is proven, it becomes a theorem. (pp. 10 and 304)

construction   The process of using a straightedge and compass to solve a problem and/or create a geometric diagram. (pp. 459 and 468)

converse   The converse of a conditional statement can be found by switching the hypothesis (the "if" part) and the conclusion (the "then" part). For example, the converse of *"If P, then Q"* is *"If Q, then P."* Knowing that a conditional statement is true does not tell you whether its converse is true. (p. 304)

convex polygon   A polygon is convex if any pair of points inside the polygon can be connected by a segment without leaving the interior of the polygon. (p. 396)

coordinate geometry   The study of geometry on a coordinate grid. (p. 395)

corresponding angles   (a) When two lines are intersected by a third line (called a transversal), angles on the same side of the two lines and on the same side of the transversal are called corresponding angles. For example, the shaded angles in the diagram at right are corresponding angles. Note that if the two lines cut by the transversal are parallel, the corresponding angles are congruent. Conversely, if the corresponding angles are congruent, then the two lines intersected by the transversal are parallel. (p. 91); (b) Angles in two figures may also correspond, as shown at right. See corresponding parts for more information. (pp. 143 and 150)

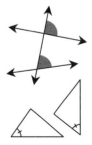

Glossary

corresponding parts Points, sides, edges, or angles in two or more figures that are images of each other with respect to a transformation. If two figures are congruent, their corresponding parts are congruent to each other. (p. 143, 150, 346)

cosine ratio In a right triangle, the cosine ratio of an acute $\measuredangle A$ is $\cos A = \frac{\text{length of adjacent side}}{\text{length of hypotenuse}}$. In the triangle at right, $\cos A = \frac{AB}{AC} = \frac{4}{5}$. (p. 241)

counter-clockwise Counter-clockwise identifies the direction of rotation shown in the diagram at right. The opposite of counter-clockwise is clockwise. (p. 34)

counterexample An example showing that a generalization has at least one exception; that is, a situation in which the statement is false. For example, the number 4 is a counterexample to the statement "All even numbers are greater than 7." (p. 238)

cross-section The intersection of a three-dimensional solid and a plane. The cross-sections of a cone (and double-cone) are called "conic sections." (pp. 538 and 578)

cube A polyhedron all of whose faces are squares. (p. 527)

cylinder (circular) A three-dimensional figure that consists of two parallel congruent circular regions (called *bases*) and a lateral surface containing segments connecting each point on the circular boundary of one base to the corresponding point on the circular boundary of the other. (pp. 30 and 451)

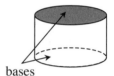

bases

decagon A polygon with ten sides. (p. 393)

delta (Δ) A Greek letter that is often used to represent a difference. Its uses include Δx and Δy, which represent the lengths of the horizontal and vertical legs of a slope triangle, respectively. (p. 47)

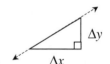

dependent events Two events are dependent if the outcome of one event affects the probability of the other event. For example, assume you will randomly select two cards, one at a time, without replacement from a deck of cards. The probability that the first card is red is $\frac{26}{52} = \frac{1}{2}$ because 26 of the 52 cards are red. However, the probability of the second card now depends on the result of the first selection. If the first card was red, there are now 25 red cards remaining in a deck of 51 cards, and the probability that the second card is red is $\frac{25}{51}$. The second event (selecting the second card) is dependent on the first event (selecting the first card). (p. 207)

diagonal In a polygon, it is a segment that connects two vertices of the polygon but is not a side of the polygon. (p. 400)

diagonal

diameter A line segment drawn through the center of a circle with both endpoints on the circle. The length of a diameter is usually denoted *d*. Note that the length of a circle's diameter is twice the length of its radius. *See* circle. (pp. 341 and 485)

626

dilation   A transformation which produces a figure similar to the original by proportionally shrinking or stretching the figure. In a dilation, a shape is stretched (or compressed) proportionally from a point, called the point of dilation. (p. 138)

dimension   (a) Flat figures have two dimensions (which can be labeled "base" and "height"); (b) Solids have three dimensions (such as "width," "height," and "depth"). (pp. 98 and 452)

directrix   A line that, along with a point (called a focus), defines a conic section (such as a parabola). (p. 574)

dissection The process of dividing a flat shape or solid into parts that have no interior points in common.

dodecahedron   A polyhedron with twelve faces. (p. 529)

double cone   Two cones placed apex to apex so that their bases are parallel. Generally we think of the cones as extending to infinity beyond their bases. (p. 588)

Double cone

edge   In three dimensions, a line segment formed by the intersection of two faces of a polyhedron. (p. 527)

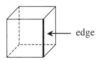

edge

ellipse   A type of conic section. An example is shown at right. The sum of the distances between each point on an ellipse and two fixed points (called the foci) is constant. (p. 585)

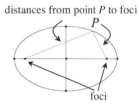

distances from point *P* to foci

foci

endpoints   See *line segment*.

enlarge   To enlarge something is to increase its size. For this course, an enlargement is a dilation of a figure whose result is similar to but larger than the original. (p. 140)

equator   If we represent the Earth as a sphere, the equator is the great circle equally distant between the North and South poles. The equator divides the Earth into two halves called the Northern and Southern hemispheres. The equator also marks a latitude of 0°. *See* latitude. (p. 549)

equilateral   A polygon is equilateral if all its sides have equal length. The word "equilateral" comes from "equi" (meaning "equal") and "lateral" (meaning "side"). Equilateral triangles not only have sides of equal length, but also angles of equal measure. However, a polygon with more than three sides can be equilateral without having congruent angles. For example, see the rhombus at right. (pp. 351 and 396)

exact answer An answer that is precisely accurate and not approximate. For example, if the length of a side of a triangle is exactly $\sqrt{10}$, the exact answer to the question "How long is that side of the triangle?" would be $\sqrt{10}$, while 3.162 would be approximate answer. (p. 123)

expected value For this course, the expected value of a game is the average amount expected to be won or lost on each play of the game if the game is played many times. The expected value need not be a value one could actually win on a single play of the game. For example, if you play a game where you roll a die and win one point for every dot on the face that comes up, the expected value of this game is $\frac{1+2+3+4+5+6}{6} = 3.5$ points for each play. (p. 516)

exterior angle When a side of a polygon is extended to form an angle with an adjacent side outside of the polygon, that angle is called an exterior angle. For example, the angles marked with letters in the diagram at right are exterior angles of the quadrilateral. Note that an exterior angle of a polygon is always adjacent and supplementary to an interior angle of that polygon. The sum of the exterior angles of a convex polygon is always 360°. (pp. 403 and 407)

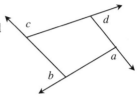

face of a polyhedron One of the flat surfaces of a polyhedron, including the base(s). (p. 527)

fair game For the purposes of this course, a fair game has an expected value of 0. Therefore, over many plays, a person would expect to neither win nor lose points or money by playing a fair game. (p. 510)

flowchart A diagram showing an argument for a conclusion from certain evidence. A flowchart uses ovals connected by arrows to show the logical structure of the argument. When each oval has a reason stated next to it showing how the evidence leads to that conclusion, the flowchart represents a proof. See the example at right. (p. 167)

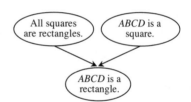

focus (plural: foci) A point that, along with a line (called a directrix), can be used to define all or part of a conic section (such as a parabola). (pp. 574 and 581)

fraction buster "Fraction busting" is a method of simplifying equations involving fractions. Fraction busting uses the Multiplicative Property of Equality to alter the equation so that no fractions remain. To use this method, multiply both sides of an equation by the common denominator of all the fractions in the equation. The result will be an equivalent equation with no fractions. For example, when given the equation $\frac{x}{7} + 2 = \frac{x}{3}$, we can multiply both sides by the "fraction buster" 21. The resulting equation, $3x + 42 = 7x$, is equivalent to the original but contains no fractions. (p. 512)

function A relation in which for each input value there is one and only one output value. For example, the relation $f(x) = x + 4$ is a function; for each input value (x) there is exactly one output value. In terms of ordered pairs (x, y), no two ordered pairs of a function have the same first member (x). (p. 589)

graph   A graph represents numerical information spatially. The numbers may come from a table, situation (pattern), rule (equation or inequality), or figure. Most of the graphs in this course show points, lines, figures, and/or curves on a two-dimensional coordinate system like the one at right.

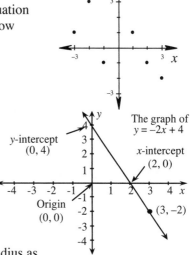

A complete graph includes all the necessary information about a line or a curve. To be complete, a graph must have the following components: (1) the *x*-axis and *y*-axis labeled, clearly showing the scale; (2) the equation of the graph written near the line or curve; (3) the line or curve extended as far as possible on the graph with arrows if the line or curve continues beyond the axes; (4) the coordinates of all special points, such as *x*- and *y*-intercepts, shown in (x, y) form. (p. 29)

y-intercept (0, 4)

The graph of $y = -2x + 4$

x-intercept (2, 0)

Origin (0, 0)

(3, -2)

great circle   A cross-section of a sphere that has the same radius as the sphere. If the Earth is represented with a sphere, the equator is an example of a great circle. (p. 549)

golden ratio   The number $\frac{1+\sqrt{5}}{2}$, which is often labeled with the Greek letter phi (ϕ), pronounced "fee." (p. 599)

height   (a) Triangle: the length of a segment that connects a vertex of the triangle to a line containing the opposite base (side) and is perpendicular to that line; (b) Trapezoid: the length of any segment that connects a point on one base of the trapezoid to the line containing the opposite base and is perpendicular to that line; (c) Parallelogram (includes rectangle, rhombus, and square): the length of any segment that connects a point on one base of the parallelogram to the line containing the opposite base and is perpendicular to that line; (d) Pyramid and cone: the length of the segment that connects the apex to a point in the plane containing the figure's base and is perpendicular to that plane; (e) Prism or cylinder: the length of a segment that connects one base of the figure to the plane containing the other base and is perpendicular to that plane. Some texts make a distinction between height and altitude, where the altitude is the segment described in the definition above and the height is its length. (pp. 106, 110, 448, 534, and 547)

hemisphere   A great circle of a sphere divides it into two congruent parts, each of which is called a hemisphere. If the Earth is represented as a sphere, the Northern hemisphere is the portion of the Earth north of (and including) the equator. Likewise, the Southern hemisphere is the portion of the Earth south of (and including) the equator. (p. 549)

heptagon   A polygon with seven sides. (p. 393)

heptahedron   A polyhedron with seven faces. (p. 529)

hexagon   A polygon with six sides. (pp. 43 and 393)

hexahedron   A polyhedron with six faces. A regular hexahedron is a cube. (p. 529)

HL ≅ (Triangle Shortcut) If the hypotenuse and one leg of one right triangle are congruent to the hypotenuse and corresponding leg of another right triangle, the two right triangles are congruent. Note that this congruence conjecture applies only to right triangles. (p. 299)

hypotenuse The longest side of a right triangle (the side opposite the right angle). (pp. 119 and 241)

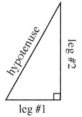

hypothesis A conjecture (or educated guess) in science.

icosahedron A polyhedron with twenty faces. (p. 529)

"If ..., then ..." statement A statement written in the form "If ..., then ...." Also known as a conditional statement. (p. 108)

image The shape that results from a transformation, such as a translation, rotation, reflection, or dilation. (p. 34)

incenter The center of the circle inscribed in a triangle. It can be found by locating the point at which the angle bisectors of a triangle intersect. The other points of concurrency studied in this course are the centroid and the circumcenter. (p. 193)

independent events If the outcome of a probabilistic event does not affect the probability of another event, the events are independent. For example, assume you plan to roll a normal six-sided die twice and want to know the probability of rolling a 1 twice. The result of the first roll does not affect the probability of rolling a 1 on the second roll. Since the probability of rolling a 1 on the first roll is $\frac{1}{6}$ and the probability of rolling a 1 on the second roll is also $\frac{1}{6}$, then the probability of rolling two 1s in a row is $\frac{1}{6} \cdot \frac{1}{6} = \frac{1}{36}$. (p. 207)

indirect proof *See* proof by contradiction.

inequality symbols The symbols < (less than), > (greater than), ≤ (less than or equal to), and ≥ (greater than or equal to).

inscribed angle An angle with its vertex on the circle and sides intersecting the circle at two distinct points. (p. 490)

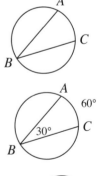

Inscribed Angle Theorem The measure of an inscribed angle is half the measure of its intercepted arc. Likewise, the measure of an intercepted arc is twice the measure of an inscribed angle whose sides pass through the endpoints of the arc. In the figure at right, if $m\overset{\frown}{AC} = 60°$, then $m\angle ABC = 30°$. (p. 494)

inscribed circle A circle is inscribed in a polygon if each side of the polygon intersects the circle at exactly one point.

$\odot B$ is inscribed in the pentagon.

$\odot C$ circumscribes the pentagon.

inscribed polygon A polygon is inscribed in a circle if each vertex of the polygon lies on the circle. (p. 461)

630

integers   The set of numbers { . . . –3, –2, –1, 0, 1, 2, 3, . . . }.

intercepted arc   The arc of a circle bounded by the points where the two sides of an inscribed angle meet the circle. In the circle at right, $\measuredangle ABC$ intercepts $\overset{\frown}{AC}$. (p. 489)

interior angle (of a polygon)   An angle formed by two consecutive sides of the polygon. The vertex of the angle is a vertex (corner) of the polygon. (p. 402)

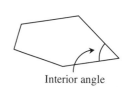
Interior angle

Interior Angle Sum Conjecture   The sum of the interior angles of a polygon with n-sides is $180°(n-2)$. (p. 407)

inverse trigonometric functions   When the ratio between the lengths of two sides of a right triangle is known, $\sin^{-1}$, $\cos^{-1}$, or $\tan^{-1}$ can be used to find the measure of one of the triangle's acute angles. (p. 248)

irrational numbers   The set of numbers that cannot be expressed in the form $\frac{a}{b}$, where a and b are integers and $b \neq 0$. For example, π and $\sqrt{2}$ are irrational numbers. (pp. 115 and 467)

isosceles trapezoid   A trapezoid with a pair of equal base angles (from the same base). Note that this will cause the non-parallel sides to be congruent. (pp. 77 and 371)

isosceles triangle   A triangle with two sides of equal length. (p. 48)

Isosceles Triangle Theorem   If a triangle is isosceles, then the base angles (which are opposite the congruent sides) are congruent. For example, if $\triangle ABC$ is isosceles with $\overline{BA} \cong \overline{BC}$, then the angles opposite these sides are congruent; that is, $\measuredangle A \cong \measuredangle C$.

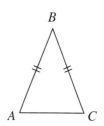

iteration   The act of repeating an action over and over. (p. 598)

justify   To give a logical reason supporting a statement or step in a proof. More generally, to use facts, definitions, rules, and/or previously proven conjectures in an organized sequence to convincingly demonstrate that your claim (or your answer) is valid (true). (p. 167)

kite   A quadrilateral with two distinct pairs of consecutive congruent sides. (p. 353)

lateral surface area   The sum of the areas of all of the faces of a prism or pyramid, with the exception of the base(s). (p. 534)

latitude   An angular measure (in degrees) that indicates how far north or south of the equator a position on the Earth is. All points on the equator have a latitude of 0°. The North Pole have a latitude of 90°, while the South Pole has a latitude of -90°. (p. 549)

Law of Cosines   For any $\triangle ABC$ with sides a, b, and c opposite $\measuredangle A$, $\measuredangle B$, and $\measuredangle C$ respectively, it is always true that $c^2 = a^2 + b^2 - 2ab \cdot \cos(C)$, $b^2 = a^2 + c^2 - 2ac \cdot \cos(B)$, and $a^2 = b^2 + c^2 - 2bc \cdot \cos(A)$. (p. 267)

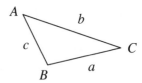

Law of Sines   For any $\triangle ABC$ with sides a, b, and c opposite $\measuredangle A$, $\measuredangle B$, and $\measuredangle C$ respectively, it is always true that $\frac{\sin A}{a} = \frac{\sin B}{b} = \frac{\sin C}{c}$. (p. 264)

legs   The two sides of a right triangle that form the right angle. Note that legs of a right triangle are always shorter than its hypotenuse. (p. 119)

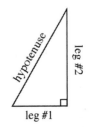

line   A line is an undefined term in geometry. It is one-dimensional and extends without end in two directions. It is made up of points and has no thickness. A line can be named with a letter (such as l), but also can be labeled using two points on the line, such as $\overleftrightarrow{AB}$ below. (p. 81)

line of symmetry   A line that divides a shape into two congruent parts that are mirror images of each other. If you fold a shape over its line of symmetry, the shapes on both sides of the line will match each other perfectly. Some shapes have more than one line of symmetry, such as the example at right. Other shapes may only have one line of symmetry or no lines of symmetry. (p. 5)

line segment   The portion of a line between two points. A line segment is named using its endpoints. For example, the line segment at right can be named either $\overline{AB}$ or $\overline{BA}$. (p. 81)

linear scale factor   (a) In the case of two similar two-dimensional figures, the linear scale factor is the ratio of the lengths of any pair of corresponding sides. This means that once it is determined that two figures are similar, all of their pairs of corresponding sides have the same ratio. The linear scale factor can also be called the *ratio of similarity*. When the ratio is comparing a figure and its image after a dilation, this ratio can also be called the *zoom factor*. (b) In the case of two similar polyhedra, the linear scale factor is the ratio of the lengths of any pair of corresponding edges. (pp. 140, 415, 457)

logical argument   A logical sequence of statements and reasons that lead to a conclusion. A logical argument can be written in a paragraph, represented with a flowchart, or documented in a two-column proof. (pp. 10, 353, and 366)

longitude   An angular measure (in degrees) that indicates how far west or east of the Prime Meridian a position on the Earth is. Lines of longitude (which are actually circles) are all great circles that pass through the North and South Poles. (pp. 549 and 551)

major arc   An arc with measure greater than 180°. Each major arc has a corresponding minor arc that has a measure that is less than 180°. A major arc is named with three letters on the arc. For example, the highlighted arc at right is the major arc $\overset{\frown}{ABC}$. (p. 485)

Geometry Connections

mat plan   A top (or bottom) view of a multiple cube solid. The number in each square is the number of cubes in that stack. For example, the mat plan at right represents the solid at the far right. (p. 443)

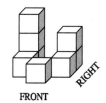

measurement   For the purposes of this course, a measurement is an indication of the size or magnitude of a geometric figure. For example, an appropriate measurement of a line segment would be its length. Appropriate measurements of a square would include not only the length of a side, but also its area and perimeter. The measure of an angle represents the number of degrees of rotation from one ray to the other about the vertex. (p. 15)

median   A line segment that connects a vertex of a triangle with the midpoint of side opposite to the vertex. For example, since D is a midpoint of $\overline{BC}$, then $\overline{AD}$ is a median of $\triangle ABC$. The three medians of a triangle intersect at a point called the centroid. (p. 471)

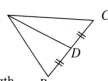

midpoint   A point that divides a segment into two segments of equal length. For example, D is the midpoint of $\overline{BC}$ in $\triangle ABC$ at right. (p. 381)

midsegment   A segment joining the midpoints of two sides of a triangle. (p. 371)

minor arc   An arc with measure less than 180°. Minor arcs are named using the endpoints of the arc. For example, the highlighted arc at right is named $\overset{\frown}{AC}$. (p. 485)

Möbius strip   A one-sided surface in a closed loop which can be formed by giving a rectangular strip of paper a half-twist and affixing its ends. (p. 8)

monomial   An expression with only one term. It can be a number, a variable, or the product of a number and one or more variables. For example, 7, $3x$, $-4ab$, and $3x^2y$ are each monomials.

n-gon   A polygon with n sides. A polygon is often referred to as an "n-gon" when we do not yet know the value of n. (p. 422)

net   A diagram that, when folded, forms the surface of a three-dimensional solid. A net is essentially the faces of a solid laid flat. It is one of several ways to represent a three-dimensional diagram. The diagram at right is a representation of the solid at the far right. Note that the shaded region of the net indicates the (bottom) base of the solid. (p. 446)

nonagon   A polygon with nine sides. (p. 393)

nonahedron   A polyhedron with nine faces. (p. 529)

non-convex polygon   A polygon that is not convex. An example of a non-convex polygon is shown at right. *See* convex. (pp. 395 and 396)

oblique   A prism (or cylinder) is oblique when its lateral surface(s) are not perpendicular to the base. A pyramid (or cone) is oblique when its apex is not directly above the center of the base. (pp. 451 and 538)

obtuse angle   Any angle that measures between (but not including) 90° and 180°. (p. 24)

octagon   A polygon with eight sides. (p. 393)

octahedron   A polyhedron with eight faces. In a regular octahedron, all of the faces are equilateral triangles. (p. 529)

opposite   In a figure, opposite means "across from." For example, in a triangle, an "opposite side" is the side that is across from a particular angle and is not a side of the angle. For example, the side $\overline{AB}$ in $\triangle ABC$ at right is opposite $\angle C$, while $\overline{AC}$ is opposite the right angle. (p. 241)

orientation   In this course, orientation refers to the placement and alignment of a figure in relation to others or an object of reference (such as coordinate axes). Unless it has rotation symmetry, the orientation of a figure changes when it is rotated (turned) less than 360°. Also, the orientation of a shape changes when it is reflected, except when reflected across a line of symmetry. (p. 241)

parabola   The graph of a quadratic function (any equation that can be written in the form $y = ax^2 + bx + c$) is a parabola. There are several other ways to find a parabola, including the intersection of a right circular cone with a flat surface parallel to an edge of the cone. (p. 574)

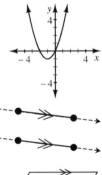

parallel lines   Two lines on a flat surface are parallel if they never intersect. Two line segments on a flat surface are parallel if the lines they lie on never intersect. There is a constant distance between two parallel lines (or line segments). Identical arrow markings are used to note parallel lines or line segments. (p. 47)

parallelogram   A quadrilateral with two pairs of parallel sides. (p. 361)

pentagon   A polygon with five sides. (p. 393)

pentahedron   A polyhedron with five faces. (p. 529)

perimeter   The distance around the exterior of a figure on a flat surface. For a polygon, the perimeter is the sum of the lengths of its sides. The perimeter of a circle is also called a circumference. (p. 15)

perpendicular   Two rays, line segments, or lines that meet (intersect) to form a right angle (90°) are called perpendicular. A line and a flat surface can also be perpendicular if the line does not lie on the flat surface but intersects it and forms a right angle with every line on the flat surface passing through the point of intersection. A small square at the point of intersection of two lines or segments indicates that the lines are perpendicular. (p. 47)

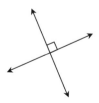

phi (ϕ)   A Greek letter, pronounced "fee," that represents the golden ratio, $\frac{1+\sqrt{5}}{2}$. (p. 599)

pi (π)   The ratio of the circumference (C) of the circle to its diameter (d). For every circle, $\pi = \frac{\text{circumference}}{\text{diameter}} = \frac{C}{d}$. Numbers such as 3.14, 3.14159, or $\frac{22}{7}$ are approximations of π. (p. 426)

plane   A plane is an undefined term in geometry. It is a two-dimensional flat surface that extends without end. It is made up of points and has no thickness.

platonic solid   A convex regular polyhedron. All faces of a platonic solid are congruent, regular polygons. There are only five possible platonic solids: tetrahedron (with four faces of equilateral triangles), cube (with six faces of squares), octahedron (with eight faces of equilateral triangles), dodecahedron (with twelve faces of regular pentagons), and icosahedron (with twenty faces of equilateral triangles). (pg. 529)

polygon   A two-dimensional closed figure of three or more line segments (sides) connected end to end. Each segment is a side and only intersects the endpoints of its two adjacent sides. Each point of intersection is a vertex. At right are two examples of polygons. (p. 42, 393, and 396)

polyhedron (plural: polyhedra)   A three-dimensional object with no holes that is bounded by polygons. The polygons are joined at their sides, forming the edges of the polyhedron. Each polygon is a face of the polyhedron. (p. 448)

postulates   Statements accepted as true without proof. Also known as "axioms."

preimage   The original figure in a transformation.

prism   A three-dimensional figure that consists of two parallel congruent polygons (called *bases*) and a lateral surface containing segments connecting each point on each side of one base to the corresponding point on the other base. The lateral surface of a prism consists of parallelograms. (p. 448)

probability   A number that represents how likely an event is to happen. When a event has a finite number of equally-likely outcomes, the probability that one of those outcomes, called A, will occur is expressed as a ratio and written as:

$$P(A) = \frac{\text{number of successful outcomes}}{\text{total number of possible outcomes}}.$$

For example, when flipping a coin, the probability of getting tails, P(tails), is $\frac{1}{2}$ because there is only one tail (successful outcome) out of the two possible equally likely outcomes (a head and a tail). Probability can be written as a ratio, decimal, or percent. A probability of 0 (or 0%) indicates that it is impossible for the event to occur, while a probability of 1 (or 100%) indicates that the event must occur. Events that "might happen" will have values somewhere between 0 and 1 (or between 0% and 100%). (p. 60)

proof   A convincing logical argument that uses definitions and previously proven conjectures in an organized sequence to show that a conjecture is true. A proof can be written in a paragraph, represented with a flowchart, or documented in a formal two-column proof. (p. 10, 353, and 366)

proof by contradiction A proof that begins by assuming that an assertion is true and then shows that this assumption leads to a contradiction of a known fact. This demonstrates that the assertion is false. Also known as an *indirect proof*. (p. 96)

proportional equation An equation stating that two ratios (fractions) are equal. (p. 145)

protractor A geometric tool used for physically measuring the number of degrees in an angle. (p. 22)

pyramid A polyhedron with a polygonal base formed by connecting each point of the base to a single given point (the **apex**) that is above or below the flat surface containing the base. Each triangular lateral face of the pyramid is formed by the segments from the apex to the endpoints of a side of the base and the side itself. A tetrahedron is a special pyramid because any face can act as its base. (p. 533)

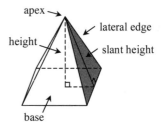

Pythagorean Theorem The statement relating the lengths of the legs of a right triangle to the length of the hypotenuse: $(\text{leg \#1})^2 + (\text{leg \#2})^2 = \text{hypotenuse}^2$. The Pythagorean Theorem is powerful because if you know the lengths of any two sides of a right triangle, you can use this relationship to find the length of the third side. (p. 123)

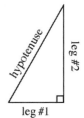

Pythagorean triple Any three positive integers a, b, and c that make the relationship $a^2 + b^2 = c^2$ true. Commonly used Pythagorean triples include 3, 4, 5 and 5, 12, 13. (p. 260)

quadratic equation An equation that can be written in the form $ax^2 + bx + c = 0$, where a, b, and c are real numbers and a is nonzero. A quadratic equation written in this form is said to be in standard form. For example, $3x^2 - 4x + 7.5 = 0$ is a quadratic equation. (p. 163)

Quadratic Formula The Quadratic Formula states that if $ax^2 + bx + c = 0$ and $a \neq 0$, then $x = \frac{-b \pm \sqrt{b^2 - 4ac}}{2a}$. For example, if $5x^2 + 9x + 3 = 0$, then $x = \frac{-9 \pm \sqrt{9^2 - 4(5)(3)}}{2(5)} = \frac{-9 \pm \sqrt{21}}{10}$. (p. 163)

quadrilateral A polygon with four sides.

radius (plural: radii) Of a circle: A line segment drawn from the center of a circle to a point on the circle. (p. 341); Of a regular polygon: A line segment that connects the center of a regular polygon with a vertex. The length of a radius is usually denoted r. (p. 410)

random An event is random if its result cannot be known (and can only be guessed) until the event is completed. For example, the flip of a fair coin is random because the coin can either land on heads or tails and the outcome cannot be known for certain until after the coin is flipped.

Geometry Connections

ratio A ratio compares two quantities by division. A ratio can be written using a colon, but is more often written as a fraction. For example, in the two similar triangles at right, a ratio can be used to compare the length of $\overline{BC}$ in $\triangle ABC$ with the length of $\overline{EF}$ in $\triangle DEF$. This ratio can be written as 5:11 or as the fraction $\frac{5}{11}$. (p. 60)

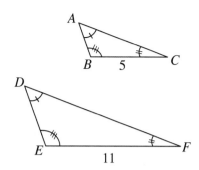

ratio of similarity The ratio of any pair of corresponding sides of two similar figures. This means that once it is determined that two figures are similar, all of their pairs of corresponding sides have the same ratio. For example, for the similar triangles $\triangle ABC$ and $\triangle DEF$ above, the ratio of similarity is $\frac{5}{11}$. The ratio of similarity can also be called the *linear scale factor*. When the ratio is comparing a figure and its image after a dilation, this ratio can also be called the *zoom factor*. (pp. 140, 415, and 457)

rationalizing the denominator Rewriting a fractional expression that has radicals in the denominator in order to eliminate them. (p. 252)

ray A ray is part of a line that starts at one point and extends without end in one direction. In the example at right, ray $\overline{AB}$ is part of $\overleftrightarrow{AB}$ that starts at A and contains all of the points of $\overleftrightarrow{AB}$ that are on the same side of A as point B, including A. Point A is the endpoint of $\overrightarrow{AB}$.

rectangle A quadrilateral with four right angles. (pp. 5 and 361)

reflection A transformation across a line that produces a mirror image of the original (preimage) shape. The reflection is called the "image" of the original figure. The line is called a "line of reflection." See the example at right. Note that a reflection is also sometimes referred to as a "flip." (pp. 34 and 38)

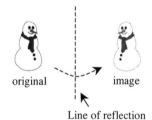

original   image

Line of reflection

Reflexive Property The Reflexive Property states that any expression is always equal to itself. That is, $a = a$. This property is often useful when proving that two triangles that share a side or an angle are congruent. For example, in the diagram at right, since $\triangle ABD$ and $\triangle CBD$ share a side ($\overline{BD}$), the Reflexive Property justifies that $BD = BD$. (p. 354)

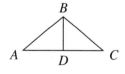

reflection symmetry *See* symmetry.

regular A polygon is regular if it is a convex polygon with congruent angles and congruent sides. For example, the shape at right is a regular hexagon. (pp. 42, 351, and 396)

REGULAR HEXAGON

relation   An equation that relates two or more variables. For example, $y = 3x - 2$ and $x^2 + y^2 = 9$ are both relations. A relation can also be thought of as a set of ordered pairs. (p. 589)

relationship   For this course, a relationship is a way that two objects (such as two line segments or two triangles) are connected. When you know that the relationship holds between two objects, learning about one object can give you information about the other. Relationships can be described in two ways: a geometric relationship (such as a pair of vertical angles or two line segments that are parallel) and a relationship between the measures (such as two angles that are complementary or two sides of a triangle that have the same length). Common geometric relationships between two figures include being similar (when two figures have the shape, but not necessarily the same size) and being congruent (when two figures have the same shape and the same size). (pp. 76, 91, 150, and 159)

reasoning   *See* logical reasoning *and* proof.

remote interior angles   If a triangle has an exterior angle, the remote interior angles are the two angles not adjacent to the exterior angle. Also called "opposite interior angles." (p. 397)

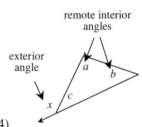

rhombus   A quadrilateral with four congruent sides. (pp. 361 and 464)

right angle   An angle that measures 90°. A small square is used to note a right angle, as shown in the example at right. (pp. 22 and 24)

right triangle   A triangle that has one right angle. The side of a right triangle opposite the right angle is called the "hypotenuse," and the two sides adjacent to the right angle are called "legs." (p. 119)

rigid motions   Movements of figures that preserve their shape and size. Also called "rigid transformations." Examples of rigid motions are reflections, rotations, and translations. Also called "isometries." (p. 34)

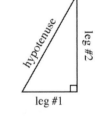

rotation   A transformation that turns all of the points in the original (preimage) figure the same number of degrees around a fixed center point (such as the origin on a graph). The result is called the "image" of the original figure. The point that the shape is rotated about is called the "center of rotation." To define a rotation, you need to state the measure of turn (in degrees), the direction the shape is turned (such as clockwise or counter-clockwise), and the center of rotation. See the example at right. Note that a rotation is also sometimes referred to as a "turn." (p. 34)

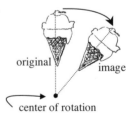

rotation symmetry   *See* symmetry.

$r : r^2 : r^3$ **Ratios of Similarity** When a two-dimensional figure is enlarged (or reduced) proportionally, its lengths and area change. If the linear scale factor is r, then all lengths (such as sides, perimeter, and heights) of the original (preimage) figure is multiplied by a factor of r while the area is multiplied by a factor of r^2.

When a polyhedron is enlarged (or reduced) proportionally, its lengths, surface area, and volume also change. If the linear scale factor is r, then the surface area is multiplied by a factor of r^2 and the volume is multiplied by a factor of r^3. Thus, if a polyhedron is enlarged proportionally by a linear scale factor of r, then the new edge lengths, surface area, and volume can be found using the relationships below. (p. 457)

New edge length = $r \cdot$ (corresponding edge length of original polyhedron)

New surface area = $r^2 \cdot$ (original surface area)

New volume = $r^3 \cdot$ (original volume)

This principle also applies to other solids such cylinders, cones, and spheres.

same-side interior angles Two angles between two lines and on the same side of a third line that intersects them (called a transversal). The shaded angles in the diagram at right are an example of a pair of same-side interior angles. Note that if the two lines that are cut by the transversal are parallel, then the two angles are supplementary (add up to 180°). (p. 91)

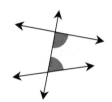

sample space The total number of outcomes that can happen in an event. The sample space for rolling a standard 6-sided die is the set {1, 2, 3, 4, 5, 6} because those are the only possible outcomes.

SAS ≅ (Triangle Congruence) Two triangles are congruent if two sides and their included angle of one triangle are congruent to the corresponding two sides and included angle of another triangle. Also referred to as "SAS Congruence" or "SAS ≅". (p. 299)

SAS ~ (Triangle Similarity) If two triangles have two pairs of corresponding sides that are proportional and have congruent included angles, then the triangles are similar. Also referred to as "SAS Similarity" or "SAS ~". (p. 171)

scale The ratio between a length of the representation (such as a map, model, or diagram) and the corresponding length of the actual object. For example, the map of a city may use one inch to represent one mile.

scalene triangle A triangle with no congruent sides. (p. 55)

secant A line that intersects a circle at two distinct points. (p. 557)

sector

sector A region formed by two radii of a central angle and the arc between their endpoints on the circle. You can think of it as a portion of a circle and its interior, resembling a piece of pizza. (p. 426)

semicircle In a circle, a semicircle is an arc with endpoints that are endpoints of any diameter of the circle. It is a half circle and has a measure of 180°. (p. 492)

side of an angle One of the two rays that form an angle.

side of a polygon *See* polygon.

similar figures Two shapes are similar if they have exactly the same shape but are not necessarily the same size. The symbol for similar is ~ . (pp. 136 and 155)

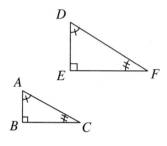

similarity statement A statement that indicates that two figures are similar. The order of the letters in the names of the shapes determine which sides and angles correspond to each other. For example, if $\triangle ABC \sim \triangle DEF$, then $\angle A$ must correspond to $\angle D$ and $\overline{AB}$ must correspond to $\overline{DE}$. (p. 150)

sine ratio In a right triangle, the sine ratio of an acute $\angle A$ is $\sin A = \frac{\text{length of opposite side}}{\text{length of hypotenuse}}$. In the triangle at right, $\sin A = \frac{BC}{AC} = \frac{3}{5}$. (p. 241)

skew lines Lines that do not lie in the same flat surface.

slant height (a) Pyramid: The height of a lateral face drawn from the apex; (b) Cone: The distance from the apex of the cone to any point on the circular boundary of the base. (p. 534)

slide See *translation.*

slope A ratio that describes how steep (or flat) a line is. Slope can be positive, negative, or even zero, but a straight line has only one slope. Slope is the ratio $\frac{\text{vertical change}}{\text{horizontal change}}$ or $\frac{\text{change in } y \text{ value}}{\text{change in } x \text{ value}}$, sometimes written $\frac{\Delta y}{\Delta x}$. When the equation of a line is written in $y = mx + b$ form, m is the slope of the line. A line has positive slope if it slopes upward from left to right on a graph, negative slope if it slopes downward from left to right, zero slope if it is horizontal, and undefined slope if it is vertical. Parallel lines have equal slopes, and the slopes of perpendicular lines are opposite reciprocals of each other (e.g. , $\frac{3}{5}$ and $-\frac{5}{3}$). (pp. 47 and 190)

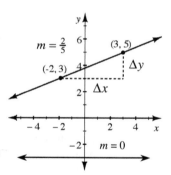

slope angle *See* slope triangle.

slope-intercept form A form of a linear equation: $y = mx + b$. In this form, m is the slope and the point $(0, b)$ is the y-intercept. (p. 123)

slope triangle A right triangle with legs (parallel to the x- and y-axes) that meet the hypotenuse at two points on a given line or line segment. The lengths of the legs of the slope triangle can be used to find the slope of the line. The angle (θ) formed in a slope triangle by the hypotenuse and the horizontal leg is called the slope angle. (pp. 47 and 190)

solid A closed three-dimensional shape and all of its interior points. Examples include regions bounded by pyramids, cylinders, and spheres. (p. 443)

Geometry Connections

solve To find all the solutions to an equation or an inequality (or a system of equations or inequalities). For example, solving the equation $x^2 = 9$ gives the solutions $x = 3$ and $x = -3$. The solution(s) may be number(s), variable(s), or an expression. (pp. 19, 87, and 163)

space The set of all points in three-dimensions.

special right triangles A right triangle with particular notable features that can be used to solve problems. Sometimes, these triangles can be recognized by the angles, such as the $45°$-$45°$-$90°$ triangle (also known as an "isosceles right triangle") and a $30°$-$60°$-$90°$ triangle. Other right triangles are special because the length of each side is an integer, such as a 3, 4, 5 triangle or a 5, 12, 13 triangle. (p. 260)

sphere The set of all points in space that are the same distance from a fixed point. The fixed point is the center of the sphere and the distance is its radius. (p. 555)

Sphere

square A quadrilateral with four right angles and four congruent sides. (p. 361)

Square

square root A number a is a square root of b if $a^2 = b$. For example, the number 9 has two square roots, 3 and –3. A negative number has no real square roots; a positive number has two; and zero has just one square root, namely, itself. In a geometric context, the principal square root of a number x (written $\sqrt{x}$) represents the length of a side of a square with area x. For example, $\sqrt{16} = 4$. Therefore if the side of a square has a length of 4 units, then its area is 16 square units. (p. 115)

SSS $\cong$ (Triangle Congruence) Two triangles are congruent if all three pairs of corresponding sides are congruent. (p. 299)

SSS ~ (Triangle Similarity) If two triangles have all three pairs of corresponding sides that are proportional (this means that the ratios of corresponding sides are equal), then the triangles are similar. (pp. 155 and 171)

straight angle An angle that measures $180°$. This occurs when the rays of the angle point in opposite directions, forming a line. (p. 24)

$180°$

straightedge A tool used as a guide to draw lines, rays, and segments. (p. 468)

statement A recording of fact to present evidence in a logical argument (proof). (p. 459)

Substitution Method A method for solving a system of equations by replacing one variable with an expression involving the remaining variable(s). For example, in the system of equations at righ t the first equation tells you that y is equal to $-3x + 5$. We can substitute $-3x + 5$ in for y in the second equation to get $2(-3x + 5) + 10x = 18$, then solve this equation to find that $x = 2$. Once we have x, we substitute that value back into either of the original equations to find that $y = -1$. (p. 87)

$$y = -3x + 5$$
$$2y + 10x = 18$$

Substitution Property The Substitution Property states that in an expression, one can replace a variable, number, or expression with something equal to it without altering the value of the whole. For example, if $x = 3$, then $5x - 7$ can be evaluated by replacing x with 3. This results in $5(3) - 7 = 8$. (p. 87)

supplementary angles Two angles a and b for which $a + b = 180°$. Each angle is called the supplement of the other. In the example at right, angles A and B are supplementary. Supplementary angles are often adjacent. For example, since $\angle LMN$ is a straight angle, then $\angle LMP$ and $\angle PMN$ are supplementary angles because $m\angle LMP + m\angle PMN = 180°$. (p. 74)

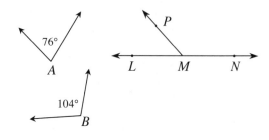

surface area The sum of all the area(s) of the surface(s) of a three-dimensional solid. For example, the surface area of a cylinder is the sum of the areas of its top base, its bottom base, and its lateral surface. (pp. 446 and 452)

symmetry (a) Rotation symmetry: A shape has rotation symmetry when it can be rotated for less than 360° and it appears not to change. For example the hexagon at right appears not to change if it is rotated 60° clockwise (or counterclockwise); (b) Reflection symmetry: A shape has reflection symmetry when it appears not to change after being reflected across a line. The regular hexagon at right also has reflection symmetry across any line drawn through opposite vertices, such as those shown in the diagram. *See* reflection, rotation, *and* line *of* symmetry. (pp. 43, 44, 45)

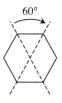

system of equations A system of equations is a set of equations with the same variables. Solving a system of equations means finding one or more solutions that make each of the equations in the system true. A solution to a system of equations gives a point of intersection of the graphs of the equations in the system. There may be zero, one, or several solutions to a system of equations. For example, (1.5, –3) is a solution to the system of linear equations at right; $x = 1.5$, $y = -3$ makes both of the equations true. Also, (1.5, –3) is a point of intersection of the graphs of these two equations. (p. 87)

$$y = 2x - 6$$
$$y = -2x$$

systematic list A list created by following a system (an orderly process). (p. 219)

tangent A line on the same flat surface as a circle that intersects the circle in exactly one point. A tangent of a circle is perpendicular to a radius of the circle at their point of intersection (also called the "point of tangency." (p. 496)

tangent ratio In a right triangle, the tangent ratio of an acute $\angle A$ is $\tan A = \frac{\text{length of opposite side}}{\text{length of adjacent side}}$. In the triangle at right, $\tan A = \frac{BC}{AB} = \frac{3}{4}$. (p. 200)

tetrahedron A polyhedron with four faces. The faces of a tetrahedron are triangles. In a regular tetrahedron, all of the faces are congruent equilateral triangles. (p. 533)

theorem   A conjecture that has been proven to be true. Some examples of theorems are the Pythagorean Theorem and the Triangle Angle Sum Theorem. (p. 367)

theta (θ)   A Greek letter that is often used to represent the measure of an angle. Other Greek letters used to represent the measure of an angle include alpha (α) and beta (β). (p. 190)

three-dimensional   An the object that has length, width, and depth. (p. 340)

transformation   This course studies four transformations: reflection, rotation, translation, and dilation. All of them preserve shape, and the first three preserve size. *See each term for its own definition.* (pp. 34 and 138)

translation   A transformation that preserves the size, shape, and orientation of a figure while sliding (moving) it to a new location. The result is called the "image" of the original figure (preimage). Note that a translation is sometimes referred to as a "slide." (p. 34)

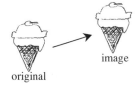

transversal   A line that intersects two or more other lines on a flat surface (plane). In this course, we often work with a transversal that intersects two parallel lines. (p. 79)

trapezoid   A quadrilateral with at least one pair of parallel sides. (pp. 107 and 361)

tree diagram   A structure used to organize the possible outcomes of two or more events. For example, the tree diagram at right represents the possible outcomes when a coin is flipped twice. (p. 219)

triangle   A polygon with three sides.

Triangle Angle Sum Theorem   The sum of the measures of the interior angles in any triangle is 180°. For example, in $\triangle ABC$ at right, $m\angle A + m\angle B + m\angle C = 180°$. (p. 99)

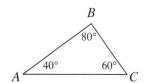

Triangle Congruence Conjectures   Conjectures that use the minimum number of congruent corresponding parts to prove that two triangles are congruent. They are: SSS $\cong$, SAS $\cong$, AAS $\cong$, ASA $\cong$, and HL $\cong$. (p. 299)

Triangle Inequality   In a triangle with side lengths a, b, and c, c must be less than the sum of a and b and greater than the difference of a and b. In the example at right, a is greater than b (that is, $a > b$), so the possible values for c are all numbers such that $c > a - b$ and $c < a + b$. (p. 119)

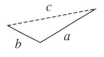

Triangle Midsegment Theorem   The segment that connects the midpoints of any two sides of a triangle measures half the length of the third side is and parallel to that side. (p. 371)

trigonometry   Literally, the "measure of triangles." In this course, this word is used to refer to the development of triangle tools such as trigonometric ratios (sine, cosine, and tangent) and the Laws of Sines and Cosines. (p. 192)

turn See *rotation*

two-column proof A form of proof in which statements are written in one column as a list and the reasons for the statements are written next to them in a second column. (p. 366)

two-dimensional A figure that that lies on a flat surface and that has length and width. *See* plane. (p. 340)

unit of measure A standard quantity (such as a centimeter, second, square foot, or gallon) that is used to measure and describe an object. A single object can be measured using different units of measure, which will usually yield different results. For example, a pencil may be 80 mm long, meaning that it is 80 times as long as a unit of 1 mm. However, the same pencil is 8 cm long, so that it is the same length as 8 cm laid end-to-end. (This is because 1 cm is the same length as 10 mm.) (p. 98)

Venn diagram A type of diagram used to classify objects. It is usually composed of two or more overlapping circles representing different conditions. An item is placed or represented in the Venn diagram in the appropriate position based on the conditions it meets. In the example of the Venn diagram at right, if an object meets one of two conditions, it is placed in region **A** or **C** but outside region **B**. If an object meets both conditions, it is placed in the intersection (**B**) of both circles. If an object does not meet either condition, it is placed outside of both circles (region **D**). (p. 51)

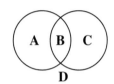

vertex (plural: vertices)   (a) For a two-dimensional geometric shape, a vertex is a point where two or more line segments or rays meet to form a "corner," such as in a polygon or angle. (b) For a three-dimensional polyhedron, a vertex is a point where the edges of the solid meet. (c) On a graph, a vertex can be used to describe the highest or lowest point on the graph of a parabola or absolute value function (depending on the graph's orientation). (pp. 74 and 527)

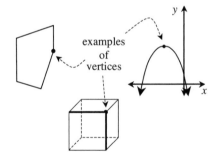

examples
of
vertices

vertical angles The two opposite (that is, non-adjacent) angles formed by two intersecting lines. "Vertical" is a relationship between pairs of angles, so one angle cannot be called vertical. Angles that form a vertical pair are always congruent. (pp. 75 and 91)

volume A measurement of the size of the three-dimensional region enclosed within an object. It is expressed as the number of 1x1x1 unit cubes (or parts of cubes) that fit inside a solid. (pp. 453 and 452)

weighted average In this course, weighted average is used to calculate expected value. The weighted average takes into account the relative probabilities of all possible outcomes and uses them to calculate the average win or loss of each game. *See* expected value *for an example.* (p. 508 and 516)

x-intercept   A point where a graph crosses the *x*-axis. A graph may have several *x*-intercepts, no *x*-intercepts, or just one.

y-intercept   A point where a graph crosses the *y*-axis. A function has at most one *y*-intercept; a relation may have several.

zoom factor   The amount each side of a figure is multiplied by when the figure is proportionally enlarged or reduced in size. It is written as the ratio of a length in the new figure (image) to a length in the original figure (preimage). For the triangles at right, the zoom factor is $\frac{8}{6}$ or $\frac{4}{3}$. See also "ratio of similarity" and "linear scale factor." (pp. 140 and 142)

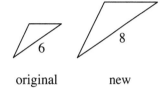

original   new

Index
Student Version

Many of the pages referenced here contain a definition or an example of the topic listed, often within the body of a Math Notes box. Others contain problems that develop or demonstrate the topic. It may be necessary to read the text on several pages to fully understand the topic. Also, some problems listed here are good examples of the topic and may not offer any explanation. The page numbers listed below reflect the pages in the Student Version. References to Math Notes boxes are bolded.

Greek Alphabet used in text

α (alpha), **190**
β (beta), **190**
Δ (delta), **47**, 599
θ (theta), **190**
φ (phi), 598, **599**

Number

30°-60°-90° triangle, **260**
30°-60°-90° triangles, 250
3-4-5 triangles, 255
45°-45°-90° triangle, **260**
45°-45°-90° triangles, 251

A

AA ~, **155**, **171**
AAS ≅, **299**
Absolute value, 453, 575
Acute angles, **24**
Adjacent leg, **241**
Adjacent side, 236
Alpha (α), **190**
Alternate interior angles, 84, **91**
Angle Notation, **190**
  alpha, **190**
  beta, **190**
  theta, **190**
Angles, 22, **24**
  acute, **24**
  alternate interior, 84, **91**
  central, 382, 394, 488, **490**
  circular, 22, **24**
  complementary, 74, **76**
  congruent, **76**
  consecutive, 370
  corresponding, **91**
  equal measure, **351**
  exterior, 358, 403
  inscribed, 489, **490**, **494**
  naming, **81**
  obtuse, **24**
  remote interior, 397
  right, 22, **24**
  same-side interior, 84, **91**
  slope, **190**
  straight, **24**
  supplementary, 74, **76**
  vertex, 74
  vertical, **91**

Apothem, **410**
Approximate, **123**, 240
Arc, 425, **426**, 430, 483, **485**, 488
  chord, 484, **485**
  intercepted, 489
  length, 488, **490**
  major, 484, **485**
  measure, 488, **490**
  minor, 484, **485**
  quarter circle, 405
  semicircle, 492
Area, **15**, 99
  of a triangle, 103
  of circle, **426**
  of parallelogram, **112**
  of regular polygons, **422**
  of trapezoid, **112**
  of triangle, **112**
  sector, **430**
  similar figures, 12
  surface, 446, **555**
Area model, 213, **219**
Argument
  proof, 365, 370
Arrow diagram, 84
ASA ≅, **299**
At Your Service Problem, 312
Average, weighted, 508

B

Base, 107
Base angles of an isosceles triangle, 90
Bases, 447, **448**
  lateral surface, 530
Beta (β), **190**
Bisect, 317
Bisector
  perpendicular, 459
Blood types, 214
Building a Kaleidoscope Problem, 21

C

Carpetmart Problem, 12
Center, **341**, **410**
  circumcenter, 500
  concentric circles, 460
  of balance, **471**
  of balance, **471**
  of circle, **490**
  radius, 410
Center of rotation, **34**
Central angle, 382, 394, 488, **490**
Centroid, **471**, **501**
Chance, 57, **60**
Choosing a strategy/tool, 283
Chords, 484, **485**
  diameter, **485**
  intersecting, 498
Circle, **341**
  arc, 425, **426**, 483, **485**, 488
  arc length, **430**
  arc measure, **490**
  area, **426**
  area of sector, **430**
  center, **341**, **490**
  central angle, 488, **490**
  chord, **485**
  circumcenter, 500
  circumference, 424, **426**
  diameter, 335, **341**, **485**
  great, 549
  inscribed angle, **490**, **494**
  length of arc, **490**
  pi (π), **426**
  quarter, 405
  radius, **341**
  secant, 496
  sector, 425, **426**
  semicircle, 492
  tangent, 496
Circles, concentric, 460
Circular angles, 22, **24**
Circumcenter, 500, **501**
Circumference, 424, **426**
Clinometer, 202
Coincide, **87**
Common ratio, 143
Compass, 460, **468**
Complementary angles, 74
Concentric circles, 460
  focus-directrix paper, 576
Concept map, 63
Concurrency, **471**, **501**
Conditional statement, 80, **108**
  arrow diagram, 84
Cone
  conic sections, 573, **578**
  ellipse, **585**
  lateral surface, **547**
  volume, **547**
Congruent, 140, **159**, 291
  angles, 75, **76**
  shapes, **159**, 291

Congruent corresponding parts, **346**
Conic sections, 573, **578**
  ellipse, **585**
  graphing, 576
  hyperbola, 584
  parabola, 574
Conjecture, **10**, 75, **304**
  proven, 354
  theorem, 367
Consecutive angles, 370
Constant diameter
  Reuleaux curves, 337
Construction, 459
  compass, 460
  ellipse, 583
  hyperbola, 583
  perpendicular bisector, 459, **468**
  straightedge, 460
Contradiction, proof by, **96**
Converse, **304**
Convex polygon, 395, **396**
Coordinate geometry
  dilation, 136
  finding a midpoint, **381**
Correspondence
  similarity statement, **150**
Corresponding angles, **91**
Corresponding parts
  congruent triangles, **346**
$\cos^{-1}$, 243, **248**
Cosine ratio (cos), 236, **241**
Counterexample, 238
Cross-sections
  conic sections, 573, **578**
  ellipse, **585**
  of three-dimensional solids, **538**
Cube, 527
Cubic units, **452**
Curves, Reuleaux, 337
Cylinder, 30, 451
  lateral surface, 530
  oblique, 451

D

Decagon, 393
Decimal, non-repeating, 467
Delta (Δ), **47**
Dependent events, 207
Designing a Quilt Problem, 3, 5
Diagonals, **400**
  of a rhombus, **364**
Diagram
  tree, 206, **219**
  Venn, 50, **51**
Diagram markings, 54
Diameter, 335, **341**, **485**
Dilation, 135, 136, **138**
  creating with rubber bands, 135
Dimensions, 98
  cubic units, **452**
Directrix, 574
Dual polyhedron, 530

E

Earth
  equator, 549
  great circle, 549
  hemisphere, 549
  latitude lines, 549
  longitude lines, 549, **551**
  prime meridian, 549
Edges, 527
  vertex, 527
Ellipse, 581, 583, **585**
  foci, **585**
Enlargement
  zoom factor, 140
Equality
  Reflexive Property, **354**
Equations
  graphing, **29**
  proportional, **145**
  Quadratic Formula, **163**
  solving linear, **19**
Equator, 549
  great circle, 549
Equiangular, **351**, 396
Equilateral, **351**
  polygon, **396**
Equilateral triangles, 11, 22
Equivalent, **512**
Euler, Leonhard, 595
Euler's Formula for Polyhedra, 595
Exact answer, **123**, 240
Examining, 129
Expected value, 507, **516**
  weighted average, 508
Experiment, 7
Exploration, **10**
Exterior angle, 358, 397, 403, **407**
  sum of, **407**

F

Faces, 527
  edge, 527
  lateral, 447, **448**, **534**
  total surface area, **452**
Facilitator, 4
Factoring, **163**
Family Fortune, The Problem, 18, 170
Flipping a shape (reflection), 27, **34**, **38**
Flowchart, 157, 167, 353
Focus (foci), 574, 581, **585**
  ellipse, 581
  hyperbola, 584
Focus-directrix paper, 576
Formula
  quadratic, **163**
Fraction busters, **512**
Function, **589**
  absolute value, 575

G

Geometry, 548
George Washington's Nose Problem, 147
Getting To Know Your Triangle Problem, 309
Golden ratio, 597, 598, **599**
  golden rectangles, 598
  golden triangles, 598
  spiral, 598, **599**
Golden rectangle, 598
Golden spiral, 598, 599
Golden triangles, 598
Graph
  focus-directrix, 576
Graphing an equation, **29**
Great circle, 549

H

Half-equilateral triangle, 40, 250
Half-square, 40
Height
  of pyramid, **534**
  slant, **534**
Hemisphere, 549
Heptagon, 393
Hexagon, 43, 393
Hinged Mirror Team Challenge, The Problem, 349
HL $\cong$, **299**
Horizontal change, **47**, **190**
How Tall Is It? Problem, 202
Hyperbola, 584
Hypotenuse, **119**, **241**

I

Icon
  Learning Log, 28
  stoplight, 88
  toolkit, 54
Image, 34
Incenter, **501**
Independent events, 207
Inscribed angle, 489, **490**
Inscribed Angle Theorem, **494**
Inscribed polygon, 461
Intercepted arc, 489
Interior angles, 402
  sum of, **407**
Interior Design Problem, 343
Intersecting chords, 498
Intersecting tangents, **560**
Intersections
  secant, 557
  tangents, **560**
Inverse, **248**
Inverse operations, 243
  trigonometric ratios, 243
Inverse trigonometric functions
  inverse cosine, **248**
  inverse sine, **248**
  inverse tangent, **248**

Investigating, 7, 66
Investigative process, **10**
Irrational number, **115**, 467
Isosceles trapezoid, 77, 371
Isosceles triangle, 38, 48
   base angles, 90
Iteration, 598

J

Justifying, 16, **167**

K

Kaleidoscope, Building a Problem, 21
Kite, 353

L

Lateral
   face, **534**
   faces, 447, **448**
   slant height, **534**
   surface area, **534**
   surface of a cone, **547**
   surface of a cylinder, 530
Latitude lines, 549
   equator, 549
Law of Cosines, 266, **267**
Law of Sines, 263, **264**
Leaning Tower of Pisa Problem, 187, 198
Learning Log, 28
Leg, **119**
Length
   of arc, 488, **490**
Lessons From Abroad Problem, 174
Light, reflection of, 85
Likelihood
   probability, **505**
Line
   naming, **81**
   of reflection, **34**, **38**
   of symmetry, **5**, 43
Line segment, 38
   midsegment, 369, **371**
   naming, **81**
Linear scale factor, 414, **415**
   Ratios of similarity, **457**
   zoom factor, 140
Lines
   latitude, 549
   longitude, 549
List, systematic, **219**
Log, Learning, 28
Longitude lines, 549, **551**
   meridian, **551**
   prime meridian, 549

M

m, **47**
Major arc, 484, **485**
Map, concept, 63
Mapping, **34**
Marking a diagram, 54
Mat plan, 443
Maximum limits, 119
Measure
   of arc, 488, **490**
Measurement
   area, **15**, 99
   dimensions, 98
   perimeter, **15**
   unit of measure, 98, 99
   unit square, 99
Medians, **471**
   centroid, **471**
Meridian, **551**
   a.m., **551**
   p.m., **551**
   prime, 549
Midnight Mystery Problem, 221
Midpoint, 40, 41, 342, **381**
   midsegment, 369, **371**
Midsegment, 369, **371**
Mighty Mascot Problem, 413
Minimum limits, 119
Minor arc, 484, **485**
Mirror, hinged, 21, 73
Möbius strip, 7, 8, 9
Models
   area, **219**
   probability, **219**

N

Nature
   golden ratio, 599
Net, 446
n-gon, 393, **422**
Nonagon, 393
Non-convex polygon, 395, **396**
Number
   irrational, 467

O

Oblique cylinder, 451
Oblique prism, 451
Oblique pyramid, 538
Obtuse angles, **24**
Octagon, 393
   regular, 402
Operations
   inverse, 243
Opposite leg, **241**
Opposite side, 236
Optimization, 343
Orientation, **241**

P

Pantograph, 135
Paper Snowflake, The Problem, 319
Parabola
    directrix, 574
    focus, 574
Parallel lines, **47**
Parallelogram, 78, **361**
    area of, 106, **112**
    diagonals, **400**
    rectangle, 360
    rhombus, **464**
Pentagon, 393
    regular, 23
Perimeter, **15**, 35
    circumference, 424, **426**
Perpendicular bisector, 459, **468**
Perpendicular lines, **47**
Phi (φ), 598, 599
pi (π), **426**
Pinwheel, 393
Plan
    mat, 443
Plato, 529
Platonic solids, 529
Point of concurrency, **471**
Point of intersection, **87**
Points
    midpoint, 342
    of concurrency, **501**
Polygonal base
    pyramid, 533
Polygons, **42**, 393, **396**
    apothem, **410**
    area of regular, **422**
    center, **410**
    convex, 395, **396**
    decagon, 393
    exterior angle, 403, **407**
    heptagon, 393
    hexagon, 393
    inscribed, 461
    n-gon, 393
    nonagon, 393
    non-convex, 395, **396**
    octagon, 393
    pentagon, 393
    quadrilateral, **361**, 393
    radius, **410**
    rectangle, 5
    regular, **42**, **351**, **396**
    regular hexagon, 21, 43
    regular octagon, 402
    sum of interior angles, **407**
    triangle, 393
Polyhedra, **448**, 528
    dual, 530
    Euler's Formula, 595
    prisms, **448**
    pyramid, 533
    regular, 528

Prediction
    expected value, 507
Prime meridian, 549
Prime notation, **34**, **81**
Prisms, 446, 447, **448**
    bases, 447, **448**
    cube, 527
    edge, 527
    face, 527
    lateral faces, 447, **448**
    lateral surface area, **534**
    oblique, 451
    rectangular, 451
    total surface area, **534**
    triangular, 451
    vertex, 527
Probability, 57, **60**, **505**
    dependent events, 207
    expected value, **516**
Probability models, **219**
    area model, 213, **219**
    systematic list, 206, **219**
    tree diagram, 206, **219**
Proof, **10**, 366, 370
    by contradiction, **96**
    flowchart, 353
    two-column, 366
Proportional equation, **145**
Protractor, 22
Proven conjecture
    theorem, 367
Pyramid, 533
    height, **534**
    lateral face, **534**
    lateral surface area, **534**
    oblique, 538
    right, 538
    total surface area, **534**
    triangular-based, 533
    truncated, 556
    volume, 543
Pythagorean Theorem, 122, **123**
Pythagorean triples, 255, **260**

Q

Quadratic equations, **163**
Quadratic Formula, **163**
Quadrilateral, **361**, 393
    parallelogram, 106, **361**
    rectangle, 360, **361**
    rhombus, 356, **361**, **464**
    special properties, **400**
    square, **361**
    trapezoid, 107, **361**
Quarter circle, 405
Quilts, 36
Quirky Quadrilaterals Problem, 591

R

$r : r^2 : r^3$ ratios, **457**
Radius, **341**, 410
Rat Race, The Problem, 204, 217
Ratio, **60**
    cosine, 236
    golden, 597, 598, 599
    Law of Sines, **264**
    linear scale factor, 414, **415**
    of similar figures, **457**
    of similar solids, **457**
    of similarity, **142**
    proportion, **145**
    sine, 236
    slope, **47**, **190**
    zoom factor, **415**
Rationalizing the denominator, **252**
Ratios of similarity, **457**
Reasoning and justifying, 16, 180
Recorder/Reporter, 4
Rectangle, 5, **361**
    diagonals, **400**
Rectangular prism, 451
Reflection, 27, **34**, **38**
    line of, **34**, **38**
    of light, 85
Reflection symmetry, 3, **5**, 43, 44
Reflexive Property of Equality, **354**
Regular hexagon, 21, 43
Regular octagon, 402
Regular pentagon, 23
Regular polygons, **42**, 349, **351**, **396**
    apothem, **410**
    area of, **422**
    center, **410**
    central angle, 382
    radius, **410**
Regular polyhedra, 528
    dual polyhedron, 530
    platonic solids, 529
Relation, **589**
Relationship
    angles, 74, **76**, 84
    Law of Cosines, **267**
    Law of Sines, **264**
    Pythagorean Theorem, 122, 123
    similarity, 136, 139, **150**
Relationships
    angles, 84
Remote interior angles, 397
Representation
    solids, 443, 444
Reuleaux curves, 337
Reversal, **304**
Rhombus, 356, **361**, **400**, **464**
    diagonals, **364**, **400**, **464**
Right angles, 22, **24**
Right pyramid, 538

Right triangle, **119**
    hypotenuse, **119**
    leg, **119**
    Pythagorean Theorem, 122
    special, **260**
Rotation, **34**
Rotation symmetry, 43, 45, 52
Rotation, center of, **34**
Rubber band, use of, 135

S

Same-side interior angles, 84, **91**
SAS $\cong$, **299**
SAS $\sim$, **171**
Secant, 496, 557
Sector, 425, **426**
Segment, 38
Semicircle, 492
Shape bucket, 50
Shape Factory, The Problem, 40, 118
Shapes
    congruent, 140, **159**, **291**
    ratio of areas, **457**
    ratio of perimeters, **457**
    similar, 136, 139
    three-dimensional, 340
    two-dimensional, 340
Side
    face, 527
Side view, 444
Sides
    congruent, **351**
Similar figures
    area of, 12, **457**
Similarity, 136, 139, **171**, **415**
    AA $\sim$, **155**
    congruent, 140, **159**, **291**
    proportional equation, **145**
    ratio, **142**
    SSS $\sim$, **155**
    SSS $\sim$, AA $\sim$, and SAS $\sim$, **171**
    statement, **150**
    zoom factor, 140, **142**
$\sin^{-1}$, 243, **248**
Sine ratio (sin), 236, **241**, **264**
Slant height, **534**
Sliding a shape (translation), **34**
Slope, 20, **47**
    negative, **47**
    of parallel lines, **47**
    of perpendicular lines, **47**
    positive, **47**
    triangle, **47**, **190**, **200**
    undefined, **47**
    zero, **47**
Slope angle, **190**, 202
    clinometer, 202
    common, **194**
Slope ratio, **200**, **241**
    common, **194**

Slope-intercept form of a line, 123
Solids, 443
   cross-section, **538**, 573, **578**
   cube, 527
   cylinder, 451
   polydedron, **448**
   prisms, 446, 447
   ratio of volumes, **457**
   representation, 443, 444
   sphere, 545
   surface area, 446
   total surface area, **452**
   views, 444
   volume, 443
Solution
   of a system of equations, **87**
Solving equations
   linear, **19**
   Quadratic Formula, **163**
Solving equations by rewriting, **512**
Solving quadratic equations, **163**
Solving systems of equations
   Substitution Method, **87**
Somebody's Watching Me Problem, 73, 94
Special right triangles, **260**
    30°-60°-90°, 250
    45°-45°-90°, 251
    Pythagorean triples, 255
Sphere, 545, **555**
   great circle, 549
   hemisphere, 549
   longitude lines, **551**
   surface area, **555**
   volume, **555**
Square, **361**
Square root, **115**
   rationalizing the denominator, **252**
SSA, 269, 270
SSS ≅, **299**
SSS ~, **155**, **171**
Statue of Liberty Problem, 202
Steepness, measurement of, **47**
Stoplight icon, 88
Straight angles, **24**
Straightedge, 459, 460, **468**
   perpendicular bisector, **468**
Stretch point, **138**
Strip, Möbius, 7
Study team expectations, 7
Substitution Method, **87**
Supplementary angles, 74
Surface area, 446
   lateral, 530, **534**
   sphere, **555**
   total, **452**, **534**
Symmetry
   line of, 43
   reflection, 3, **5**, 43, 44
   rotation, 43, 45, 52
   translation, 45
Systematic list, 206, **219**
Systems of equations, **87**

T

Table, **29**
Take A Shot Problem, 345
Take It To The Bank Problem, 306
$\tan^{-1}$, 243, **248**
Tangent ratio (tan), 199, **200**, **241**
Tangents, 496, **560**
Task Manager, 4
Tetrahedron
   triangular-based pyramid, 533
Theorem, 354, 367
   Inscribed Angle, **494**
   Pythagorean, 122
   Triangle Angle Sum, **99**
   Triangle Midsegment, **371**
Theta (θ), **190**
Three-dimensional solids, 340, 443
   cross-section, **538**
   net, 446
   polyhedra, 528
   sphere, **555**
   volume, 443, **452**
Tiling, 78
Time zones, **551**
Toolkit
   angle relationships, 86
   area, 111
   shapes, 54
   triangle, 237
Top view, 444
Total surface area, **452**
Tracing paper
   construction, 459
   perpendicular bisector, **468**
Transformations
   dilation, 135, 136, **138**
   prime notation, **34**, **81**
   reflection, 27, **34**, **38**
   rotation, **34**
   translation, **34**
Translation symmetry, 45
Translations, **34**
Transversal, 79
Trapezoid, 107, **361**
   area of, 107, **112**
   bases of, 107
   isosceles, 77, 371, **400**
Tree diagram, 206, **219**
Trial of the Century Problem, 17

Triangle, 393
    area of, 103, **112**
    centroid, **471**
    circumcenter, 500
    congruent corresponding parts, **346**
    equilateral, 11, 22
    half-equilateral, 40
    half-square, 40
    hypotenuse, **119**
    isosceles, 38, 48
    leg, **119**
    medians, **471**
    minimum and maximum side lengths, 119
    points of concurrency, **501**
    right, **119**
    right isosceles, **260**
    special right, **260**
    sum of the angles, 90, **99**
Triangle ambiguity, 270
Triangle Angle Sum Theorem, 90, **99**
Triangle congruence conjectures, **299**
    AAS ≅, **299**
    ASA ≅, **299**
    HL ≅, **299**
    SAS ≅, **299**
    SSS ≅, **299**
Triangle Inequality, 118, 119
Triangle Midsegment Theorem, **371**
Triangle similarity, **171**
    AA ~, **155**
    SSS ~, **155**
Triangle toolkit, 237
    Law of Cosines, 266, **267**
    Law of Sines, 263
Triangular prism, 451
Triangular-based pyramid, 533
Trig table, 193, **194**
Trigonometric ratios, **241**
    adjacent side, 236
    cosine, 236, **241**
    inverse, 243
    opposite side, 236
    sine, 236, **241**
    tangent, **241**
Truncated pyramid, 556
Turning a shape (rotation), **34**
Two-column proof, 366
Two-dimensional shapes, 340
    polygon, 393, **396**

U

Unit of measure, 98, 99
Unit square, 99

V

Value
    expected, 507
Venn diagrams, 50, **51**
Vertex, 74, 527
Vertical angles, 75, **91**
Vertical change, **47, 190**
Views of a solid, 444
Visualizing, 26, 229
Volume, 443, **452**
    cone, **547**
    cubic units, **452**
    pyramid, 543
    sphere, **555**

W

Ways of Thinking, 129, 180, 283, 329
    choosing a strategy/tool, 283
    examining, 129
    investigating, 7, 66
    reasoning and justifying, 16, 180
    visualizing, 26, 229
Weighted average, 508, **516**
Writing a flowchart, 167

Y

You Are Getting Sleepy Problem, 173

Z

Zero Product Property, **163**
Zoe and the Poison Weed Problem, 601
Zoom factor, 140, **142**
    linear scale factor, **415**

Index of Symbols

$\geq$	"greater than or equal to"	$\angle A$	angle with vertex A
$\leq$	"less than or equal to"	$m\angle A$	measure of angle A
$>$	"greater than"	$36°$	"36 degrees"
$<$	"less than"	$\cong$	"congruent to"
$=$	"equal to"	$\tan\theta$	"tangent of theta"
$\sqrt{}$	square root or radical	$\sin\theta$	"sine of theta"
$\triangle ABC$	a triangle with vertices A, B, and C	$\cos\theta$	"cosine of theta"
		$\sim$	"similar to"
$X'Y'Z'$	"X prime, Y prime, Z prime"	$P:Q$	ratio of P to Q
$ABCD$	quadrilateral $ABCD$	$P(A)$	probability of event A
cm^2	square centimeter	π	"pi"
in^2	square inch	$\odot C$	circle C
$//$	"parallel to"	$\overset{\frown}{ABC}$	arc ABC
$\perp$	"perpendicular to"	$m\overset{\frown}{ABC}$	measure of arc ABC
AB	length of line segment $\overline{AB}$	α	alpha
$\overleftrightarrow{AB}$	line through points A and B	β	beta
$\overline{AB}$	line segment with endpoints at A and B	Δ	delta
		θ	theta
$\overrightarrow{AB}$	ray starting at A and passing through B	ϕ	phi

Equation Reference

$y = mx + b$	Slope-intercept form of a line	$x^2 + y^2 = r^2$	Equation of a circle with center at the origin $(0, 0)$, with radius r
$y = ax^2 + bx + c$	Standard form of a quadratic equation		